ELECTRONICS Made Simple

The Made Simple series
has been created
primarily for self-education
but can equally well
be used as
an aid to group study.
However complex the subject,
the reader is taken
step by step,
clearly and methodically
through the course. Each volume
has been prepared by
experts,
using throughout the
Made Simple technique of teaching.
Consequently the gaining
of knowledge now becomes
an experience to be enjoyed.

In the same series

Accounting
Acting and Stagecraft
Additional Mathematics
Advertising
Anthropology
Applied Economics
Applied Mathematics
Applied Mechanics
Art Appreciation
Art of Speaking
Art of Writing
Biology
Book-keeping
British Constitution
Calculus
Chemistry
Childcare
Commerce
Commercial Law
Company Administration
Computer Programming
Cookery
Cost and Management Accounting
Data Processing
Dressmaking
Economic History
Economic and Social Geography
Economics
Electricity
Electronic Computers
Electronics
English
English Literature
Export
French
Geology
German
Human Anatomy
Italian
Journalism
Latin
Law
Management
Marketing
Mathematics
Modern Electronics
Modern European History
New Mathematics
Office Practice
Organic Chemistry
Philosophy
Photography
Physical Geography
Physics
Pottery
Psychology
Rapid Reading
Retailing
Russian
Salesmanship
Secretarial Practice
Soft Furnishing
Spanish
Statistics
Transport and Distribution
Typing
Woodwork

ELECTRONICS Made Simple

Henry Jacobowitz

Advisory editor
Leslie Basford, B.Sc.

Made Simple Books
W. H. ALLEN London
A Howard & Wyndham Company

Made and printed in Great Britain
by Richard Clay (The Chaucer Press), Ltd., Bungay, Suffolk
for the publishers W. H. Allen & Co. Ltd.
44 Hill Street, London W1X 8LB

First edition April 1967
Reprinted May 1969
Reprinted September 1970
Reprinted February 1972
Reprinted October 1972
Reprinted October 1973
Reprinted September 1974
Revised and reprinted September 1975
Reprinted April 1977
Reprinted July 1978

ISBN 0 491 01822 3 Paperback

Foreword

It would be stating the obvious to preface this book with a reminder of the importance of electronics today. But it is worth bearing in mind that electronics is still a young science (the word itself got into English dictionaries during the Second World War), and one that continues to grow. At the same time, more and more people are taking a serious interest in this subject, either as a career or as an absorbing hobby.

Here is a book for every reader who wants to gain a sound understanding of the principles of electronics. It assumes no more than a prior knowledge of elementary electricity, and seeks to avoid mathematics as far as possible.

Although it is not a formal text, since it is deliberately written in everyday language, it should be of use to students following A-level G.C.E. and O.N.C. courses in schools and technical colleges. The systematic arrangement of the principles and applications of electronics will be particularly appreciated by the reader who is studying alone.

A logical approach to the subject requires a deliberate progression from fundamental electronic principles and devices to their application in circuits, which, in turn, form the building blocks for complex systems such as radio transmitters and receivers. Following this approach, the book falls naturally into three major sections. The first, covering Chapters 1–7, explains basic principles, valves, and transistors. The second, covering Chapters 8–14, delves into a variety of electronic circuits, including amplifiers, oscillators, power supplies, modulation, and detection. The remaining chapters are devoted to some of the major electronic systems of general interest, such as radio, television, radar, and high-fidelity reproduction.

L.B.

Table of Contents

CHAPTER ONE

ATOMIC STRUCTURE

Electronics—the word immediately conjures up pictures of television receivers, radio circuits, computers, radar systems and all the other products of a very large branch of engineering—is in fact concerned with the movement of incredibly small electrical charges that have managed to detach themselves from atoms, the building bricks of all matter. According to simple theory, an atom consists of a central nucleus of positive charge around which tiny, negatively charged particles, called *electrons*, revolve in fixed orbits, just as the planets revolve around the Sun. In each type of atom, the negative charge of all the orbital electrons just balances the positive charge of the nucleus, thus making the combination electrically neutral.

The positively charged nucleus, in turn, reveals a complex structure, but for the purpose of understanding electronics a vastly simplified picture is adequate. According to this simplified picture, the nucleus of the atom is made up of two fundamental particles, known as the *proton* and the *neutron*. The proton is a relatively heavy particle (1,840 times heavier than the electron) with a positive (+) charge, while the neutron has about the same mass as the proton, but has no charge at all.

The positive charge on each proton is equal to the negative charge residing on each electron. Since atoms are ordinarily electrically neutral, the number of positive charges equals the number of negative charges, that is, the number of protons in the nucleus is equal to the number of electrons revolving around the nucleus.

Practically the entire weight of the atom is made up by the protons and neutrons in the nucleus, the weight of the orbital electrons around the nucleus being negligible in comparison. Lest you should think that substantial weights are involved, let us hasten to say that the mass of an electron is only about $9{\cdot}11 \times 10^{-28}$ grams (a number with 27 zeros after the decimal point), while that of the proton is only about 1,840 times as much, which is still fantastically little. To give you an idea how small this really is, you might consider that an electron is about as small compared to a standard ping-pong ball, as a ping-pong ball is compared to the 186,000,000-mile diameter of the Earth's orbit.

We have seen that all atoms are made up of protons, neutrons, and electrons. The difference between various types of elements is the number and arrangement of the protons, neutrons, and electrons within their atoms. The orbits of the electrons are arranged in 'shells' about the nucleus, each shell having a definite maximum capacity of electrons. It is the outermost shell which determines the chemical valence of an atom and its principal physical characteristics. The outermost shell is also the most important for electronics, since it is the only one from which electrons are relatively easily dislodged to become 'free' electrons capable of carrying a current in a valve or conductor. The electrons in the inner shells cannot be easily forced out from their orbits and, hence, are said to be 'bound' to the atom.

While atoms are the smallest bits of matter in each element, it may be well to keep in mind that most materials in the world are compounds of various elements, formed by combinations of different atoms. These smallest combinations are called molecules.

IONS AND IONIZATION

An atom (or molecule) that has become electrically unbalanced by the loss or gain of one or more electrons is called an ion. An atom that has lost an electron is called a positive ion, while an atom that has gained an electron is a negative ion. The reason is clear. When an atom loses an electron, its remaining orbital electrons no longer balance the positive charge of the nucleus, and the atom acquires a charge of +1. Similarly, when an atom gains an electron in some way, it acquires an excess negative charge of −1. The process of producing ions is called ionization.

Ionization does not change the chemical properties of an atom, but it does produce an electrical change. It can be brought about in a number of ways. As we have seen, the electrons in the outermost shell of an atom are rather loosely held and they can be dislodged entirely by collision with other electrons or atoms, or by exposure to X-rays.

Ionization is important in electronic valves. Although valves are evacuated, there is always a trace of gas left. An electron moving rapidly through a valve may collide with a gas atom, resulting in several possible effects. The electron may become attached to the gas atom, thus changing it into a negative ion. The electron may simply bounce off the atom. Finally, the impact of a rapidly moving electron may knock out one or more of the outer-shell electrons in the atom, thus producing a positive ion. If one electron is knocked off, the gas is said to be singly ionized, if two are dislodged, it is doubly ionized, and so on. Positive ions may adversely affect the operation of the valve.

FREE ELECTRONS

Electrons that have become dislodged from the outer shell of an atom are known as free electrons. These electrons can exist by themselves outside of the atom, and it is these free electrons which are responsible for most electrical and electronic phenomena. Free electrons carry the current in ordinary conductors (wires), as well as in valves and transistors. The motion of free electrons in aerials gives rise to electromagnetic radiations (radio waves). Free electrons also constitute the so-called cathode rays and beta rays, with which we shall become familiar in Chapter 6.

Most substances normally contain a number of free electrons, which are capable of moving freely from atom to atom. Materials such as silver, copper, or aluminium, which contain relatively many free electrons capable of carrying an electric current, are called conductors; materials that contain relatively few free electrons are called insulators. Materials that have an intermediate number of available free electrons are classed as semiconductors. Actually, there are no perfect conductors and no perfect insulators. The more free electrons a material contains, the better it will conduct. All substances can be arranged in a series according to their conductivity, that is, in accordance with their relative number of available free electrons.

The free electrons in a conductor are ordinarily in a state of chaotic motion in all possible directions. When a battery is connected across a conductor, however, the free electrons will move in an orderly fashion, atom to atom, from the negative terminal of the battery, through the wire, and to the positive terminal of the battery. This orderly drifting motion of free electrons is said to constitute an electric current. Although the motion is rather slow, the impulse is transmitted almost at the speed of light.

When free electrons are introduced in an evacuated tube across which a voltage (electric force) has been applied, they will move very rapidly towards the positive terminal of the tube. The drifting motion is speeded up, since an evacuated tube has very few atoms to impede the progress of the electrons. The greater the voltage across the tube, the greater is the attraction of the (negatively charged) electrons to the positive terminal and, hence, the greater is the speed of the electron motion.

In the following chapter we shall learn how free electrons may be introduced into such an evacuated tube.

CHAPTER TWO

ELECTRON EMISSION

The radio valve depends for its action on a stream of electrons that act as current carriers. To produce this stream of electrons a special metal electrode (emitter) is present in every valve. But at ordinary room temperatures the 'free' electrons in the metallic emitter cannot leave its surface because of certain restraining forces that act as a barrier. These attractive surface forces tend to keep the electrons within the emitter substance, except for a small portion that happens to have sufficient kinetic energy (energy of motion) to break through the barrier. The majority of electrons move too slowly for this to happen.

To escape from the surface of the emitter the electrons must perform a certain amount of work to overcome the restraining surface forces. To do this work the electrons must have sufficient energy imparted to them from some external source of energy, since their own kinetic energy is inadequate. This external energy may come from a variety of sources, such as heat energy, light energy, energy stored in electric or magnetic fields, or the kinetic energy of electric charges bombarding the metal surface. Accordingly, we can classify the four principal methods of obtaining electron emission from the surface of a metallic conductor, as follows:

1. *Thermionic (primary) emission.* In this method the emitter metal is heated, resulting in increased thermal (or kinetic) energy of the unbound electrons. Thus, a greater number of electrons will attain sufficient speed and energy to escape from the surface of the emitter. The number of electrons released depends on the temperature.

2. *Photoelectric emission.* In this process the energy of the light radiation falling upon the metal surface is transferred to the free electrons within the metal and speeds them up sufficiently to enable them to leave the surface. The number of electrons emitted depends upon the intensity (brightness) of the light beam falling upon the emitter surface.

3. *Field emission (cold-cathode emission).* The application of a strong electric field (i.e. a high positive voltage) outside the emitter surface will literally 'yank' the electrons out of the emitter surface, because of the attraction of the positive field. The stronger the field, the greater the field emission from the cold emitter surface.

4. *Secondary emission.* When high-speed electrons suddenly strike a metallic surface they give up their kinetic energy to the electrons and atoms which they strike. Some of the bombarding electrons collide directly with free electrons on the metal surface and may knock them out from the surface, similar to a billiard-ball collision. The electrons freed in this way are known as secondary-emission electrons, since primary electrons from some other emitter must be available to bombard the secondary electron-emitting surface.

Of these four methods of emission, thermionic emission is the most important and the one most commonly used in electronic valves.

THERMIONIC EMISSION

Electron emission from a heated metal surface is very similar to the evaporation of a liquid from its surface. When a liquid is heated, an increasing number of molecules acquire sufficient energy to overcome the restraining forces of the liquid surface and are evaporated. The number of molecules evaporated increases rapidly as the temperature is raised. Similarly, when a metallic body is heated, a progressively larger number of electrons overcome the restraining surface barrier and are literally 'boiled out' from the metal, like steam from a kettle.

The number of electrons 'evaporated' per unit area of an emitting surface is related to the absolute temperature T (°A) of the emitter and a quantity b that is a measure of the work an electron must perform when escaping the emitter surface, according to an equation derived by O. W. Richardson about 50 years ago:

$$\text{emission current } I = A\,T^2 e^{-b/T} \text{ amps per square cm}$$

where $e = 2{\cdot}7183$ (base of natural logs) and A is a constant which has a value of about 60 for pure metals, such as tungsten or tantalum, but varies widely for other practical emitters.

The combination of the squared and exponential terms in the emission equation makes it extremely sensitive to small changes in temperature.

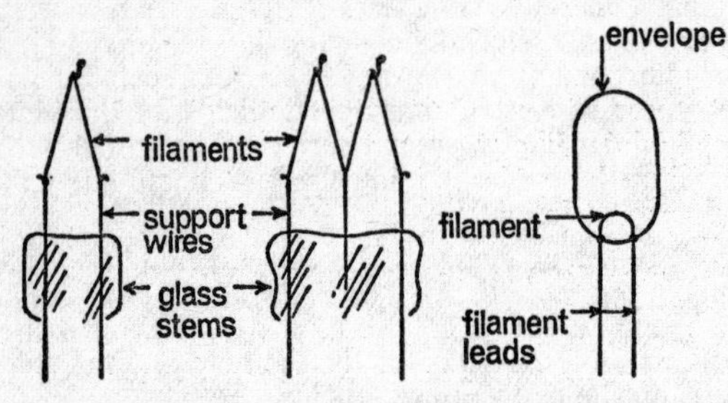

Fig. 1. Directly heated filaments and schematic symbol

Doubling the temperature of an emitter may increase electron emission by more than 10,000,000 times. For example, the emission from pure tungsten metal is about a millionth ampere per sq. cm at 1600° A (1327° C), but rises to the enormous value of about a 100 amps per sq. cm when the temperature is raised to 3200° A (2927° C).

The most commonly used emitters in radio valves are tungsten, heated to temperatures between 2200 and 3000° A, thoriated tungsten (thorium and carbon added to tungsten), operated at about 1900° A (1627° C) and oxide-coated cathodes (barium-strontium oxide), operated at about 1000 to 1150° A (727 to 877° C).

METHODS OF HEATING

Electron emitters are heated electrically, either directly or indirectly. In the direct method, the electric current is applied directly to a wire, called the filament, that also serves as electron emitter (Fig. 1). In the indirect method, the electric current is applied to a separate heater element, located inside a cylindrical sleeve (cathode) that is coated with the emitting material (Fig. 2). The cathode is thus heated indirectly through heat transfer from the heater element.

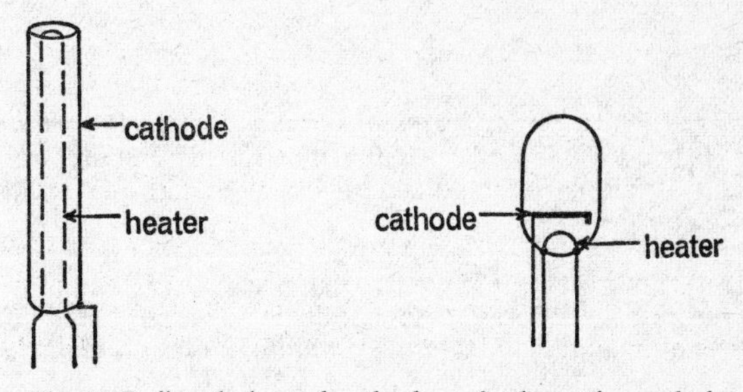

Fig. 2. Indirectly heated cathode and schematic symbol

All the emitters previously described (tungsten, thoriated tungsten, or oxide-coated) can be used for directly heated filaments, but indirectly heated cathodes always use oxide-coated emitters. The valves in a radio are usually indirectly heated.

SUMMARY

Valves depend for their action on a stream of electrons that act as current carriers. Free electrons may be produced from a metal electrode, called the emitter. The process is called emission. Emission does not occur at room temperature because of restraining forces at the surface of the emitter.

To escape from the surface of an emitter, electrons must be given sufficient kinetic energy from an external energy source to overcome the surface barrier.

When electrons are emitted from a surface because of heat energy supplied to it, the process is called thermionic emission.

In photoelectric emission light energy falling upon the emitter is transferred to free electrons and ejects them from the surface.

In field or cold-cathode emission the electric field set up by a high positive voltage 'yanks' free electrons from the emitter.

When primary electrons strike a metal surface at high speed, some will collide with electrons on the metal surface and project them outward like billiard balls. This is called secondary emission.

Thermionic emission from a metal surface is similar to the evaporation of molecules from the surface of a liquid when it is heated.

The number of electrons 'boiled out' from the emitter, called emission current, increases extremely rapidly when the emitter temperature is increased.

In direct heating, electric current is applied to a filament wire that serves as an emitter. In indirect heating, electric current is applied to a separate heater element, located inside a cylindrical cathode that serves as the emitter.

CHAPTER THREE

DIODES

The simplest combination of elements constituting an electronic valve is the diode ('two electrodes'). It consists of a cathode, which serves as an emitter of electrons, and an anode or plate surrounding the cathode, which acts as a collector of electrons (see Fig. 3). Both electrodes are enclosed in a highly evacuated envelope of glass or metal. The emitter may be either directly or indirectly heated. The size of diode valves varies from tiny metal tubes to large-sized glass-envelope rectifiers. The anode is generally a hollow metallic cylinder made of nickel, molybdenum, graphite, tantalum, monel, or iron.

OPERATION

A basic law of electricity states that like charges repel each other and unlike charges attract each other. Electrons emitted from the cathode of an electron tube are negative electric charges. These charges may be either attracted to or repelled from the anode of a diode valve, depending on whether the anode is

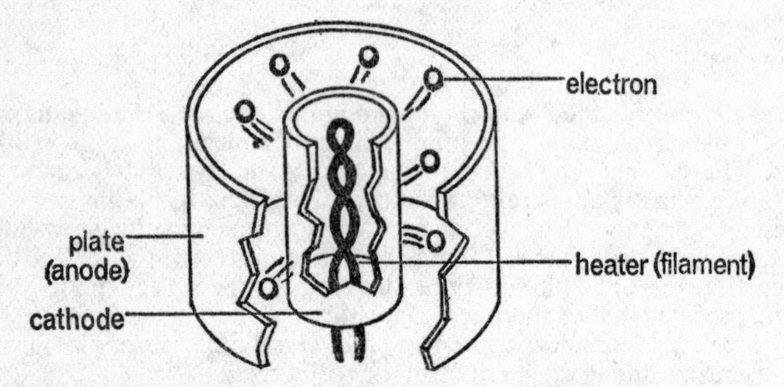

Fig. 3. Construction of a diode valve

positively or negatively charged. Actually, by applying a potential difference (voltage) from a battery or other source between the anode and cathode of a diode, an electric field is established within the valve. The lines of force of this field always extend from the negatively charged element to the positively charged element. Electrons, being negative electric charges, follow the direction of the lines of force in an electric field. Note that some older text-books use the *opposite* convention, whereby lines of force are regarded as running from positive to negative, in the same way that a 'conventional' current is imagined as a flow of positive charges instead of electrons.

Fig. 4 is a simplified schematic of a diode circuit to illustrate its basic action. A battery has been connected between anode and cathode of a diode, so as to make the anode negative with respect to the cathode. The lines of force of the field established within the valve thus extend from the anode to the cathode. When a voltage is now applied to the heater element (*H*), the cathode will emit a copious flow of electrons. However, these electrons are strongly repelled from the negatively charged anode and tend to fill the interelectrode space between cathode and anode. (Some of the electrons actually fall back into the cathode.) Since no electrons actually reach the anode, the valve acts like an open circuit, and the meter connected externally between anode and cathode indicates no current.

In Fig. 5, the battery connexion has been reversed so as to make the anode positive with respect to the cathode. The lines of force of the electric field now extend in a direction from the cathode to the anode. Again, applying a heater voltage results in copious emission of electrons from the cathode. Now, however, the electrons follow the lines of force to the positive anode and strike it at high speed. Since moving charges comprise an electric current, the stream of electrons to the anode is an electric current, called the anode current. Upon

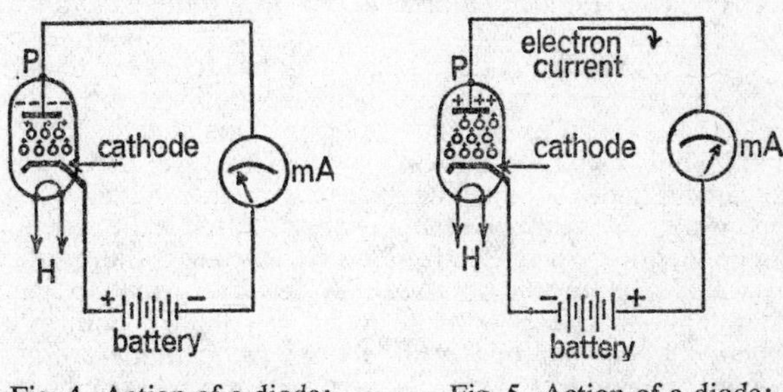

Fig. 4. Action of a diode: negative anode

Fig. 5. Action of a diode: positive anode

reaching the anode the electron current continues to flow through the external circuit made up of the connecting wires, meter, and the battery. The arriving electrons are absorbed into the positive terminal of the battery, and an equal number of electrons flow out from the negative battery terminal and return to the cathode, thus replenishing the supply of electrons lost by emission. The flow of electrons from the anode to the cathode through the external circuit registers on the meter, as shown in Fig. 5. As long as the cathode of the valve is maintained at emitting temperatures and the anode remains positive, electrons will continue to flow from the cathode to the anode within the valve and from the anode back to the cathode through the external circuit.

The following conclusions may be drawn from this simplified picture of diode valve operation:

1. Electron current (anode current) flows in the diode only when the anode is made positive with respect to the cathode. No current can flow when the anode is negative with respect to the cathode.
2. Electron flow within a diode takes place only from the cathode to the anode, never from the anode to the cathode. This is known as unidirectional or unilateral conduction.
3. Because of its unidirectional characteristic, a diode can be made to act like a switch or valve, automatically starting or stopping the anode current, depending on whether the anode is positive or negative with respect to the cathode. This ability permits diodes to change alternating current to direct current, or rectify it. We shall become acquainted with the operation of rectifiers in Chapter 14.

SPACE CHARGE

The total number of electrons emitted by the cathode of a diode is always the same at a given operating temperature and is determined by the Richardson emission equation, as we saw on p. 4. The anode voltage (voltage between anode and cathode) has no effect, therefore, on the amount of electrons emitted from the cathode. Whether or not these electrons actually reach the anode, however, is determined by the anode-to-cathode voltage, as well as by a phenomenon known as space charge. The term space charge is applied to the cloud of electrons that is formed in the interelectrode space between cathode and anode. Since it is made up of negatively charged electrons, this cloud constitutes a negative charge in the interelectrode space, and has a repelling effect on the electrons being emitted from the cathode. The effect of this negative space charge alone, therefore, is to force a considerable portion of the emitted electrons back into the cathode and prevent others from reaching the anode.

The space charge, however, does not act alone. It is counteracted by the electric field from the positive anode, which reaches through the space charge to attract electrons and thus partially overcomes its effects. At low positive anode voltages, only electrons nearest to the anode are attracted to it and constitute a small anode current. The space charge then has a strong effect on limiting the number of electrons reaching the anode. As the anode voltage is increased, a greater number of electrons are attracted to the anode through the negative space charge and correspondingly fewer are repelled back to the cathode. If the anode voltage is made sufficiently high, a point is eventually reached where all the electrons emitted from the cathode are attracted to the anode and the effect of the space charge is completely overcome. Further increases in the anode voltage cannot increase the anode current through the valve, and the emission from the cathode limits the maximum current flow.

DIODE CHARACTERISTICS

The relation between the anode current in a diode and the anode-to-cathode voltage just discussed can be represented by a characteristic curve, obtained by plotting the anode-current (I_b) values for different values of the applied anode voltage (E_b). The diode characteristics for a typical diode valve at various cathode operating temperatures are shown in Fig. 6.

It can be seen from Fig. 6 that all the curves are the same at low anode voltages, where the negative space charge is most effective in limiting the flow of electrons. The anode current in the low anode-voltage region is completely controlled by the voltage at the anode and is independent of the emitter (cathode) temperature. Under these conditions the anode current is said to be space-charge limited.

As the anode voltage is made progressively higher, an increasingly greater portion of the total supply of emitted electrons is attracted to the anode and the effect of the space charge is eventually completely overcome. This is seen by the flattening of the characteristic curves as the anode voltage is increased. When the entire supply of emitted electrons (at a given cathode temperature) is attracted to the anode, the anode current becomes independent of the anode voltage and reaches a constant value equal to the total emission current. Emission saturation takes place and the anode current is said to be emission-limited in the high anode-voltage region. This is seen best by the flattening of the dotted graphs in Fig. 6, which represent tungsten or thoriated-tungsten cathode emitters. For each constant operating temperature, the anode current at high anode voltages reaches its specific saturation value, equal to the total emission current determined from Richardson's emission equation. Oxide-coated emitters, represented by the solid lines in Fig. 6, do not have such a specific emission saturation value, and the anode current—though tapering off at high anode voltages—never becomes completely independent of the

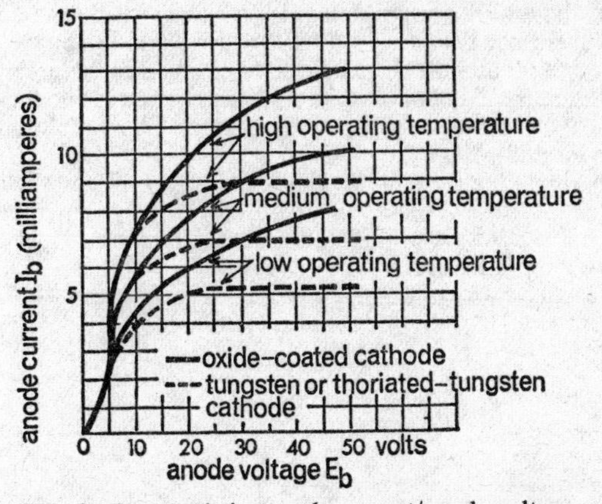

Fig. 6. Diode characteristic: anode-current/anode-voltage curves

anode voltage. At very high anode voltages, oxide-coated cathodes may become damaged because of the abnormally large emission. In general, thermionic valves are operated in the space-charge limited (low E_b) region.

CHILD'S LAW FOR DIODES

An interesting relation holds for valves that are operated in the region where space charge limits the value of the anode current (see Fig. 6). It has been found that the anode current (I_b) in this region is approximately proportional to the 3/2 (three-halves) power of the voltage between the anode and the cathode (E_b). This may be expressed mathematically by the so-called Child's law:

$$\text{anode current } I_b = K\, E_b^{3/2}$$

where K is a constant that depends on the shape of the electrodes and the geometry of the valve. Although this law is a guide, it is not entirely accurate in practice.

SUMMARY

A diode is a two-element valve that consists of a cathode, serving as an electron emitter, and an anode or plate acting as an electron collector.

By applying a positive voltage to the anode of a diode, an electric field is established between cathode and anode that attracts electrons emitted from the cathode to the anode.

Anode current is the flow of electrons from cathode to anode and their return to the cathode through the external circuit.

Anode current flows in a diode when the anode is made positive with respect to the cathode. No current flows when the anode is negative with respect to the cathode.

Current within a diode flows from cathode to anode, never from anode to cathode. This is called unidirectional conduction.

The cloud of electrons formed in the space between cathode and anode is called space charge. The space charge is negative and hence has a repelling effect on electrons emitted from the cathode.

The size of the anode current depends on the space charge and the relative strength of the electric field set up by the positive anode voltage.

At low anode voltages the negative space charge limits the flow of electrons and the anode current is completely controlled by the anode voltage and is independent of emitter (cathode) temperature. The anode current thus is space-charge limited.

At high anode voltages the space charge is drawn off and the anode current reaches saturation at a value equal to the total emission current. It is then independent of the anode voltage and is said to be emission-limited for a specific cathode temperature.

The anode current in the space-charge limited region is approximately proportional to the 3/2 power of the anode voltage.

CHAPTER FOUR

TRIODES

In 1907 Lee De Forest added a third element—the control grid—between the cathode and anode of a diode and so provided the resulting triode valve with the ability to amplify tiny radio signals. This led to the sensational development of radio communication, broadcasting and electronics in general, with which we are all familiar.

The construction of a typical triode is shown in Fig. 7. The control grid in this valve is a circular helix (spiral) of fine wire that completely surrounds the cathode. Because of its open construction, the grid does not directly hinder the flow of electrons to the anode; but a voltage placed on the grid has a profound effect on the electric field between cathode and anode and, hence, on the total electron flow.

Grid structures take many other forms besides a circular helix, such as flattened or elliptical helixes, ladders, etc. Different sizes and spacings of the control grid wires are employed, depending on the desired field configuration and design of the valve. Metals used for grids include nichrome, molybdenum, iron, nickel, tungsten, tantalum, and various alloys. Triode valves differ widely in size and electrode spacing, depending on their power rating and on the purpose for which they are intended.

CONTROL-GRID ACTION

Since the control grid is nearer to the cathode than the anode, a voltage placed on the grid has a much larger effect on the electric field within the valve —and hence upon the anode current—than the same voltage placed on the anode. The grid thus has a controlling effect on the anode current. A triode requires three operating voltages, one on each electrode, to operate correctly. The anode is normally connected to a high positive voltage (called 'B+') to attract the stream of electrons. A relatively low 'A' voltage (a.c. or d.c.) is connected to the filament or heater to bring the cathode to its proper emitting temperature and thus make available a supply of electrons. Finally, a voltage is placed on the control grid to govern the flow of anode current. This voltage generally consists of two components. One is a fixed direct voltage, called the bias ('C– ') which is normally a few volts negative with respect to the cathode. Its purpose is to operate or 'bias' the valve on a definite point on its characteristic curve so that a certain amount of anode current is always flowing. Superimposed upon the bias voltage is a varying voltage, usually called the signal voltage. The purpose of this voltage is to vary the current through the valve in

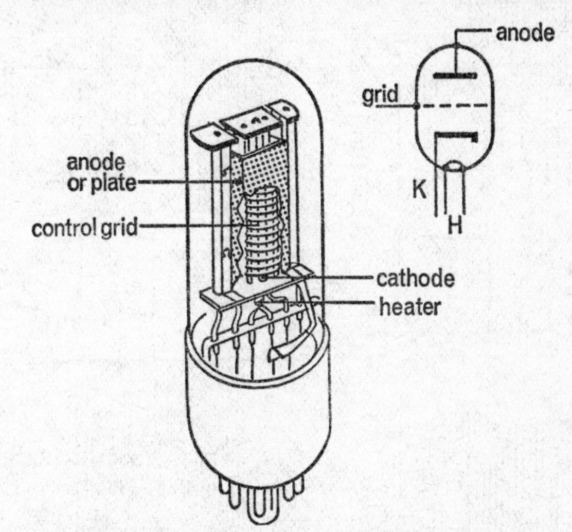

Fig. 7. Triode construction and schematic symbol

strict accordance with the signal variations, so as to make the anode current an amplified replica of the signal voltage. Amplification takes place, since a small variation of the signal voltage on the grid results in a large variation of the anode current.

Fig. 8 presents a simplified picture of the action of the grid in a triode. Instead of applying a bias and a signal voltage to the grid, we have varied the value of the bias voltage in each part of the figure, thus simulating the effect of a varying or signal voltage. The d.c. supply voltages (obtained from batteries here) are measured with respect to the cathode, and a milliampere meter has been inserted into the anode-to-cathode return circuit to measure the amount of anode current.

Cut-off Bias. Fig. 8a shows a cross-section of the cathode, grid wires and anode in a simplified schematic manner. A voltage (from the 'A' battery) has been applied to the heater, and the cathode is emitting a normal supply of electrons. The anode is at a high positive voltage and would normally attract a large number of electrons from the space charge, if it were not for the large negative bias voltage applied to the grid from the 'C' battery. Because of this large negative bias voltage, the electrostatic field normally existing between anode and cathode cannot penetrate to the cathode and actually terminates on the grid wires. This is shown by the lines of force extending from the grid to the anode. Under these conditions the grid entirely neutralizes the electrostatic field and, hence, the attraction of the anode. Since there is no electro-

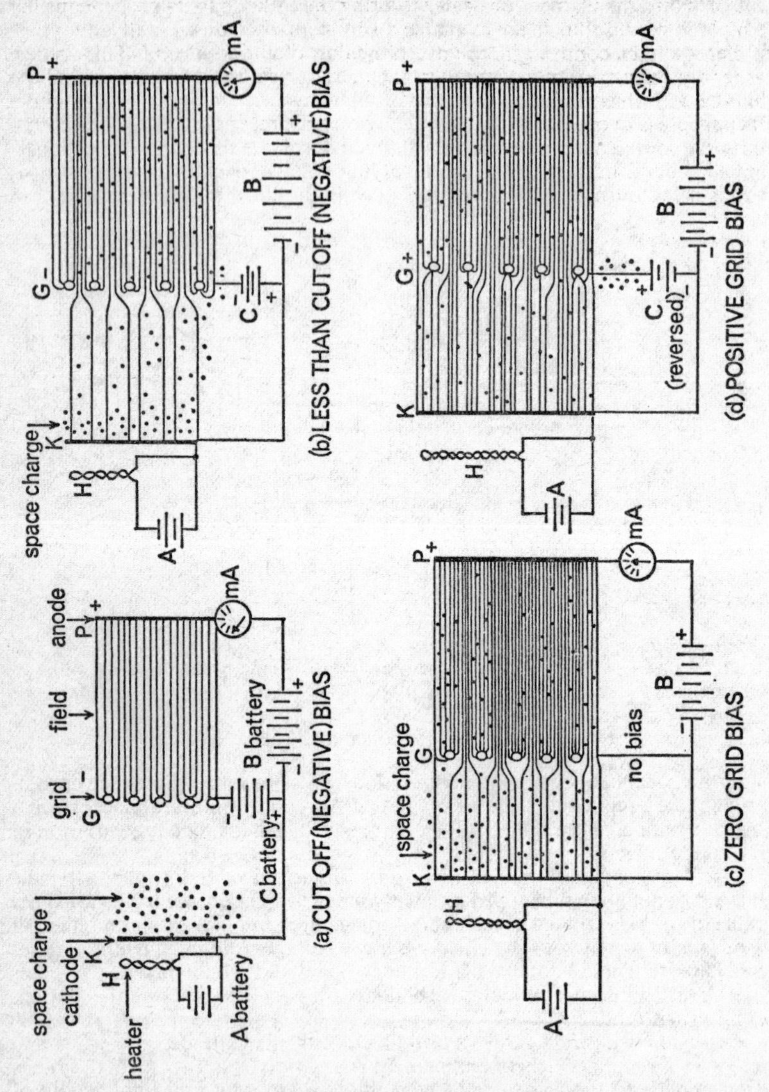

Fig. 8. Action of control grid in a triode

static field near the cathode to draw away the electrons, the current through the valve is zero (as indicated by the zero reading of the milliampere meter) and a large space charge accumulates in the region between cathode and grid. The smallest negative voltage between grid and cathode that is just capable of cutting off the anode current is called the cut-off bias. Bias voltages that are more negative than this cut-off value have no effect on the action of the tube.

Less Than Cut-off Bias. In Fig. 8b everything has been left unchanged, except that the negative bias voltage has been reduced to a value less than cut-off. The grid is now no longer capable of neutralizing the filed between anode and cathode completely and some of the lines of force penetrate between the grid wires to the cathode, as shown. Consequently, some electrons are attracted away from the space charge and move between the grid wires towards the positive anode. This results in a moderate anode current, as indicated on the milliampere meter. As the negative grid voltage is further reduced (that is, made less negative), progressively more electrons are able to pass between the grid wires to the anode, and the anode current continues to increase. Note, however, that electrons are not attracted to the grid itself, as long as it is maintained at a negative bias voltage with respect to the cathode.

Zero Grid Bias. When the 'C' battery is removed and the grid voltage is zero (Fig. 8c), the positive voltage on the anode produces a substantial electric field at the cathode, and large numbers of electrons are attracted through the grid wires to the anode, resulting in a fairly large anode current. The action is similar to that of a diode, except that the grid still has some retarding effect on the electrons because of its shielding action, and hence the anode current is somewhat less than it would be with the grid removed entirely. Again, electrons are not attracted to the grid itself, since it is at zero voltage with respect to the cathode.

Positive Grid Bias. In Fig. 8d, the 'C' battery has been reversed in polarity, thus making the grid positive with respect to the cathode. The grid potential now aids the anode voltage and produces a very strong electrostatic field at the cathode, resulting in a large anode current through the valve. If the grid is made sufficiently positive with respect to the cathode, a point will be reached when the electrons are attracted to the anode as fast as they can be emitted from the cathode. As shown in the illustration, no space charge can accumulate under these conditions and the anode current reaches its saturation value. Still further increases in either the grid or the anode voltage cannot cause an increase in the anode current.

Fig. 8d also shows that some of the lines of force of the electric field actually terminate on the grid itself because of its positive voltage. As a result, some of the electrons are attracted to the positive grid and cause a grid current to flow between grid and cathode (through the 'C' battery). Under these conditions power is dissipated in the grid circuit. To avoid this power consumption and also the large saturation current, which eventually can damage the valve, valves are generally operated at negative grid voltages with respect to the cathode. We shall see later that the shape and relative linearity of the characteristic curves are another important reason for operating valves at negative grid voltages. (A big exception occurs in the so-called pulse circuits, to be discussed in Chapter 17.)

Amplification. As we have seen, the anode current in a triode is determined by the electrostatic field resulting from the combined action of the grid and anode voltages on the space charge near the cathode. Since the grid is placed closer to the space charge than the anode, the grid voltage has a greater effect on the anode current than has the anode voltage. Thus, if the grid voltage is made more negative by a fixed number of volts, the anode current is reduced far more than the same decrease in anode voltage would produce. When a load resistance is placed in series with the anode circuit, the voltage drop produced across this resistance is a function of the anode current and, hence, is controlled by the grid voltage. Thus a tiny change in the grid (or signal) voltage can cause a large change in the anode current and in the resulting voltage across the load resistance. In other words, the signal voltage appearing at the grid is amplified in the anode circuit of the valve. This amplification takes place without any grid current or power consumption in the grid circuit, as long as the grid voltage is negative with respect to the cathode. We shall see later (p. 18) how the amplification of a triode can be defined more precisely in quantitative terms.

TRIODE CHARACTERISTIC CURVES

The relationships between the anode voltage, grid voltage, and anode current in a triode, which we have explored in the last few paragraphs, can be (as in the case of the diode) conveniently summarized in the triode's characteristic performance curves. Actually, a three-dimensional surface model is required to represent the relation between all three quantities at the same time, but for convenience two-dimensional cuts through this surface will give the relation between any two quantities, while the third is held constant. Thus, we can plot a curve that shows the values of the anode current (I_b) as a function of varying anode voltages (E_b), when the grid voltage (E_c) is held at some fixed value. This is known as the anode-current/anode-voltage ($I_b - E_b$) characteristic. Or we can show graphically the effect on the anode current caused by varying the grid voltage (i.e. the bias), while holding the anode voltage at a constant value. This is called the anode-current/grid-voltage ($I_b - E_c$) characteristic of the triode. We can, of course, obtain a whole set of either of these characteristics by assuming different values for the constant quantity (either anode or grid voltage) and plotting a curve between the remaining quantities ($I_b - E_b$, or $I_b - E_c$) for each of these conditions. Such a set of characteristic curves is known as a family of static triode characteristics. The term static denotes that the characteristics are obtained when various steady voltages are applied to the electrodes. In Chapter 8 we shall become acquainted with the dynamic characteristics of the triode, i.e. the characteristics obtained under actual operating conditions, with a signal voltage applied and a load resistance inserted in the anode circuit to extract power from the valve.

We have neglected to mention the filament or heater voltage, which also substantially affects the anode current. Sets of curves can be plotted showing the effect of varying heater voltages (or emission) on the anode current for different anode voltages and grid voltages. However, the heater voltage is usually fixed at a value to provide sufficient emission from the cathode for normal operation and, hence, we need not concern ourselves with these characteristics.

TRIODE CHARACTERISTICS

A circuit for obtaining the static characteristics of a triode is illustrated in Fig. 9. Variable voltage dividers (potentiometers) are connected across the anode-voltage and grid-voltage supplies (denoted by E_{bb} and E_{cc}, respectively) to permit ascertaining the effect of varying either voltage on the valve's anode current, while the remaining voltage is held constant. The voltages at the electrodes, and the anode and grid currents resulting, are then measured by suitable voltmeters and ammeters, inserted in the grid and anode circuits. Note that resistors cannot be inserted in series with either grid or anode to obtain the static characteristics of the triode. There are no voltage drops, therefore, external to the valve and thus the grid-to-cathode voltage (E_c) equals the grid supply voltage (E_{cc}) and the anode-to-cathode voltage (E_b) equals the anode supply voltage (E_{bb}). This is not true for an actually operating (dynamic) amplifier circuit, as we shall see on p. 76.

A grid family of characteristic curves for a typical triode is shown in Fig. 10. (This is also known as the static transfer characteristic of the valve.) In

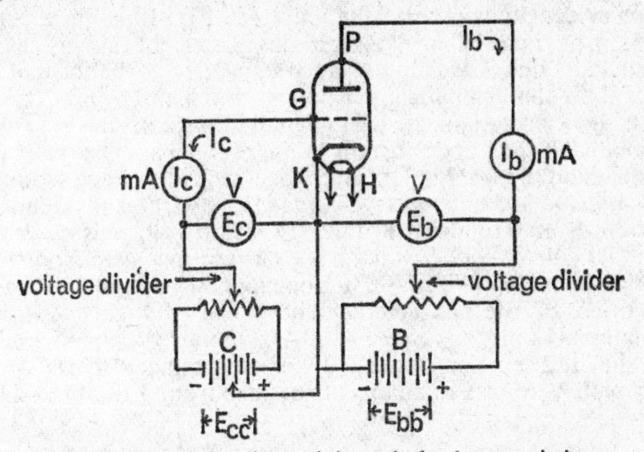

Fig. 9. Circuit for determining triode characteristics

these curves, the anode current (I_b) has been plotted as a function of the grid voltage (E_c) for various constant values of the anode voltage (E_b). The shape of the curves is typical of most receiving triodes. Note that each curve intersects the grid-voltage axis at a specific point that indicates the value of the negative grid voltage required to stop the anode current, at the fixed value of the applied anode voltage. This is the so-called cut-off bias. As the anode voltage is increased, it may be seen that the negative bias required to cut off the anode current also increases.

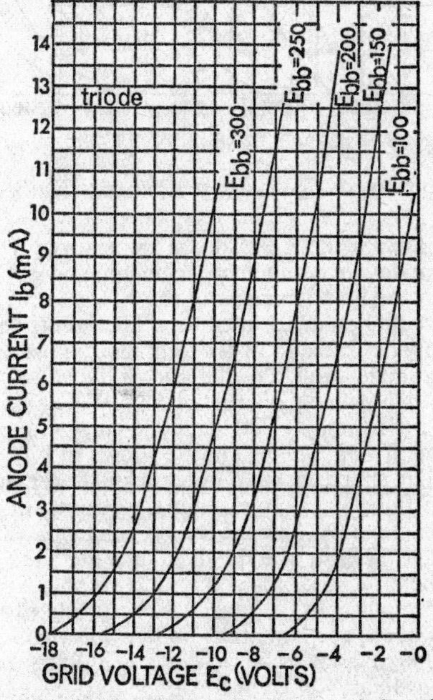

Fig. 10. Anode-current/grid-voltage static characteristics of a typical triode

It is also evident that each of the graphs in Fig. 10 is quite curved in the lower portion, near cut-off, while it is almost a straight line in the central and upper portions. Triodes are almost always operated in the straight-line, linear portions of their characteristic and rarely in the curved or non-linear portion. This is because increases in the grid (signal) voltage do not result in proportional increases of the anode current in the non-linear portion of the characteristic. For example, in Fig. 10, a change in grid voltage from −18 V to −15 V on the 300 V anode voltage curve (E_{bb}=300) results in an increase in anode current from 0 to about 1½ mA. If the grid voltage is reduced again by 3 V, from −15 V to −12 V, the anode current increases from 1½ to about 5¾ mA, or an increase of 4¼ mA. Thus equal (3 V) grid-voltage changes in the curved portion of the characteristic have caused highly unequal anode-current changes (1½ mA as against 4¼ mA). On the other hand, a grid-voltage change in the linear portion of the 250 V anode-voltage curve (E_{bb}=250) from −11 V to −10 V results in a current increase from 3 to about 4¾ mA, or a

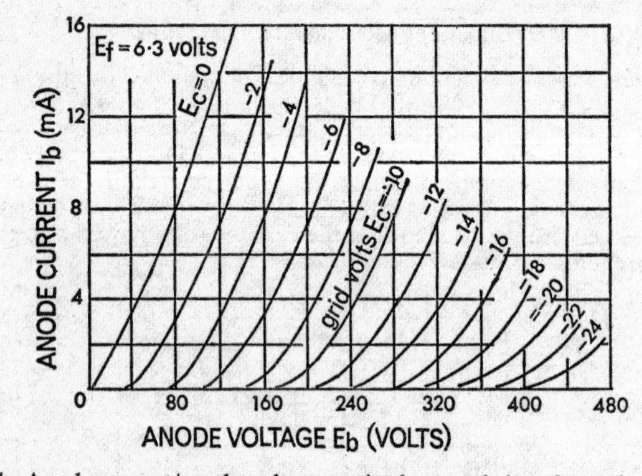

Fig. 11. Anode-current/anode-voltage static characteristics of a typical triode

change of 1¾ mA. Decreasing the grid voltage again by 1 V, from −10 V to −9 V, causes an anode-current increase from 4¾ to 6½ mA, or again a change of 1¾ mA. Thus, in the linear portion of the characteristic, equal changes in grid voltage cause equal changes in the anode current. As we shall see later (p. 77), this is highly desirable in order to obtain undistorted, linear amplification of a signal voltage applied to the grid of the valve.

Anode-current/anode-voltage characteristics. We can use the circuit of Fig. 9 to determine the family of static anode-current/anode-voltage characteristics for the same triode. The results are shown in Fig. 11. Here again, the curves have a similar shape, being curved in the lower portion near anode-current cut-off and fairly linear in the upper portions. Each curve shows the effect on the anode current when the anode voltage is varied over a certain range, while holding the grid voltage fixed at a definite value. A new curve results each time the grid voltage is changed to another fixed value, and when this is done over a representative range of constant grid voltages, the entire family of curves is obtained. Note that no new information can be obtained from the anode family of static characteristics; it offers the same data as the grid family, but in a slightly different form.

Fig. 12 illustrates another anode family of static characteristics for a twin triode (two triodes in one envelope). The effect of positive grid voltages on the $I_b - E_b$ curves may be seen clearly. Even for low positive values of the grid voltage, the anode current is seen to increase very rapidly with small increases in anode voltage. In addition, the grid circuit of the valve draws a grid current (I_c) for positive grid voltages. This grid current evidently decreases as the

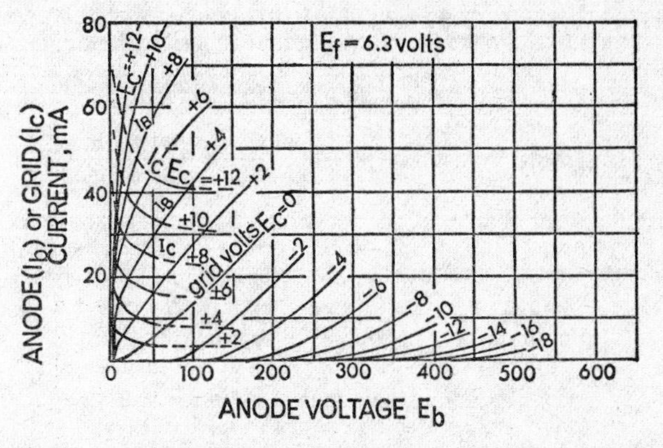

Fig. 12. Anode family of static characteristics of a twin triode

anode voltage is made large relative to the grid voltage, and the electrons are rapidly attracted to the anode.

VALVE CONSTANTS

The families of characteristics, which show the characteristic performance of each type of valve, are not a result of accident. Rather they represent the outcome of purposeful design to make each valve behave in a certain manner. Valve design takes into account the geometric configuration of the electrodes and the spacing between them, the power dissipation capabilities and many other factors. It is these factors that determine the maximum voltages applied to the electrodes, the maximum permissible anode current through the valve, anode-current cut-off, the amplification, etc. The design factors are summarized by a series of numbers, called the valve constants. All major valve manufacturers publish manuals in which these constants are listed along with other data and the characteristic curves for each type of valve. The three most important valve constants are the amplification factor, the a.c. anode resistance, and the transconductance.

AMPLIFICATION FACTOR

The amplification factor of a triode is a measure of the relative effectiveness of the control grid in overcoming the electrostatic field produced by the anode. Numerically, it expresses how much more effective the grid voltage is in producing an electrostatic field at the cathode surface (or space charge) than the field set up by the anode voltage. The amplification factor depends to a large extent on the spacing between the grid and anode, and thus can be predicted for each type of valve. Still, it is more practicable and accurate to determine the amplification factor from the actual performance of the valve, as expressed in the characteristic curves. The characteristic curves reflect the effectiveness of the control-grid voltage in determining the anode current, compared with that of the anode voltage—which is another way of defining the amplification factor. All that is necessary, therefore, to determine the amplification factor is to change the anode voltage by a certain amount, record the change in anode current, and then change the grid voltage (in the opposite direction) by an amount just sufficient to restore the previous anode-current value. By comparing the anode-voltage change to the grid-voltage change for the same change in anode current (essentially holding it constant), we can determine their relative effectiveness, which is the amplification factor. Accordingly, we simply define the amplification factor by the formula:

$$\text{amplification factor} = \frac{\text{small change in anode voltage}}{\text{small change in grid voltage}}$$

to produce the same change in anode current. The changes have to be small since the non-linear characteristics lead to errors for large changes.

We can abbreviate this formula by the following symbols:

$$\text{amplification factor } \mu = \frac{\Delta E_b}{\Delta E_c} \text{ (with } I_b \text{ constant)}$$

where ΔE_b = a small change in anode voltage
and ΔE_c = a small change in grid voltage

Thus, if the amplification factor, μ, of a valve is 20, a change in grid (or signal) voltage will be 20 times as effective in changing the anode current as the same change in the anode voltage. To illustrate further, suppose an anode-current change of 1 mA is produced by an anode-voltage change of 50 V and

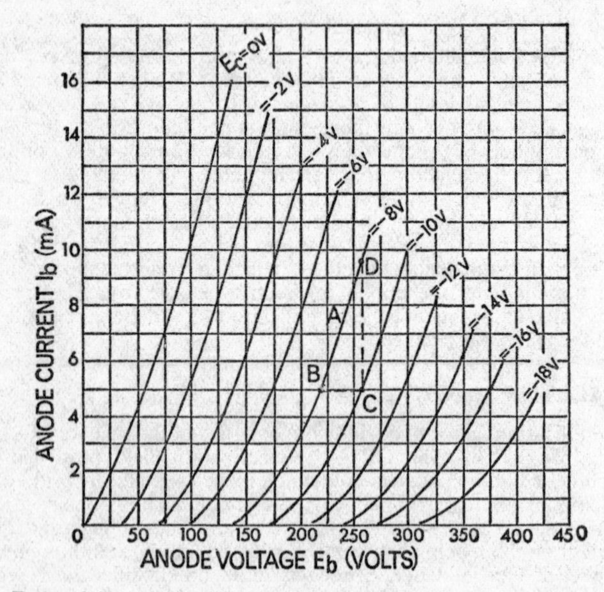

Fig. 13. Determining the amplification factor and anode resistance from the anode family of characteristics

that a grid-voltage change of only 2 V results in the same 1 mA anode-current change. The amplification factor then is

$$\mu=\frac{\Delta E_b}{\Delta E_c}=\frac{50}{2}=25$$

In practice we need not perform any measurements to ascertain the amplification factor, but we simply make a paper experiment on the valve characteristics. As an example, let us determine μ from the family of static characteristics shown in Fig. 11. Fig. 13 shows the same family, somewhat enlarged for greater accuracy, and provided with a simple construction for determining μ. Suppose we want to find the amplification factor near point *A* in the illustration (μ changes slightly depending on the chosen point of operation). The grid voltage at the operating point *A* is − 8 V, the anode current is 7 mA and the anode voltage is about 235 V. First follow the −8 V curve down to a convenient point *B*. Draw a horizontal line along a constant anode-current value of 5 mA to the next grid-voltage curve (− 10 V) until it intersects at point *C*. Now draw a vertical line upward for a constant anode voltage of 255 V, until it intersects the −8 V grid-voltage curve again at point *D*.

This construction has the following significance: moving from point *B* to point *D* along a constant grid voltage of −8 V, the anode current has increased from 5 mA to about 9·6 mA, or by 4·6 mA. To produce this increase, the anode voltage has to be increased from about 215 V (at *B*) to about 255 V (at *D*), or an increase of 40 V. Thus an anode-voltage change of 40 V, is necessary to produce an anode-current increase of 4·6 mA. We now reduce the anode current to its previous value of 5 mA, by moving vertically down from point *D* to point *C*, along a constant anode voltage of 255 V. To reduce the anode current to its former value, the grid voltage must be increased from − 8 V (at *D*) to −10 V (at *C*), or an increase of 2 V. A change of 2 V in the negative grid voltage thus has produced the same anode-current change as a 40 V change in the anode voltage. The amplification factor is therefore

$$\mu=\frac{\Delta E_b}{\Delta E_c}=\frac{40}{2}=20$$

This, incidentally, is the value given by the manufacturer for this particular triode. The amplification factor could have been equally well determined from the grid family of characteristics (Fig. 10), but the anode family is more frequently available in manufacturer's valve-data manuals.

ANODE RESISTANCE

The anode resistance describes the internal resistance or opposition of the valve to the flow of electrons from cathode to anode. This resistance is not the same for the flow of direct current as for alternating current. Accordingly, the d.c. anode resistance is defined as the resistance of the path between cathode and anode to the flow of direct current, i.e. when steady values of voltage are applied to the electrodes. This d.c. resistance can be found from the anode family of characteristics by a simple application of Ohm's law. Thus,

$$\text{d.c. anode resistance, } R_p=\frac{E_b}{I_b}\text{ ohms}$$

For example, in Fig. 13, the anode current at point *A* is 7 mA (0·007 amp) and the anode voltage is 235 V. Hence, the d.c. anode resistance is

$$R_p=\frac{235\text{ V}}{0{\cdot}007\text{ A}}=33{,}600\text{ ohms}$$

More significant for the valve's performance as an amplifier is the a.c. anode resistance, which is a measure of the opposition encountered by electrons when varying or alternating voltages are applied to the electrodes. The a.c. anode resistance is defined as the ratio of a small change in anode voltage to the change in anode current produced thereby, when the grid voltage is kept at a constant value. Expressing this definition in equation form, as before, the

$$\text{a.c. anode resistance, } r_p = \frac{\Delta E_b}{\Delta I_b} \text{ (with } E_c \text{ constant)}$$

The anode resistance is usually determined from the linear part of the anode characteristics. Returning to our example (Fig. 13), let us determine the anode resistance of the triode near operating point *A*. We have previously found that a change in anode voltage from 215 V (point *B*) to 255 V (point *D*) on the −8 V grid-voltage curve produced a change in anode current from 5 mA (at *B*) to 9·6 mA (at *D*). Thus, the a.c. anode resistance near point *A* is

$$r_p = \frac{\Delta E_b}{\Delta I_b} = \frac{(255-215) \text{ V}}{(9{\cdot}6-5) \text{ mA}} = \frac{40 \text{ V}}{0{\cdot}0046 \text{ A}} = 8{,}700 \text{ ohms}$$

The valve manual lists a value of 7,700 ohms for the a.c. anode resistance, but the discrepancy is not too serious and must be expected, since the a.c. anode resistance is a function of the operating point along any particular grid-voltage characteristic. We might also point out that the triangle *BCD* in Fig. 13 could have been made smaller (it is not necessary to draw a line to the next grid-voltage curve, as in the case of μ), with probably greater accuracy resulting.

TRANSCONDUCTANCE

A third constant used in describing the properties of valves is the control-grid/anode transconductance (sometimes called mutual conductance) designated by the symbol g_m. The transconductance is the most important of the three valve constants, since it reveals the effectiveness of the control grid in securing changes in the anode current and, hence, in the signal output of the valve. It is defined as the ratio of a small change in anode current to the small change in control-grid voltage producing it, when the anode voltage is held constant. Expressing this in equation form,

$$\text{transconductance, } g_m = \frac{\Delta I_b}{\Delta E_c} \text{ (with } E_b \text{ constant)}$$

Since g_m is the ratio of a current to a voltage, it has the form of a conductance and is expressed in units of *mho* (also known as the *reciprocal ohm* and the *siemens*). This unit is, however, too large for valve usage and hence one-millionth part of a mho, the micromho (μmho) is generally used. Thus, a valve in which a 1 V change in grid voltage produces a 2 mA (0·002 amp) change in anode current has a transconductance of

$$\frac{0{\cdot}002}{1} \times 10^6 = 2{,}000\ \mu\text{mho}$$

Transconductance is easily found from the anode family of characteristics. Again returning to our original triode, we found from Fig. 13 that a change in grid voltage from −10 V to −8 V (from point *C* to point *D*) produced a change in anode current from 5 mA to 9·6 mA. Hence the transconductance

$$g_m = \frac{\Delta I_b}{\Delta E_c} = \frac{(9{\cdot}6-5)\ \text{mA}}{(10-8)\ \text{V}} = \frac{0{\cdot}0046\ \text{A}}{2\ \text{V}} = 2{,}300\ \mu\text{mho}$$

The valve-data manual lists the transconductance near this operating point as 2,600 μmho, which is in fair agreement with the above.

Relation between μ, r_p, *and* g_m. The three valve constants are interrelated, in accordance with their definition, as is easily shown by dividing the relation defining the amplification factor by that defining the anode resistance. Thus,

$$\frac{\mu}{r_p} = \frac{\Delta E_b / \Delta E_c}{\Delta E_b / \Delta I_b} = \frac{\Delta I_b}{\Delta E_c} = g_m$$

Hence $g_m = \dfrac{\mu}{r_p}$ and equivalently $\mu = g_m \times r_p$

$$\text{and } r_p = \frac{\mu}{g_m}$$

If two of the valve constants are known, therefore, the third may be obtained from the relations above. The transconductance is frequently used as a figure of merit for comparing different valves in the same general classification. Transconductance expresses the ratio of the amplification factor to the anode resistance of the valve. A high amplification factor is desired to obtain

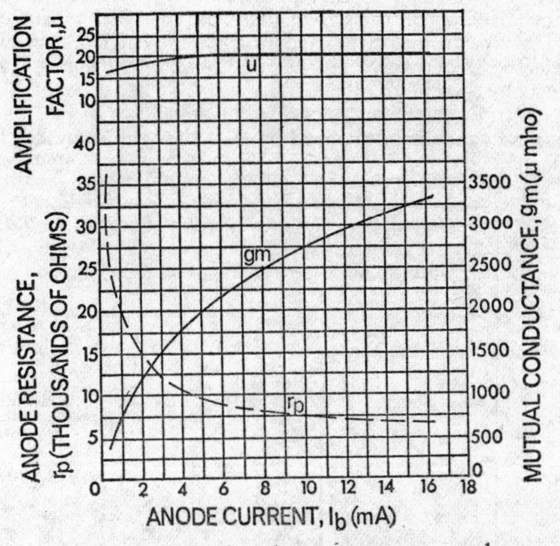

Fig. 14. Variation in valve constants μ, g_m, and r_p

large output voltages from small input signals. A low anode resistance, on the other hand, permits the flow of a large anode current, resulting in large output power. Transconductance, thus, is a measure of the voltage and power amplification possible in any specific valve. But it is not easy for the designer to raise the value of the transconductance, since the anode resistance increases as the amplification factor is made larger. Besides serving as a figure of merit, the transconductance is the factor most frequently measured when comparing the performance of the valve with its original ratings.

Although we have stated that the valve constants depend to a large degree on the spacing and geometrical configuration of the original design, we have also seen from the examples that these so-called 'constants' vary to some degree depending on the choice of operating voltages on the electrodes. To show this variation with operating conditions, we have plotted the 'constants' for a constant anode voltage of 250 V in Fig. 14. (To keep the anode voltage constant, we had to decrease the negative grid voltage, as the anode current increases.) Each of the constants has its own vertical scale, but all have the same horizontal scale, i.e. the anode current in mA. Note that the amplification factor, μ, is essentially constant, while the anode resistance decreases with increasing anode currents and the transconductance, g_m, increases correspondingly—as expected.

INTERELECTRODE CAPACITANCES

Among the important design factors are the so-called interelectrode capacitances. You will remember that various electrostatic fields exist between the charged electrodes of a triode, such as the field between anode and cathode, between anode and grid, and between grid and cathode. From elementary electricity you will also recall that an electrostatic field between any two charged metal plates is the equivalent of an electric capacitor. Capacitance exists between any two pieces of metal separated by a dielectric. The amount of capacitance depends on the size of the metal pieces, the distance between them, and the type of dielectric. Accordingly, the capacitances existing between the electrodes of a valve depends on their size and spacing, and on the dielectric (usually a vacuum).

The interelectrode capacitances of a triode are shown in Fig. 15. The grid-to-cathode capacitance is symbolized by C_{gk}, the grid-to-anode capacitance

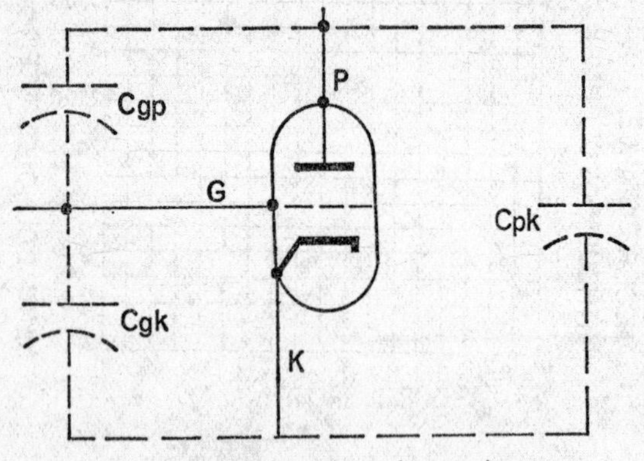

Fig. 15. Triode interelectrode capacitances

by C_{gp}, and the anode-to-cathode capacitance by C_{pk}. Although these capacitances are generally small (from 2 to 10 micromicrofarads), their reactance at the higher radio frequencies becomes low and leads to undesirable coupling effects. The grid-to-anode capacitance, C_{gp}, especially has the property of feeding back energy from the anode (output) circuit to the grid (input) circuit, which may lead to instability and oscillations. This property is generally undesirable, although it is sometimes utilized to produce oscillations in suitable circuits, as we shall see in Chapter 11.

A reduction of the interelectrode capacitances can be achieved by additional shielding electrodes, and leads to the design of multielectrode valves, which will be described in the next chapter. However, at the ultra-high frequencies (u.h.f.) used in radar and similar circuitry, interelectrode capacitances become so objectionable that ordinary vacuum valves can no longer be used. Special u.h.f. valves of very tiny physical dimensions and with closely spaced electrodes have been designed and are discussed in Chapter 6.

SUMMARY

A triode is a three-element valve, consisting of a cathode (emitter), an anode (collector), and a control grid (spiral of fine wire), placed between cathode and anode.

The voltage placed on the control grid usually is made up of a fixed, negative, direct voltage, called the bias, and a varying or alternating voltage, called the signal.

The anode current in a triode is determined by the combined effect of the electric fields set up by the anode and control-grid voltages. The control grid, being nearer to the cathode than the anode, has a greater effect on the anode current than has the anode voltage. This permits amplification of an input signal.

When a large negative bias voltage is placed on the control grid, the grid neutralizes the electrostatic field of the anode, and no anode current flows. The smallest negative grid voltage capable of cutting off the anode current is called the cut-off bias. Larger bias values have no effect.

As the negative grid bias is reduced, the grid is no longer capable of neutralizing the electric field of the plate completely, and the anode current increases in proportion to the reduction of the negative grid bias. When the bias is made positive, the field of the grid aids that of the anode and a point will be reached where the anode current reaches its saturation value, equal to the total supply of electrons emitted. Grid current will also flow for positive grid voltages. This results in power consumption in the grid circuit and is undesirable.

The functional relations between anode voltage, grid voltage, and anode current are called the triode characteristic curves.

When a number of these curves are plotted on graph paper, expressing the relation between any two variables while a third is held constant (at a different value for each curve), a family of triode characteristic curves results.

Static characteristics are obtained when steady (d.c.) voltages are applied to the electrodes; dynamic characteristics are obtained under actual operating conditions, i.e. with a varying signal voltage applied to the grid and a load resistance inserted into the anode circuit.

In the linear portion of each characteristic equal changes in grid or anode voltage cause equal changes in the anode current. This is necessary for undistorted amplification.

The most important design factors of a valve, the valve constants, are the amplification factor, the a.c. anode resistance, the mutual conductance or transconductance, and the interelectrode capacitances.

The amplification factor is a measure of the relative effectiveness of the control grid in overcoming the electrostatic field of the anode. It is defined as the ratio of a small change in anode voltage to the small change in grid voltage required to produce the same change in the anode current.

The a.c. or variational anode resistance is the internal opposition of the valve to the flow of anode current, when alternating voltages are applied to the electrodes. It is defined as the ratio of a small change in anode voltage to the corresponding change in anode current produced thereby, when the grid voltage is kept constant.

The transconductance or mutual conductance, is the most important valve constant, since it reveals the effectiveness of the control grid in securing changes in the anode current and, hence, signal output of the valve. It is defined as the ratio of a small change in anode current to the small change in control-grid voltage producing it, when the anode voltage is held constant.

The transconductance is also equal to the ratio of the amplification factor to the a.c. anode resistance of the valve. It sometimes serves as a figure of merit for comparing valves.

Interelectrode capacitances exist between cathode, anode, and grid, because of the electric field present between these charged electrodes. They are designated as the grid-to-cathode capacitance, C_{gk}, the grid-to-anode capacitance, C_{gp}, and the anode-to-cathode capacitance, C_{pk}.

The grid-to-anode capacitance feeds back energy from the anode (output) circuit to the grid (input) circuit. This may lead to instability and oscillations at radio frequencies.

CHAPTER FIVE

MULTIELECTRODE VALVES

Multielectrode valves are valves having more than one grid, which makes it possible to achieve many desirable characteristics not possible with the triode. Among the most common multielectrode valves are the tetrodes, which have four electrodes (two grids), and the pentodes, which have five electrodes (three grids).

TETRODES

The development of the tetrode is a direct outcome of the undesirable grid-to-anode capacitance in triodes, which we have previously discussed. This capacitance leads to coupling effects and instability in radio-frequency amplifiers that cannot be eliminated, except by costly 'neutralization circuits'. The most effective answer to the feedback problem caused by the grid-to-anode capacitance is to insert an additional shielding electrode, called the screen grid, between the control grid and anode of a triode. This additional grid almost completely encloses the anode and thus acts as an effective electrostatic shield between it and the control grid. The grid-to-anode capacitance, which usually runs from 2 to 5 micromicrofarads in a triode, is thereby reduced to values as low as 0·01 micromicrofarad in a tetrode. This effectively cancels out the feedback action and resulting instability in radio-frequency amplifiers. The tetrode produces some undesirable distortion of the valve characteristics, however, as we shall see shortly. Its successor, the pentode, overcomes this disadvantage while retaining the tetrode's desirable characteristics and has for this reason almost replaced the tetrode. Nevertheless, it is instructive to study the basic action of the tetrode, since the pentode's action is very similar.

The physical arrangement of the four electrodes in a tetrode is illustrated in Fig. 16, along with the tetrode schematic symbol. The cathode, anode, and control grid are substantially the same as in a triode. The anode is separated from the control grid either by a single wire screen on one side, or it may be

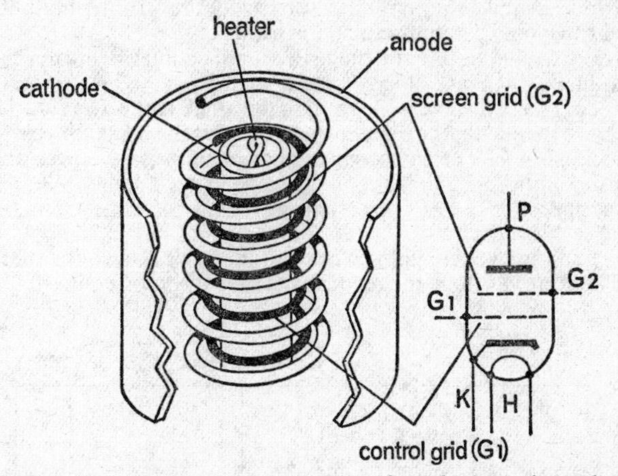

Fig. 16. Simplified construction of tetrode and schematic symbol

completely enclosed by the screen grid. The latter is similar to the control grid, except for a somewhat coarser mesh.

TETRODE CHARACTERISTICS

The tetrode is operated similarly to the triode, with the cathode near earth potential, the control grid at a small negative (bias) voltage, and the anode at fairly high (several hundred volts) positive voltage. The screen grid is also placed at a high positive potential with respect to the cathode, but somewhat lower than the anode voltage. The positive voltage on the screen grid helps to accelerate the electrons on their way from the cathode to the anode and thus aids the electrostatic field of the anode. Some of the electrons strike the screen grid and produce a screen current, which is not useful, but most of the electrons pass through the open mesh of the screen towards the anode to produce the anode current.

Because of the shielding effect of the screen grid, the electrostatic field of the anode has little effect on the electron space charge near the cathode. As a consequence, variations in anode voltage have little effect on the anode current and a much greater change in the plate voltage is required to produce the same change in anode current than would be necessary in a triode. Since the a.c. anode resistance is defined as the ratio of a change in anode voltage to the corresponding change produced in the anode current, the anode resistance of a tetrode is far greater than that of a triode. (The tetrode anode resistance is in the order of 0·5 to 1 megohm.)

The control grid of a tetrode has about the same effect in governing the flow of anode current as in a triode, since it is not shielded from the space charge by the screen grid. Thus, a small change in control-grid voltage will produce a large change in the anode current, just as in a triode. On the other hand, the tetrode requires a very large change in anode voltage to produce a small change in anode current, as we have just seen. Since the amplification factor (μ) is defined as the ratio of a small change in anode voltage to the small change in grid voltage that will produce the same change in anode current, the amplification factor of a tetrode, clearly, will be far higher than that of a comparable triode. Amplification factors for tetrodes run from about 400 to 800, compared to values of about 5 to 50 for triodes.

The price we have to pay for the high amplification factor of a tetrode is a correspondingly high anode resistance, as we have seen. Since the transconductance (g_m) of a valve is the ratio of the amplification factor to the anode resistance, we would expect this ratio to be about the same for tetrodes as for triodes—and so it is. The transconductance of tetrodes averages about 1,000 to 1,500 micromho, which is about the same as for triodes. There are, however, some special types of power tetrodes, with g_m values as high as 4,500 micromho.

A family of anode characteristic curves for a typical tetrode is illustrated in Fig. 17. These curves show the relation between the anode current (I_b) and the

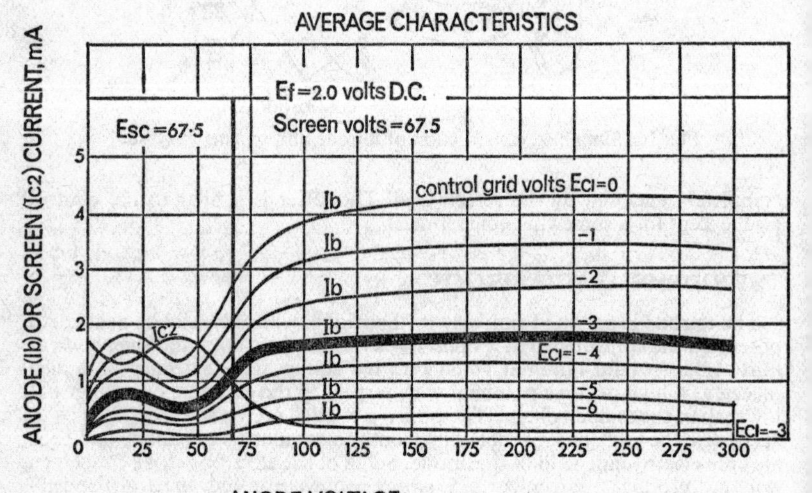

Fig. 17. Anode family of characteristic curves for a tetrode

anode voltage (E_b) for various values of the control-grid voltage (E_{c1}). The screen voltage (E_{c2}) is fixed at 67·5 V. One curve, showing the variation in the screen-grid current (I_{c2}) with anode voltage, for a control-grid voltage of -3 V, is also shown.

The distorted shape of the characteristics in the region of low anode voltages may be explained by considering a single $I_b - E_b$ curve, such as the one for a control grid voltage of -3 V (Fig. 17). Since a positive potential of $+67{\cdot}5$ V is applied to the screen, an electrostatic field exists between screen and cathode, even when the anode voltage is zero. Consequently, the electrons are attracted towards the screen (through the slightly negative control grid) and a small screen current (I_{c2}) of about 2 mA flows, when the anode voltage is zero. As the anode voltage is increased, some of the electrons passing through the screen are diverted to the anode, with a consequent increase in anode current and a decrease in the screen current. The anode current continues to increase and the screen current decreases until the anode voltage is about +12·5 V. The total space current (sum of anode plus screen currents) also is seen to increase slightly.

As the anode voltage is further increased, however, the anode current suddenly begins to fall off, while the screen current increases correspondingly. The reason for this strange behaviour is the phenomenon of secondary emission, which we mentioned in Chapter 2 (p. 3). When the anode voltage is raised above a few volts, the electrons are speeded up sufficiently to dislodge loosely held electrons within the anode material and project them as secondary electrons into the region between anode and screen grid. These additional, secondary electrons are immediately collected by the screen grid because its voltage is more positive than that of the anode. Secondary emission also takes place in the triode, but since the anode in a triode is the only positive electrode, the secondary electrons are recollected by the anode. In the tetrode, however, these secondary electrons are diverted from the anode to the screen and are effectively subtracted from the anode current. (The flow of secondary electrons to the screen is in the opposite direction to the normal anode-current flow.) The net anode current then is the number of primary electrons from the cathode received by the anode minus the number of secondary electrons lost to the screen grid. When each primary electron striking the anode produces on the average more than one secondary electron, the screen will receive more electrons from the anode than the anode from the cathode, and the net anode current will then be negative. This is shown by the negative slope of the $I_b - E_b$ characteristic between anode-voltage values of 12·5 V and 50 V. The screen current (I_{c2}) in the same region is seen to increase in accordance with the anode-current decrease. (The total space current remains about the same in this region, but its division between anode and screen continually changes.)

As the anode voltage is still further increased and begins to approach the value of the screen voltage, the force of attraction exerted by the anode on the secondary electrons becomes greater than that exerted by the screen grid, and the secondary electrons return to the anode. Consequently, the anode current begins to rise sharply and the screen current decreases correspondingly. When the anode voltage is made substantially higher than the screen voltage, the anode collects practically all the electrons from the cathode and, in addition, a small number of secondary electrons from the screen grid. The screen grid collects only a few electrons whose paths happen to be intercepted by it. The anode current now stabilizes at a high value, almost equal to the total space current, while the screen current drops to a low, constant value. Because of the shielding effect of the screen, further increases in anode voltage have little effect and the anode current becomes almost independent of the anode voltage, as is evident from the flattening out of the curves.

Negative Resistance. The negative slope of the anode-current/anode-voltage characteristic in the low-voltage region signifies that the anode current decreases for increasing anode voltages, contrary to Ohm's law. This negative current characteristic is termed negative resistance and the valve in this region may be interpreted as a source of power rather than a consumer of it. As we shall see in Chapter 10, this behaviour leads to instability and oscillations of the anode current under certain conditions. For this reason, the anode voltage in tetrodes must be made sufficiently high so that it never drops to the negative resistance region for any signal voltage on the grid. This disadvantage is overcome in pentodes, to be discussed next. Incidentally, the negative resistance characteristic of tetrodes is utilized in special oscillator circuits.

PENTODES

The insertion of an additional grid, called the suppressor grid, between the anode and screen grid of a tetrode overcomes the undesirable effects of secondary emission and the resulting negative-resistance characteristic at low anode voltages. By adding this electrode the valve becomes a five-electrode unit, or pentode. While eliminating the distortion of the anode family of characteristics, pentodes retain all the advantages of tetrodes, such as low grid-to-anode capacitance, high amplification factor, and high power output.

Fig. 18 illustrates a typical pentode valve, along with its schematic symbol. The pentode contains an emitter (cathode), three grids and an anode. The grid closest to the cathode, *G*1, is the control grid, next is the screen grid, *G*2, and the third is the suppressor grid, *G*3, located between screen grid and anode.

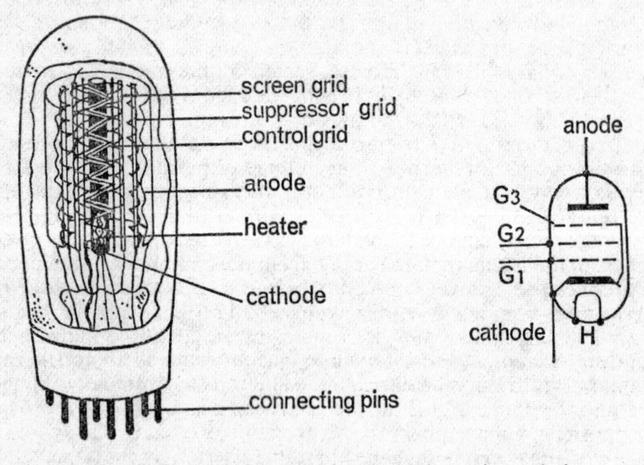

Fig. 18. Pentode construction and schematic symbol

Pentodes resemble tetrodes in external appearance. The connexions to all electrodes are generally made from pins in the base of the valve, but in some of the older types the control grid is connected via a cap on top of the envelope.

ACTION OF SUPPRESSOR GRID

The suppressor grid is usually connected directly to the cathode and is thus at a substantial negative voltage with respect to the anode. As you might suspect, the secondary electrons emitted by the anode are repelled by the negative suppressor grid and are driven back to the anode. Thus, while not preventing secondary emission by the anode, the suppressor grid does eliminate the effects of secondary emission. As a result, the anode current rises smoothly with increasing voltage, from zero up to the maximum value for each control-grid voltage.

The presence of the suppressor grid further increases the shielding action between anode and control grid and thus further reduces the grid-to-anode capacitance. For the same reason, the anode current is even more independent of the anode voltage than in the tetrode, as demonstrated by the further 'flattening' (low slope) of the anode characteristics (Fig. 19). With the anode

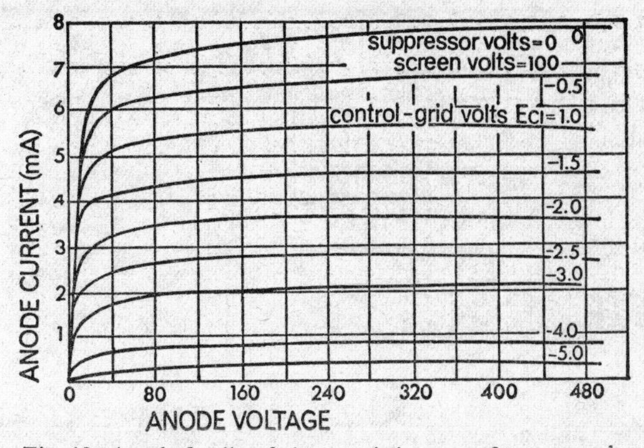

Fig. 19. Anode family of characteristic curves for a pentode

current little affected by the anode voltage because of the high screening, both the amplification factor and the anode resistances of the pentode become extremely high. The amplification factor in pentodes may run as high as 1,000 to 1,500, and the anode resistance is usually in the order of about one megohm (one million ohms). As we shall see later (in the chapter on audio amplifiers) the high anode resistance makes it impossible to use more than about one-tenth of the pentode amplification factor. The undistorted anode characteristics of pentodes, however, permit large variations in the input signal voltage at the grid, resulting in large available power output. As is the case for the tetrode, the combination of a high amplification factor and high anode resistance results in only average values of the transconductance.

BEAM-POWER TETRODE

A useful hybrid between a tetrode and a pentode is the so-called beam-power tetrode. It has two grids, like a tetrode, but yet is capable of suppressing the effects of secondary emission and thus operate as a pentode. The effect of the pentode suppressor grid is obtained by special electrodes and by arranging the valve elements in such a way as to produce a negative 'space charge' near the anode. The action of this space charge repels the secondary electrons back to the anode, just as the suppressor grid does in a pentode.

The action of the beam-power tetrode is attained through its structural features, which are illustrated in Fig. 20. As illustrated, the valve produces a dense beam of electrons between screen grid and anode. This beam is formed by a combination of two features. First, the control grid and screen grid are of the same pitch, and the grid wires are so aligned that the screen grid lies in the 'shadow' of the control grid. The 'shadowed' screen grid intercepts few electrons, resulting in low screen current and high anode-current density. Secondly, special beam-confining electrodes, electrically connected to the cathode, further assist in producing dense electron beams in the region between the screen and the anode. This bunching up of electrons leads to a voltage minimum and a resulting space-charge effect between the anode and screen, which effectively repels the secondary electrons from the anode. Moreover, the distance between screen and anode is made larger than in a tetrode (or pentode) to make sure that there will be a large supply of electrons in the region, which further assists in increasing the space-charge effect.

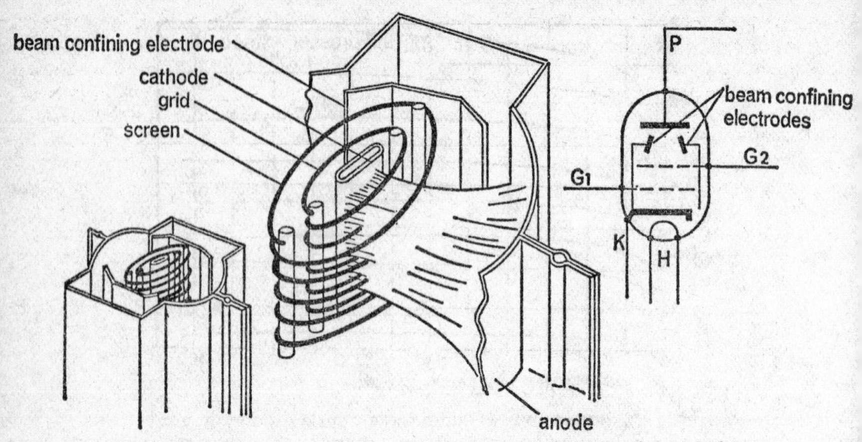

Fig. 20. Structure of beam-power tetrode and schematic symbol

As a matter of fact, the suppressor action of the space charge in a beam-power tetrode is superior to that of a pentode, resulting in anode-current characteristics that are even less distorted and have a sharper 'knee' (at low anode voltages). Because of this low distortion, low screen current, and large effective cathode and anode areas, the beam-power tetrode has the advantages of exceptionally high power output for small input signals and a high efficiency. It is often preferred, therefore, in the output stages of amplifiers.

VARIABLE-MU (REMOTE CUT-OFF) VALVES

The amplification factor (μ) depends to a large extent on the spacing of the control-grid wires. In conventional valves the turns are uniformly spaced throughout the length of the control grid, resulting in a constant amplification factor for most of the anode-current and grid-voltage ranges. Furthermore, when making the grid voltage more negative, all parts of the control-grid structure begin to cut off the anode current at the same time, producing a sharp cut-off characteristic. (See Fig. 21.)

It is sometimes desirable to produce a more gradual, or remote cut-off characteristic to accommodate large-amplitude signals without the distortion accompanying anode-current cut-off. This is accomplished by spacing the turns of the control-grid wire in tetrodes or pentodes non-uniformly, winding

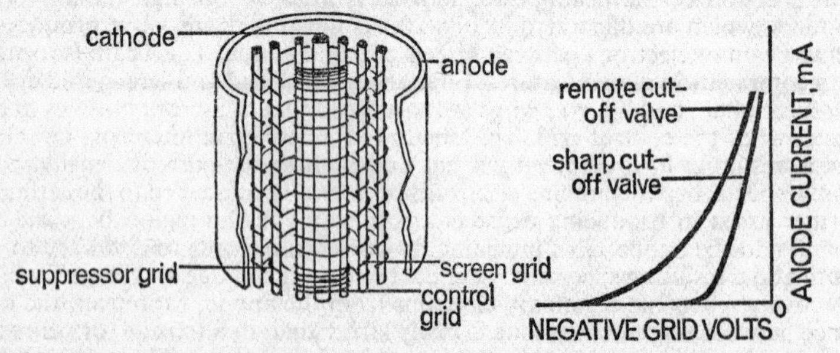

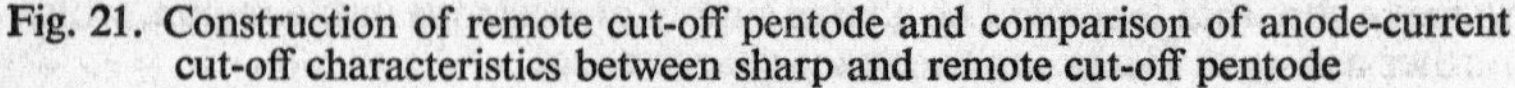

Fig. 21. Construction of remote cut-off pentode and comparison of anode-current cut-off characteristics between sharp and remote cut-off pentode

them closer near the ends of the structure than at the centre (Fig. 21). This effectively gives the valve a variable mu (μ), low at the centre and high at the ends.

For low values of the grid bias and high values of anode current, the valve operates normally. As the control grid is made more negative, however, the closely spaced (high-mu) regions at the ends of the grid reach anode-current cut-off first, while the centre (low-mu) portions of the grid still permit anode current to flow. As operation is progressively forced into the more central (low-mu) portions of the grid, the amplification factor decreases considerably and the negative control-grid voltage must be made very high to achieve complete anode-current cut-off. Because of these features, such valves are called either variable-mu or remote cut-off valves. Since the amplification depends on the value of the control-grid bias, they are very useful in automatic volume-control (a.v.c.) circuits, where it is required to vary the amplification automatically by changing the grid bias obtained from the signal amplitude.

SUMMARY

Multielectrode-valves have more than one grid. The tetrode has four electrodes (two grids) and the pentode has five electrodes (three grids).

The tetrode eliminates the feedback of energy, caused by the grid-to-anode capacitance, by the insertion of a screen grid between control grid and anode of the valve.

The anode-current/anode-voltage characteristics of the tetrode are distorted in the region of low anode voltages, resulting in a negative resistance characteristic and instability in this region.

Negative resistance and distortion at low anode voltages is caused by secondary emission from the tetrode anode. The secondary electrons are attracted to the highly positive screen and subtract from the anode current. The net anode current will be negative, when the number of secondary electrons flowing to the screen exceeds the number of primary electrons arriving at the anode.

At anode voltages higher than the screen voltage, the anode collects all primary and secondary electrons and, because of the shielding action of the screen, the anode current becomes practically independent of the anode voltage.

Since variations in anode voltage have little effect on the anode current (at high anode voltages), both the a.c. anode resistance and the amplification factor of the tetrode are *far greater* than those of the triode. The transconductance, which is the ratio of the amplification factor to the anode resistance, is about the same for tetrodes as for triodes.

In the pentode, a suppressor grid is inserted between the anode and screen grid. When connected to the cathode, the suppressor grid is highly negative with respect to the anode and is able to overcome the effects of secondary emission and thus the distortion of the anode characteristics.

The suppressor grid also increases the shielding action between anode and control grid. As a result, both the anode resistance and the amplification factor of pentodes are extremely high. The transconductance is about the same as that of triodes or tetrodes.

The pentode permits higher power output and efficiency than either the triode or the tetrode.

A beam-power tetrode is a hybrid between a tetrode and a pentode. It has two grids, like a tetrode, but is also equipped with special beam-confining electrodes capable of overcoming, as does a pentode, the effects of secondary emission.

In the beam-power tetrode dense electron beams produce an electron space charge between anode and screen that repels secondary electrons and returns them to the anode.

A variable-mu or remote cut-off valve is a pentode that has a gradual anode-current cut-off when the negative bias is increased. This is achieved by non-uniform spacing of the control-grid wires, resulting in an amplification factor that is low at the centre and high at the ends of the grid structure. Thus, the amplification of the valve may be changed by varying the grid bias.

CHAPTER SIX

SPECIAL VALVES

Standard diodes, triodes, tetrodes, and pentodes are the heart of practically all electronic circuits, with which we shall become intimately acquainted later on. Now let us turn to a group of specialized valves which are capable of doing various jobs that ordinary valves cannot perform and which are especially useful in industrial, control, and ultra-high-frequency (u.h.f.) circuits. Although numerous special-purpose valves have sprung up, we shall consider here only four of the most important generic types. These are gas-filled valves, phototubes and photomultipliers, cathode-ray tubes, and special u.h.f. valves.

GAS-FILLED VALVES

In the manufacture of ordinary valves as much air as possible is removed from the envelope to prevent significant ionization of residual gases and the resulting large, uncontrolled currents. You will remember that the secret of the triode is the fine control of free electrons within the valve by the electrostatic fields of the grid and anode. Ionized gas molecules would interfere with this control and make the device useless as an amplifier. When fine control is of less importance, than the efficient handling and turning on and off of heavy currents in industrial applications, gases such as nitrogen, helium, neon, argon, or mercury vapour may be purposely introduced into the valve envelope.

As we saw in Chapter 1, ionization within a gas-filled valve consists of the removal of one or more electrons from a normal gas molecule, leaving the resulting ion with a positive charge. Ionization of molecules may occur when electrons travelling from cathode to anode collide with gas molecules, or when gas molecules collide with other gas molecules. In either case, additional electrons are freed which join the original electron stream and may be capable of liberating still more electrons by colliding with other gas molecules. This process is cumulative, as in an avalanche, and results in a sudden, sharp increase in the electron flow towards the anode. At the same time the heavier, positively charged ions (ionized gas molecules) slowly drift towards the negative cathode; during the journey they attract electrons from the cathode and recombine with them to form gas molecules.

The energy required to produce ionization of gas molecules by collision with high-speed electrons is supplied by means of the anode-to-cathode voltage. A definite voltage value exists for each type of gas at which ionization suddenly begins. Before this point is reached, the anode current is about the same as for a vacuum valve at the same anode voltage. When ionization occurs, however, the anode current increases dramatically to large values and the anode-to-cathode voltage (voltage drop across the valve) drops to a relatively low value, which remains constant regardless of the anode current. The voltage at which ionization occurs is known as ionization potential, striking voltage, or firing point.

Once ionization has started, the action maintains itself at anode-to-cathode voltages considerably lower than the ionization potential. However, a minimum voltage, called de-ionizing or extinction potential, exists below which ionization cannot be maintained; the gas then de-ionizes and conduction stops. Because of this relay or switching action, the valve can be used as an electronic switch, which closes at the striking potential, permitting a large current to flow, and opens at the de-ionization voltage, blocking the current flow.

Because of their unidirectional characteristics (current will only flow from cathode to anode), gas-filled valves are useful for rectification of heavy alternating currents used in industry. However, caution must be exercised when placing gas valves in a.c. circuits, where the polarity at the anode continuously reverses. It takes a certain time, known as de-ionization time, for the gas ions to recombine with free electrons and thus stop the ionization current. If the anode voltage is made negative before de-ionization is completed, the gas ions will flow to the negative anode, constituting an inverse current flow, or arc-back. The valve thus carries a heavy current on both a.c. half-cycles, which might destroy it. Arc-back becomes more severe as the operating frequency is increased and less time is available for de-ionization to be completed. It is important, therefore, to know the inverse-voltage rating of the valve at the operating frequency.

Gas-filled valves may be classified, according to the type of electron emission employed, into cold-cathode types (usually diodes), hot-cathode (thermionic) types, available as diodes, triodes, and tetrodes, and mercury-pool valves, generally triodes known as ignitrons. Cold-cathode types obtain field emission of electrons from an unheated cathode (see Chapter 2). Gas-filled thermionic valves have oxide-coated, heated cathodes, just as conventional vacuum valves, while mercury-pool valves obtain electron emission from a pool of liquid mercury.

The construction of gas-filled valves is similar to that of high-vacuum valves, except that the cathodes, grids, and anodes are usually larger in order to carry the heavier currents. As we shall see later, grid control in gas-filled valves is limited to starting conduction and cannot be used to stop it, as in high-vacuum types.

Schematic symbols for four types of gas-filled valves are shown in Fig. 22. A small dot within the circle generally indicates that a valve is gas-filled.

COLD-CATHODE GAS-FILLED DIODES

Usually filled with neon gas in combination with other gases, cold-cathode diodes utilize field emission to obtain ionization of the gas. Since there are no emitted electrons to help the process, the striking voltage for cold-cathode valves is higher than for the hot-cathode types and it is also somewhat erratic. Two types of neon valve may be distinguished. The cathode may have the same shape and size as the anode, such as the neon-glow lamp shown in Fig. 22d, in which case the valve can conduct in either direction, depending on the values of the applied electrode potentials. Since the negative electrode (cathode) is surrounded by a characteristic glow (usually orange), the valve is useful to indicate the polarity of a direct voltage. When an alternating voltage is applied, both electrodes are surrounded by a glow discharge. A strong radio-frequency (r.f.) field is capable of ionizing the gas without direct connexion to the valve, and neon-glow lamps are, therefore, frequently used to indicate the presence of an r.f. field.

More frequently the cathode of a cold-cathode valve is larger than the anode, in which case the valve permits conduction in one direction only (see Fig. 22c). This makes the valve useful as a rectifier, to handle large currents at a relatively low voltage drop. These valves are also employed as voltage regulators, since the voltage drop remains nearly constant over a wide range of current values. Various types are available with different voltage and current ratings.

HOT-CATHODE GAS-FILLED DIODES

Designed for use as a rectifier, the chief type of hot-cathode gas diode is the so-called mercury-vapour valve, which consists of a thermionic (hot) cathode, an anode, and a small amount of liquid mercury that vaporizes when the cathode is heated (Fig. 22a). When the anode voltage is applied, the vapour ionizes and sustains a heavy current (several times as high as that of an

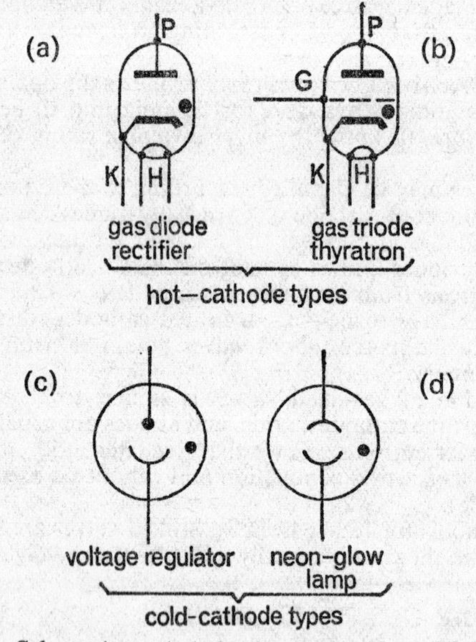

Fig. 22. Schematic symbols for four types of gas-filled valves

equivalent high-vacuum type) at a low, constant voltage drop of about 15 V. This means that most of the available supply voltage will appear as rectified output voltage and little will be wasted in the voltage drop across the valve. The cathodes of mercury-vapour valves must be preheated for one to two minutes before the anode voltage is applied, to permit the mercury to be completely vaporized. Only then is the valve capable of carrying its rated anode current.

GAS-FILLED TRIODES AND TETRODES (THYRATRONS)

A triode or tetrode to which a small amount of gas (argon, neon, or mercury vapour) has been added is known as a thyratron. (See Fig. 22b.) Although the electrode structure is basically the same as in conventional valves, the

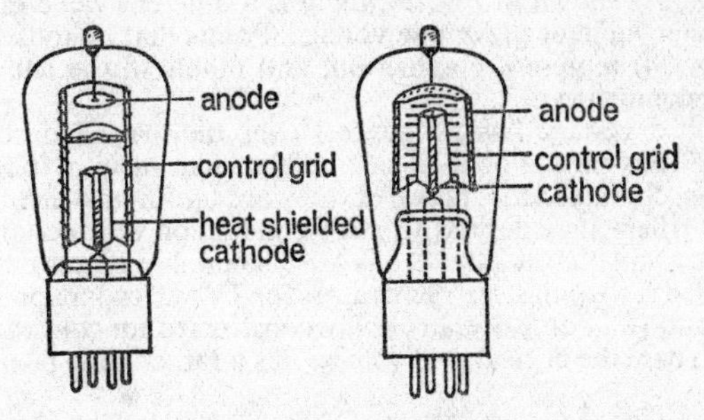

Fig. 23. Cross-section of gas-filled triodes (thyratrons)

characteristics of gas triodes and tetrodes are entirely different. The grid in a thyratron is used only to start conduction of the anode current by ionization; it cannot be employed to control the amount of anode current or stop it. The action of a thyratron, thus, is essentially that of a trigger which starts the anode current. To stop it, the anode voltage must be removed from the valve.

The construction of two types of thyratron is illustrated in Fig. 23; the grid voltage necessary to start the anode current for different anode voltages is shown in Fig. 24.

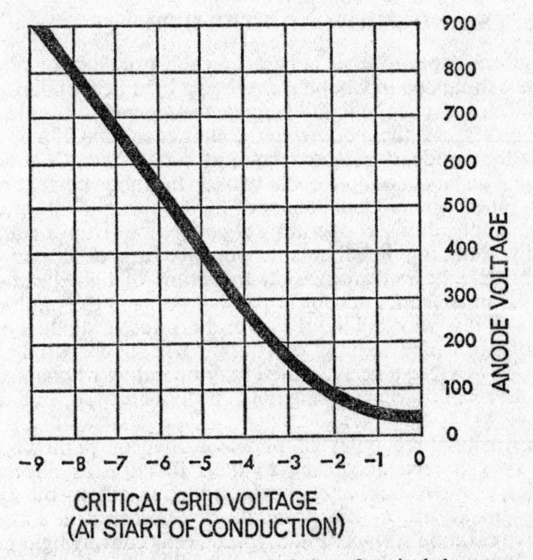

Fig. 24. Grid control characteristics of typical thyratron

When the grid voltage is made sufficiently negative the electrons emitted from the cathode do not acquire the necessary velocity to ionize the gas, and the anode current is substantially zero. As the negative grid voltage is reduced, the electrons acquire more speed and energy and a point—called the critical grid voltage—is reached, where ionization occurs and a large anode current flows. After conduction has started, the grid has no further control over the anode current. As shown in Fig. 24, there is a different value of critical grid voltage for each value of the anode voltage. Points that lie to the right of the curve (in Fig. 24) represent conduction, and points to the left of the curve represent nonconduction.

Since the grid voltage has no control over the magnitude of the anode current, thyratrons cannot be used as amplifiers like vacuum triodes. Because of their trigger characteristic, however, they are useful in switching and relay applications, where it is desired to start conduction at a certain instant by control of the grid voltage. Thyratrons are also used in motor-control circuits and in so-called sawtooth sweep generators for TV and radar applications, (see Chapters 17 and 18). Some thyratrons are constructed for cold-cathode operation, in which case the critical grid voltage has a rather large positive value.

MERCURY-POOL VALVES (IGNITRONS)

Used primarily for heavy-duty, industrial rectifier service, mercury-pool valves are one of the oldest types of gas-filled valves. In the modern version, called an ignitron, a pointed electrode (called the ignitor) dipping into a mercury pool is utilized to trigger the mercury-vapour discharge in the valve at the desired instant. The mercury pool is used itself as a cathode and requires no heating power. In practice, a current is passed through the ignitor, which generates sufficient heat to produce a 'cathode spot' at its tip. Electrons are emitted from this bright spot and are immediately ionized. Ignitrons are capable of carrying currents from 10 A to 5,000 A.

PHOTOTUBES AND PHOTOMULTIPLIERS

The operation of phototubes is based on the principle of photoemission, which we have discussed in Chapter 2. When a light beam falls upon a photo-emissive cathode, part of the light energy is transferred to the electrons within the material and 'kicks' them out from the emitter surface. If a positive voltage is placed on the anode, the electrons are attracted towards it and an anode current results, just as in a thermionic valve. The anode current may be made to operate a relay, whenever the beam of light falls on the phototube. This is used for such applications as operating a garage door from a car's headlights, automatically dimming headlights when two cars encounter each other, telegraph and telephone transmission by means of light beams, and many others. On the other hand, an object passing between the phototube and the light source will interrupt the lightbeam and produce a shadow on the photo-emissive cathode, which reduces or cuts off the anode current. Again, this effect can open or close a relay, which may operate a mechanical register to count the number of objects passing in front of the tube, open a door, or start an escalator.

There are really three types of photosensitive or photoelectric devices, which may be classified according to their function as photoemissive (or photoelectric), photovoltaic, and photoconductive. Photoemissive or photo-electric tubes make use of the principle of photoelectric emission from a photosensitive cathode surface. Photovoltaic cells convert light energy falling upon them directly into electric current. Finally, photoconductive cells contain a semiconducting solid material whose resistance decreases as the amount of light falling on it increases. All three devices depend on the same basic fact—namely, the liberation of electrons from a photosensitive surface when light strikes it.

BASIC LAWS

Regardless of type, all photosensitive surfaces act in accordance with the following empirical laws, when light energy falls upon them:

1. The number of electrons emitted per second from a photosensitive surface is proportional to the intensity of the incident light.
2. The maximum kinetic energy of each released electron is independent of the light intensity, but is directly proportional to the frequency (number of vibrations per second) of the light. This is equivalent to saying that the initial

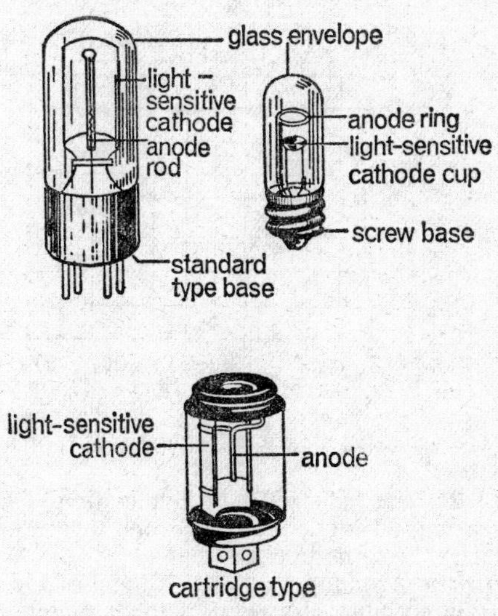

Fig. 25. Three typical forms of phototube

velocity of the emitted electrons is proportional to the square root of the frequency of the incident light. You will remember from basic physics that light frequencies are measured in Ångstroms (one Å $= 10^{-8}$ cm) and that the entire visible light spectrum extends from about 4,000 Å (0·00004 cm) for violet to about 7,400 Å for red.

3. From the second law it is also apparent that there must be some frequency, called the threshold frequency, below which the kinetic energy of the electrons is insufficient to liberate them from the photoemissive surface. This is indeed so, and each type of material has its characteristic threshold frequency.

PHOTOEMISSIVE TUBES

Photoemissive tubes (or phototubes, for short) consist of cathodes and anodes, similar to conventional diodes. The cathode (emitter) is usually large in area and arranged so that it exposes as much surface area to incident light as possible. The anode, or collector of electrons, is frequently a metal rod that is surrounded by the cylindrical cathode. Fig. 25 illustrates three typical forms of phototubes.

Cathodes and anodes are enclosed in a sealed glass envelope, which is either highly evacuated or filled with an inert gas at low pressure. The gas-filled tubes, being more sensitive than the vacuum types, have been used as light-operated switches and in the sound-reproduction heads of cinema projectors. However the response of the gas-filled tube is non-linear and its behaviour tends to be more erratic than the vacuum type, which is therefore preferred for high-precision laboratory work.

Anode Characteristics. A typical anode-voltage/anode-current characteristic family at various light intensities for a vacuum phototube is shown in Fig. 26. As is apparent from the curves, the anode current saturates for anode voltages above about 20 V to a constant value, determined by the total cathode

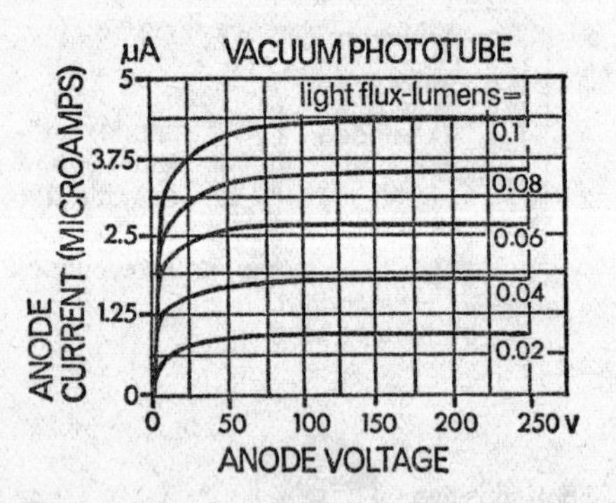

Fig. 26. Anode characteristic of vacuum phototube at various light intensities

emission for the particular value of light illumination. There is little point, therefore, in increasing the anode voltage beyond that required to overcome the space charge near the cathode.

Spectral Sensitivity Characteristics. The cathode of the phototube determines its two most important characteristics, the luminous sensitivity and the spectral response. The luminous sensitivity, expressed in microamperes per lumen, is a measure of the amount of current emitted from the cathode for a given amount of light falling on it. The spectral response of a photocathode expresses the relative amount of photoelectric current produced by light of different colours.

Phototube cathodes utilize a light-sensitive surface coating, consisting of some form of an alkali metal or alkaline-earth metal. Among these are caesium, lithium, potassium, rubidium, sodium, barium, calcium, and strontium. The tube is usually designated by the name of the material that serves as light-sensitive surface. Among the most important is the caesium-silver-oxide-silver tube, which has a luminous sensitivity of about 30 microamperes per lumen. This tube is often simply referred to as a 'caesium tube' and is prominent in TV and sound reproduction.

The spectral response of a phototube is influenced not only by the type of photosensitive cathode material, but also by the technique used to prepare the surface and by the type of glass employed for the envelope. The spectral response of most phototubes generally resembles that of the human eye. The

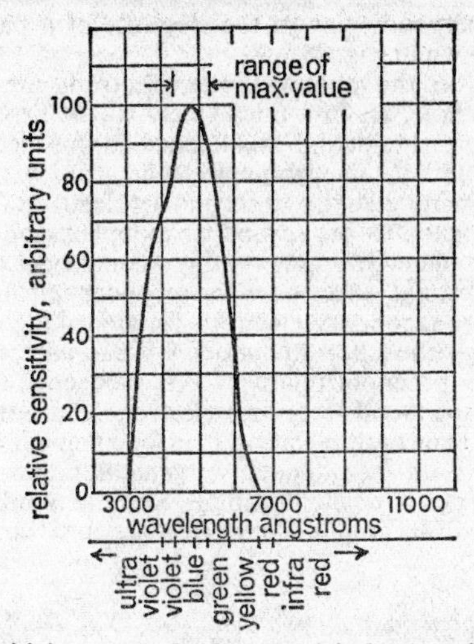

Fig. 27. Spectral sensitivity curve of a commercial photosensitive cathode material

spectral response of all phototubes reaches a peak (Fig. 27) at some specific wavelength. In some tubes, the peak response is made to coincide with that of the human eye, at about 5,500 Ångstroms. In other tubes the peak is set at some other value for special applications, and the spectral response may be extended into the ultra-violet and near-infrared regions, which are both invisible to the human eye.

PHOTOMULTIPLIERS (ELECTRON MULTIPLIERS)

The drawback of the conventional phototube is its tiny anode current for relatively large values of the illumination and, hence, the need for additional amplification of the current. Photomultipliers (sometimes called electron

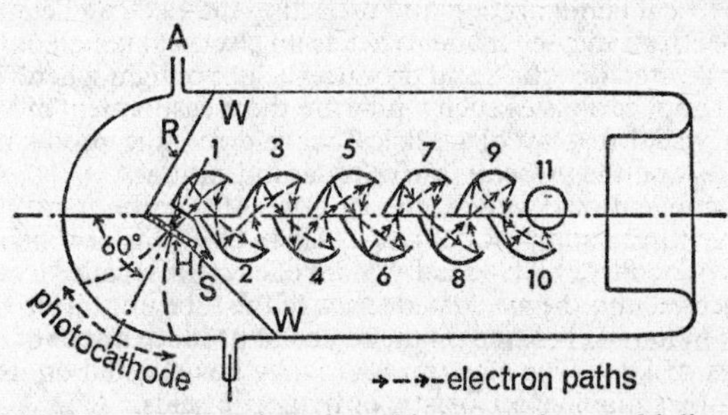

Fig. 28. Cross-section of partition-type photomultiplier

multipliers) overcome this disadvantage by utilizing the principle of secondary emission to multiply the electron current by a tremendous value. A photomultiplier tube consists of a light-sensitive photocathode, a system of nine or ten secondary-emission electrodes, called dynodes, and a collector anode. A cross-section view perpendicular to the elements of a partition-type photomultiplier tube is shown in Fig. 28.

When light falls on the photocathode, electrons are liberated and are attracted through a hole, *H*, in a mica shield to the first positive electrode, dynode 1. An electrostatic shield, *S*, is attached to this electrode, which has a voltage of about +150 V. The photoelectrons striking the concave side of dynode 1 at high velocity give rise to secondary electrons (by secondary emission) which are attracted to the second dynode, electrode 2, which is again 150 V more positive than dynode 1. Again, secondary electrons are added to the initial stream of primary electrons. The secondary electrons from dynode 2 give rise to still more secondary electrons upon striking dynode 3, which is again 150 V positive with respect to dynode 2. Thus, each successive dynode is set at a fixed higher potential relative to its predecessor and each electron striking it gives rise to several secondary electrons. Since the action, clearly, is cumulative, the electron multiplication may be as much as 2·5 million after nine stages, for 150 V on each dynode. In general, if each dynode gives off r secondary electrons per electron striking it, and the number of dynodes is n, the output current will be amplified r^{n-1} over the initial photoelectric current from the cathode.

CATHODE-RAY TUBES

We are all familiar with one type of cathode-ray tube, namely the television picture tube which converts the complex 'video signal' into a visual representation of a distant scene. But cathode-ray tubes come in many types, sizes, and shapes and about the only thing they have in common is that they all convert an electrical signal (voltage) into a visual one.

BASIC ACTION

The radio valves that we have mentioned so far make use of the amount and intensity of the electron stream flowing from cathode to anode to produce a useful anode current. Electrons may have been emitted in various ways, but the 'payoff' was always the production of the largest possible anode current. The cathode-ray tube is in a class by itself, since it is less concerned with the intensity of the electron stream than with its direction. By giving the tube a certain geometrical configuration and deflecting the electron beam by various means, the electrons may be made to act as an electrical pencil of light, which moves in any desired direction and produces a spot of light wherever it strikes. The principal applications of such a tube are the measurement of voltages and currents, the visual display of electrical waveforms, the production of television pictures, and the presentation of radar indications.

Just as in conventional valves, the cathode of a cathode-ray tube makes available a plentiful supply of electrons. These electrons are then formed by various electrodes into a high-velocity beam of electrons (called a cathode ray), which is projected into the evacuated space of the tube until it strikes a screen. A fluorescent material is coated on the screen to produce a spot of light wherever electrons strike it. The electron beam may be deflected on its journey in any direction by suitable electrostatic or magnetic fields.

MAGIC-EYE INDICATOR

Let us first look at a very simple type of cathode-ray tube, called the magic-eye indicator. This is really a combination of two sets of elements, one being an ordinary triode amplifier valve, while the other is a cathode-ray indicator

(Fig. 29). The anode of the triode section is internally connected to the ray-control electrode, so that the voltage on this electrode varies in the same manner as the anode voltage with the applied signal voltage. The ray-control electrode is a metal cylinder (Fig. 29), so placed between the cathode and the fluorescent-coated anode (or target) that it deflects a portion of the electrons emitted from the cathode.

All electrons that strike the target cause it to fluoresce with a greenish glow. The part of the target that lies in the 'shadow' of the ray-control electrode is not struck by electrons and, hence, remains dark. The size or angle of this wedge-shaped shadow depends on the electric field produced by the ray-control electrode, that is, the voltage placed on it. When the ray-control

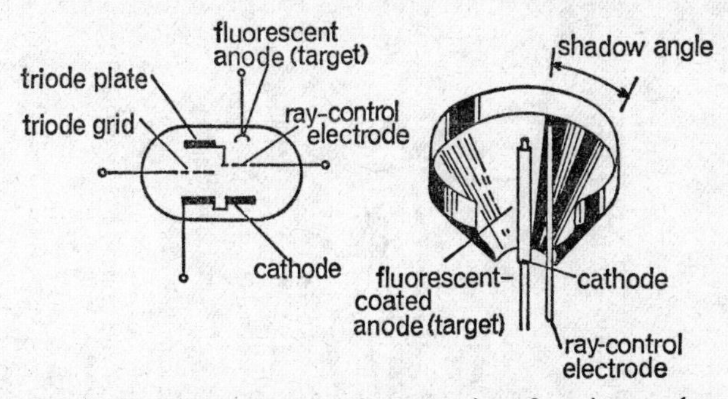

Fig. 29. Schematic symbol and construction of magic-eye tube

electrode is at the same potential as the fluorescent anode, the shadow disappears entirely. As the ray-control electrode is made negative (less positive) with respect to the anode, a shadow appears. The width of the shadow is proportional to the voltage on the ray-control electrode and, thus, can be used for rough voltage measurements. The principal uses of magic-eye tubes are as tuning indicators in radio receivers and as balance indicators in electrical bridge circuits (Wheatstone bridges).

OPERATION OF CATHODE-RAY TUBE

Now let us turn to those cathode-ray tubes that provide a visual representation of voltage and current waveforms. These tubes are usually meant when the term 'cathode-ray tube' is used, rather than the magic-eye tube discussed above. Cathode-ray tubes consist of three basic components:

1. The electron gun, which produces, accelerates and focuses the emitted electrons into a narrow beam.
2. A deflection system, which deflects the electron beam either electrostatically or magnetically in accordance with the phenomenon (voltage or current waveform) to be displayed.
3. A fluorescent screen, upon which the beam of electrons impinges to produce a spot of visible light.

These essential parts of a cathode-ray tube are mounted inside a highly evacuated, funnel-shaped glass envelope, the large end of which has the fluorescent screen coated on the inside surface. The construction of an electrostatic-deflection type cathode-ray tube is illustrated in Fig. 30. The electron gun consists of the group of electrodes to the left of the vertical deflecting plates. Commercial tubes range from 10 to 20 inches in length and have

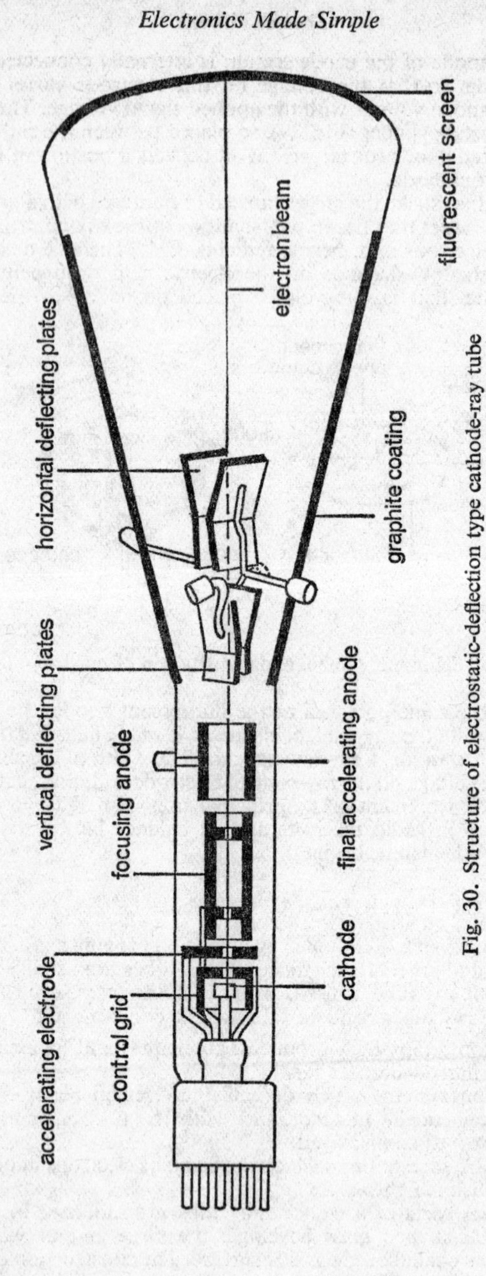

Fig. 30. Structure of electrostatic-deflection type cathode-ray tube

fluorescent screens varying from one inch in diameter for radio servicing to giant 24-inch diameter television picture tubes. Beam-accelerating potentials vary from 800 V for small tubes to about 12,000 V for large ones.

The Electron Gun. The electron gun has the job of producing and focusing the electrons into a narrow beam so that it makes a tiny spot when impinging on the fluorescent screen. A broad beam is of little use for accurate display of waveforms or TV pictures. An electron gun of the electrostatic type (Fig. 30) consists of an indirectly heated cathode, a control grid, an accelerating electrode or grid, a focusing anode, and a final accelerating anode. These electrodes are in the form of cylinders surrounding the cathode. Connexions to the electrodes are brought to pins in the base.

The cathode, control grid, and accelerating electrode perform essentially the same function as the cathode, control grid, and screen grid in a tetrode or pentode. The cathode emits the electrons, the control grid determines the amount of electron flow by means of the negative bias placed on it, and the accelerating electrode—being highly positive with respect to the cathode—speeds up the electrons passing through it.

Electrons passing through the openings in the accelerating electrode are focused into a sharp electron beam by the combined effect of the focusing anode and the final accelerating anode, which are in effect a multiple electronic lens system. Both the accelerating and focusing anodes are at a positive potential with respect to the cathode, but the voltage on the focusing anode is always considerably lower (by several hundred volts) than that on the final accelerating anode.

To understand the focusing action that takes place between the focusing and (final) accelerating anodes, refer to Fig. 31. As shown in the figure, a strong

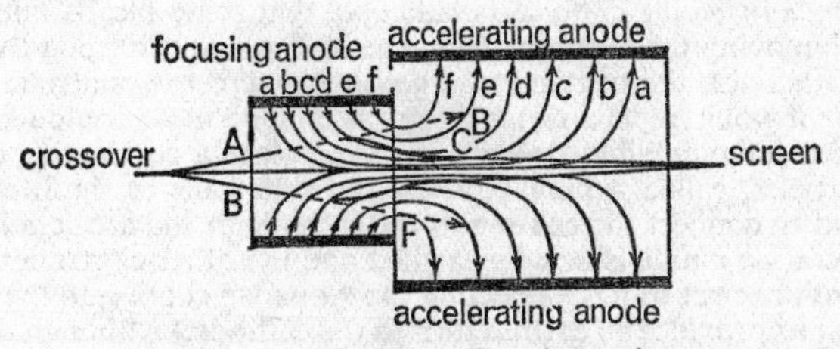

Fig. 31. Electrostatic focusing action

electrostatic field exists between the focusing and accelerating anode cylinders. The field lines (lines of force) are semicircular or radial in shape because of the geometry of the two anodes and they extend from the focusing anode to the accelerating anode, because the latter is at a high positive voltage with respect to the focusing anode. Electrons enter this 'lens system' from a point near the control grid, called the crossover, where they have been brought to an initial focus by the combined effect of the control-grid aperture and the negative grid bias. From this first crossover point, the electrons diverge rapidly outward and would eventually strike the sides of the tube, as shown by the dotted rays in Fig. 31, if it were not for the action of the radial field between the two anodes.

As mentioned on page 6, electrons tend to follow the lines of force of an electric field. Accordingly, the electrons entering the field between the focusing and accelerating anodes, tend to follow the field lines (shown by arrows) and would normally fall into the accelerating-anode cylinder, if they were moving slowly enough. However, because of the rapid forward acceleration imparted to the electrons by the attraction of the accelerating electrodes, electrons are not sufficiently long in the space between the anodes to be completely pulled off course. Rather, they move tangentially along the lines of force and then out of the field, converging into a narrow spot on the screen.

Thus, the radial field lines just have sufficient strength to guide the electrons (along their tangents) to a smoothly converging beam, but they are not strong enough to pull them away from their forward movement towards the sides of the anode cylinder. We might mention in passing that a similar focusing action can be attained by means of a magnetic field produced by focus coils around the neck of the tube.

The Fluorescent Screen. If the electron beam leaving the electron gun were not deflected, it would produce a luminous spot at the centre of the fluorescent screen. The intensity of this spot can be controlled by adjusting the control-grid bias, but it is also dependent on the fluorescent material, called phosphor, coated on the inside of the tube. Among commonly used phosphors are Willemite (zinc silicate), which gives off predominantly green light; zinc oxide, which gives off a blue colour; zinc beryllium silicate or zinc sulphide, which glow yellow; and combination phosphors, which can be selected to give off nearly white light.

Another important consideration in the choice of phosphors is their afterglow or persistence of glow after the electrons have ceased to bombard a spot on the screen. If the afterglow is less than 0·1 second, the screen is said to have short persistence; if it is 1 second or more, it has a long persistence. (Between these limits the persistence is medium.) Willemite has a short persistence. Short persistence is desirable for rapidly changing images, such as those displayed in television. Long persistence is occasionally of advantage, such as in radar presentations and waveform comparisons.

The Aquadag Coating. You can readily see that some means must be provided for removing the electrons from the screen and returning them to the cathode. Otherwise the negative charge on the screen would build up to a point where it would repel arriving electrons, and no more could reach it. The method used for removing the electrons is to place a conducting coating of carbon particles, called Aquadag, along the side walls of the tube (not the screen), and to connect the coating to the cathode or the accelerating anode. (The accelerating anode is usually earthed and the other electrodes are made negative with respect to it. Connecting the Aquadag coating to the accelerating anode thus provides an earth return to the cathode.) Although the coating is not directly connected to the screen, the electrons are removed by means of secondary emission from the screen and no pile-up occurs. The secondary electrons are collected by the coating and then returned to the cathode.

Electrostatic Deflection. We have mentioned that the electron beam may be deflected either electrostatically or magnetically. Let us first consider electrostatic deflection. An electrostatic field can be produced between two metal plates (electrodes) simply by applying a voltage between them. The lines of force of this field are straight from the negative to the positive plate, except near the edges, where they bulge outwards, as shown in Fig. 32. Electrons entering the field between the deflecting plates will be bent in the direction of the lines of force, that is, they will be attracted towards the positive plate. The resultant path of the beam will be the net effect (vector sum) between its forward velocity and—in this case—its upward deflection, or transverse velocity. This path is parabolic between the deflecting plates (inside the field) and becomes a straight line after the beam leaves the field. As a result the electron spot is deflected upward (in this case) from its central position on the screen. The spot may be deflected downward by reversing the polarity of the voltage on the plates.

Note in Fig. 30 that two pairs of deflecting plates at right angles to each other are set into the path of the electron beam. The vertical deflecting plates move the beam vertically up and down, while the horizontal deflecting plates move the beam sideways, to the left or right of centre. The terms 'vertical' and 'horizontal' thus refer to the direction in which the beam is deflected and not to the position of the plates within the tube. If a direct voltage is applied to the vertical plates, the electron spot on the screen will be displaced up or down from centre, depending on the polarity of the deflection voltage, and by an amount that depends on the relative magnitude of the voltage. The same is true for sideways (left or right) deflection by the horizontal plates.

Suppose now that an alternating voltage is applied to the vertical plates. The spot will now move up and down on the screen at the frequency of the alternating voltage. Because of the persistence of the screen and that of the eye, the moving spot will actually appear as a continuous, luminous vertical line. (At very low frequencies the movement of the spot may actually be seen.) Similarly, by placing an alternating voltage on the horizontal deflecting plates, a continuous, luminous horizontal line can be produced. Finally, if alternating voltages are applied simultaneously to the two sets of deflecting plates,

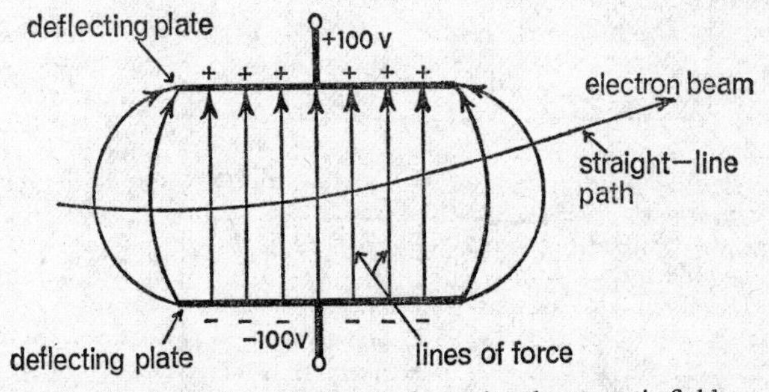

Fig. 32. Deflection of electron beam by electrostatic field

various patterns are formed on the screen, depending on the relative magnitudes, frequency, phase and waveform of the deflecting voltages. These patterns are called Lissajous figures.

Fig. 33 illustrates some typical, simple Lissajous figures. Part (a) of the figure illustrates the 45° inclined line formed when alternating voltages of equal magnitude, phase and frequency are applied to the two sets of deflecting plates. In (b) the voltages are equal in magnitude and frequency, but 90 degrees out of phase with each other; as a result a circle appears. In (c), the magnitudes of the deflecting voltages are equal, but the frequency of the vertical deflecting voltage is twice that of the horizontal deflecting voltage. In (d) the vertical and horizontal deflecting voltages have been interchanged. Finally, in (e), the frequency of the vertical voltage is three times that of the horizontal deflecting voltage. In each case the Lissajous pattern that appears on the screen represents the vector sum of the horizontal and vertical deflecting voltages, and thus may be predicted by plotting it on graph paper.

Deflection Sensitivity. The deflection sensitivity of an electrostatic-deflection cathode-ray tube is the amount of spot deflection on the screen (measured in inches) per volt applied to the deflecting plates. An average value is 50 V per inch, which means that a positive voltage of 50 V applied to the upper vertical deflection plate will deflect the electron spot upward by one inch on the screen.

If an alternating voltage with a peak amplitude of 50 V is applied to the vertical plates the spot will trace out a vertical line two inches high, one inch up from centre and one inch down from centre.

Note that the deflection sensitivity is inversely proportional to the forward velocity of the electrons to which they are accelerated by the final anode voltage. It is also directly proportional to the length of the deflecting plates and the length of the beam from the centre of the deflecting plates to the screen and inversely proportional to the distance between the deflecting plates (at right angles to the beam).

Magnetic Deflection. Magnetic deflection of the electron beam is employed wherever it is inconvenient or impossible to obtain suitable electrostatic

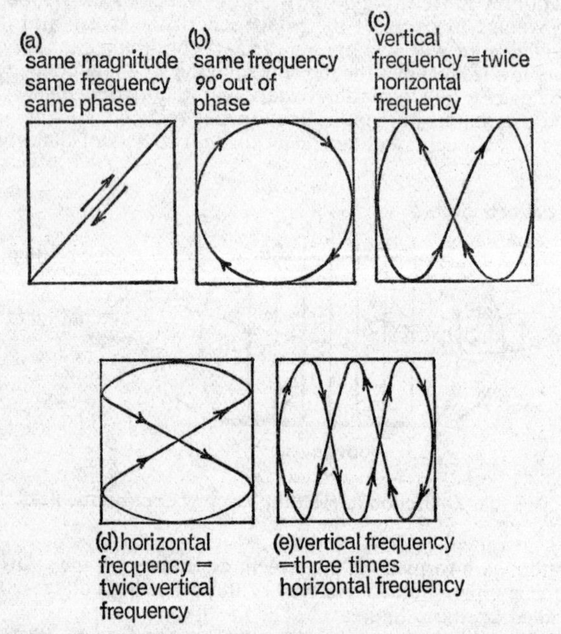

Fig. 33. Typical Lissajous figures on screen of cathode-ray tube

deflection voltages. Radar indicators and television picture tubes use it. A big advantage of magnetic deflection is that it obviates the need for deflecting plates inside the tube and thus simplifies the tube construction. In magnetic deflection, the deflecting force is obtained by a magnetic field set up by coils placed around the neck of the tube (Fig. 35).

To understand how the electron beam can be deflected by a magnetic field, you must consider that the stream of electrons travelling in one direction constitutes an electric current. It is immaterial whether the electrons flow in a wire conductor or in free space. One of the basic laws of electricity states that a magnetic field surrounds every electric current. When a current-carrying wire is placed in a uniform magnetic field, its own field interacts with the external magnetic field and the wire experiences a force that depends on the relative direction of the fields and the strength of the current. The same is true for a beam of electrons, which is an electric current, although not carried in a wire.

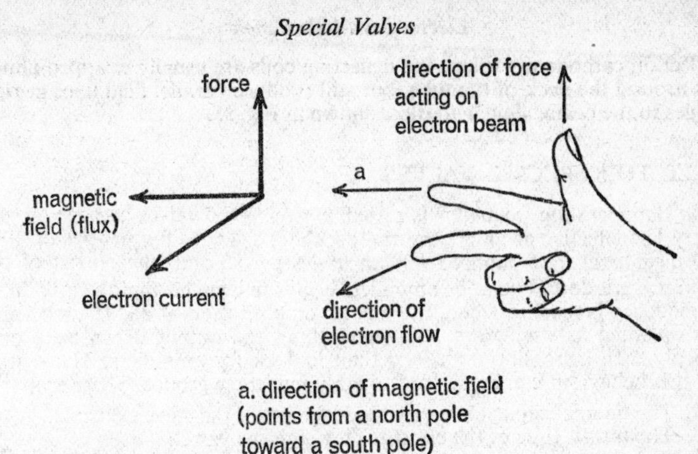

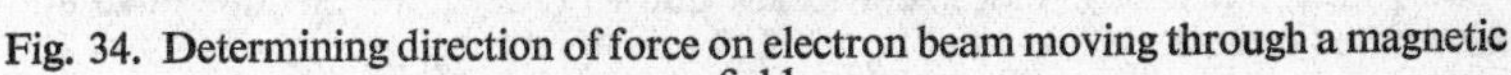

Fig. 34. Determining direction of force on electron beam moving through a magnetic field

A simple way of determining the direction in which an electron current (beam or wire conductor) will be deflected is by the application of the so-called right-hand rule (Fig. 34). If the thumb, index finger, and middle finger are held at right angles to each other, and the index finger is pointed in the direction of the magnetic field (from north pole to south pole), while the middle finger points in the direction the electron beam is travelling, then the thumb will indicate the direction of the force on the electron beam.

Fig. 35 illustrates the action of the vertical deflecting coils in a magnetic-deflection type cathode-ray tube. For horizontal beam deflection another pair of deflecting coils must be mounted alongside the neck of the tube at right angles to the vertical coils. The amount of deflection produced is roughly proportional to the product of the number of turns in each coil and the number of amperes flowing through the coil (i.e. the ampere-turns). By applying the right-hand rule, you can see that the beam illustrated in Fig. 35 will be deflected upward by the action of the magnetic field. In actual magnetic-

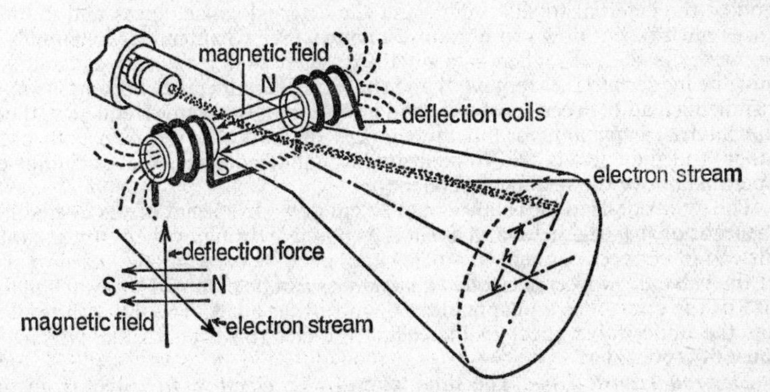

Fig. 35. Action of vertical deflecting coils on electron beam in magnetic-deflection cathode-ray tube

deflection cathode-ray tubes, the deflecting coils are usually wrapped almost flat around the neck of the tube, but still produce parallel field lines at right angles to the beam, similar to those shown in Fig. 35.

NEED FOR SPECIAL VALVES

As the operating frequency is raised, conventional valves become progressively less effective as amplifiers and oscillators. Above frequencies of about 100 megahertz (one-hundred million cycles per second) the ouptut of conventional triode and pentode amplifiers begins to drop off rapidly until finally a frequency is reached where there is no amplification at all. If such valves are operated as oscillators (a.c. generators), the output drops even more sharply with increasing frequency than it does for amplifiers. The reasons for this behaviour can be divided roughly into three groups. These are:

1. The internal capacitances and inductances of the valve elements.
2. The transit time of the electrons through the valve.
3. Radio-frequency losses and losses due to radiation from the valve and associated circuit.

Internal Capacitances and Inductances. We have already become familiar with the interelectrode capacitances of conventional valves. You will remember these capacitances exist between the grid, anode, and cathode, and are labelled C_{gp}, C_{gk}, and C_{pk}, respectively. As the frequency goes up, the reactance (opposition to the flow of a.c.) of these capacitances goes down and begins to short out part of the input and output signal voltages, as well as feed back increasingly more energy from the anode (output) to the grid (input). All this cuts down the amplification.

Furthermore, the leads between the various electrodes and the connecting base pins have a definite, though small, inductance. The reactance of this inductance goes up with increased frequency. This means that the voltage applied to the base pins will not actually appear at the electrodes, but will be reduced by the amount wasted in the series (inductive) reactance of the leads. Again, this cuts down the available gain from the valve.

Finally, you must consider that at radio-frequencies valve electrodes are connected to external tuned or resonant circuits, consisting of inductances and capacitances. As the frequency is raised, the effect of the valve's internal capacitances and inductances begins to become progressively greater in relation to the external tuned circuit, and the external capacitances and inductances must be cut down to obtain resonance. (See Chapter 13.) Eventually a frequency is reached, where—to obtain resonance—the external capacitance must be made zero (i.e. removed) and the external inductance becomes simply a shorting lead between the anode and grid terminals. At this frequency, then, the internal capacitances and inductances of the valve resonate with each other, and the valve is said to be at its resonant frequency. Valves cannot be operated above the resonant frequency.

The electrode-lead inductances can be cut down in special valves by making the leads of large diameter and as short as possible (bringing them straight out, instead of connecting them to a base) and, also, by reducing the physical size of the valve. Low interelectrode capacitances can be obtained by reducing the size of the electrodes and spreading them further apart. The latter, however, has the undesirable effect of increasing the electron transit time, which we must now consider.

Electron Transit Time. The time taken by an electron to travel from the cathode to the anode is known as the transit time. This time represents a negligible part of one cycle at operating frequencies lower than 100 megahertz and, hence, the output anode current may be assumed to respond instantaneously to changes in the control-grid voltage. But as the operating frequency is raised into the u.h.f. region (300–3,000 megahertz), the transit time of the electrons from cathode to anode becomes an appreciable part of the a.c. cycle. At a 1,000 megahertz, for example, one cycle lasts only 0·001 microsecond, which is considerably less than the transit time in conventional valves. Thus, a change in the control-grid (or other) voltage will not affect the anode current immediately, but there will be a definite time-lag before the anode current can respond to the change in control-grid voltage. This time-lag may be thought to be effectively the same as the lagging of the current behind the applied voltage in an inductance. (You will remember that this is expressed by a lagging phase angle.) Thus, the transconductance of the valve (i.e. the ratio of a change in anode current to a change in grid voltage) has a lagging phase angle at very high frequencies.

A more serious consequence of the finite transit time is that the grid of the valve absorbs power at high frequencies, even when the grid bias is made negative and no grid current flows. The mechanism by which this happens is somewhat too complex to be discussed here, but the fact is that power is actually absorbed by the grid circuit and this power consumption increases with the square of the frequency.

The transit time can be reduced appreciably by decreasing the physical size of the valve and, particularly, the spacing between cathode and control grid. Increasing the positive anode voltage increases the electron velocity and thus also assists in cutting down the electron transit time.

Radio-Frequency and Radiation Losses. Finally, there are certain power losses associated with the valve and its circuit which all tend to increase with frequency. At ultra-high frequencies all currents flow in thin surface layers on the conductor, a phenomenon that is called the skin effect. Skin effect causes increased resistance. The skin effect and, hence, the conductor resistance and associated power loss, increase with frequency.

Electric fields produce molecular movements in glass and other insulating supports used in valves, leading to heat and power losses. These dielectric-hysteresis losses, as they are called, go up directly with the frequency.

As we shall learn later (Chapter 15), any piece of wire (such as an electrode lead) will radiate radio-frequency power, if its dimensions are comparable to the wavelengths of the radio-frequency current through it.

$$\text{Wavelength (metres)} = \frac{3 \times 10^8}{\text{frequency (hertz)}}$$

As a consequence, the valve and its associated circuit will have appreciable radiation losses at ultra-high frequencies. All the factors discussed above generally decrease the valve efficiency as the frequency goes up.

The skin effect can be reduced and the consequent resistance losses may be made lower by increasing the surface area of the current-carrying conductors. Dielectric losses can be reduced by properly positioning the glass insulators with respect to the electric field. Radiation losses can be reduced by shielding the valve and its circuit in an enclosure, or by using 'concentric lines' in the construction to confine the electric fields of the valve and its associated circuit.

TYPICAL U.H.F. VALVES

The factors we have discussed above, which make conventional valves unsuitable for operation at ultra-high frequencies, have led to the design of special u.h.f. valves that minimize internal capacitances and inductances, cut down transit time, and reduce radiation and radio-frequency losses. This is done by reducing the physical size of the valves, making the electrodes quite small and spacing them very close together, and by making special lead arrangements. Because of these constructional features, the power handling ability of these u.h.f. valves is smaller than that of conventional valves used at lower frequencies. The valves are, therefore, usually operated at high cathode-emission currents and relatively heavy anode currents and anode voltages (large anode dissipation). Ultra-high-frequency valves frequently have no base; connexions to the electrodes are made through pins protruding through the envelope at the shortest possible distance from each electrode.

Some typical examples of commercial high-frequency tubes that incorporate these features, are illustrated in Fig. 36. The acorn, doorknob, pencil, and lighthouse valves are so named because of their characteristic shapes and sizes. Acorn and doorknob valves are available as diodes, triodes, and pentodes, while the pencil and lighthouse types come as triodes.

Acorn valves are characterized by very small size (about 1½ inches in height

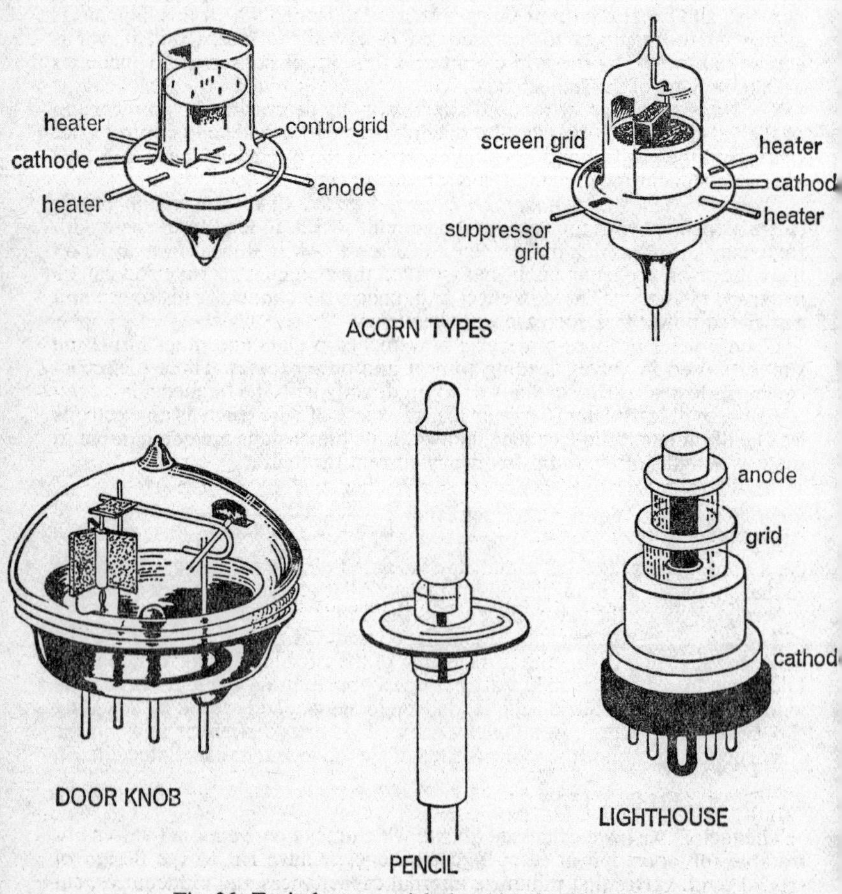

Fig. 36. Typical commercial ultra-high-frequency valves

and diameter), close spacing of the electrodes, and leads that come out through a ring seal instead of a base, thus minimizing their lengths and the capacitance between them. The resonant frequency of acorn valves is about 1,500 megahertz and they operate at frequencies anywhere from 400 to 1,500 megahertz.

The doorknob valve is an enlarged version of the acorn type and has an upper frequency limit of about 700 megahertz, when used as an oscillator. Note that the leads are brought out directly through the glass envelope and are widely spaced to reduce the capacitance between them. The electrodes are small and closely spaced. The valve shown in Fig. 36 is about two-thirds of actual size and has a plate dissipation of 30 W.

The pencil triode is used as amplifier and oscillator at frequencies up to 1,700 megahertz. It is characterized by a closely spaced cylindrical structure that is designed for operation with coaxial and cavity resonators, rather than ordinary tuned circuit elements (Chapter 11).

The lighthouse type represents a different approach to the problem of losses at ultra-high frequencies. The active parts of the cathode, grid, and anode in the lighthouse valve are parallel planes and the leads are metal discs, as illustrated. Such a structure can be fitted into a system of concentric lines, used to form the tuned (resonant) circuit at ultra-high frequencies. By connecting the valve directly to its tuned circuit in this manner, the losses due to lead inductance of the connecting wires are practically eliminated. In addition, the parallel-plane construction permits extremely close spacing of the electrodes. (The spacing between grid and cathode is about one-thousandth of an inch.)

CERAMIC VALVES

The planar structure of the lighthouse valves has led to the design of tiny ceramic u.h.f. valves, measuring approximately $\frac{1}{4} \times \frac{1}{2}$ inch overall. Constructed of alternating layers of titanium electrodes and special ceramic insulators, the ultra-small ceramic valves have proved themselves capable of high-quality performance up to several thousand megacycles under severe environmental conditions, such as are encountered in space applications. They have exceptional tolerance to shock and vibration, and function equally well in the icy regions of outer space and in high-temperature and high nuclear-radiation environments. Except for being much smaller in size, the appearance of ceramic valves is similar to the lighthouse valve illustrated opposite.

MICROWAVE VALVES

Frequencies higher than about 2,000 megahertz are usually referred to as microwave frequencies. The u.h.f. valves just discussed will not operate well at these frequencies. Special microwave valves operating on different principles, such as magnetrons and klystrons, have been developed for this frequency region. These will be discussed in Chapter 11.

SUMMARY

In gas-filled valves gases such as nitrogen, helium, neon, argon, or mercury vapour are purposely introduced into the valve envelope. This gives them the ability to handle heavy currents and makes them useful for rectifier, relay, and switching applications.

The voltage at which ionization occurs in these valves is known as the ionization potential, striking voltage, or firing point. Anode conduction begins with the firing point, and the valve voltage drop falls to a low, constant value.

Below the de-ionizing or extinction potential, ionization cannot be maintained and conduction stops.

Grid control in gas-filled valves is limited to starting conduction, but cannot be used to stop it. To stop conduction the anode voltage must be removed.

Cold-cathode, gas-filled valves utilize field emission from an unheated cathode; hot-cathode or thermionic types have conventional heater-type cathodes; and ignitrons obtain electron emission from a pool of liquid mercury.

Cold-cathode gas diodes are used to indicate polarity of a.c., the presence of r.f. fields, and serve as voltage regulators.

Mercury-vapour valves are hot-cathode gas diodes that are chiefly used as heavy-duty rectifiers.

Gas-filled triodes and tetrodes are known as thyratrons and are used for relay applications and for saw-tooth sweep generators.

Critical grid voltage in thyratrons is the value of the negative bias to start conduction at a certain anode voltage.

Ignitrons utilize pointed ignitor electrodes, dipped into a pool of mercury, to trigger off the mercury-vapour discharge and thus start rectification.

Photosensitive devices can be classified as photoemissive (or photoelectric), photovoltaic, and photoconductive.

Photoemissive or photoelectric tubes utilize the principle of photoelectric emission from a light-sensitive cathode surface.

Photovoltaic cells convert incident light energy directly into electric current.

Photoconductive cells contain a semiconducting, solid material whose resistance decreases in proportion to the light falling on it.

The number of electrons emitted from a photosensitive surface (regardless of type) is proportional to the intensity of the incident light.

The maximum kinetic energy of the released electrons is independent of the light intensity, but is proportional to the frequency of the incident light. The initial electron velocity is proportional to the square root of the incident light frequency.

Electrons are not emitted below the characteristic threshold frequency of the photosensitive material.

Gas-filled photoemissive valves are more sensitive than high-vacuum types, but are more erratic in operation and their characteristic is non-linear. They do not respond well to very rapid changes in illumination.

The photocathode determines luminous sensitivity (microamperes per lumen) and spectral (colour) response.

Photomultipliers (or electron multipliers) utilize the principle of secondary emission to multiply the electron current.

Cathode-ray tubes convert an electrical signal (current or voltage) into a visual one by shooting a beam of electrons at a fluorescent screen and deflecting the beam in accordance with the variations of the electrical signal.

Cathode-ray tubes consist of an electron gun, a deflection system, and a fluorescent screen, all housed in a glass envelope.

The electron gun—usually made up of a heater-type cathode, a control grid, an accelerating electrode or grid, a focusing anode, and a final accelerating anode—produces, accelerates, and focuses the emitted electrons into a narrow beam. It acts as an electron emitter and electronic lens system.

The fluorescent screen is coated with a phosphor (Willemite, zinc oxide, zinc sulphide) that produces visible light when electrons impinge on it.

The deflection system deflects the electron beam either electrostatically, by means of deflecting plates, or magnetically, by coils placed around the neck of the tube, in accordance with the voltage or current waveform to be displayed.

Electrostatic deflection sensitivity, measured in volts per inch deflection, is inversely proportional to the velocity of the electrons (i.e. the final anode voltage) and to the spacing between the deflecting plates, and is directly proportional to the length of the deflecting plates and the length of the beam from the centre of the deflecting plates to the screen.

Magnetic deflection is proportional to the ampere-turns of the deflecting coils.

Conventional valves become inoperative at ultra-high frequencies because of their internal capacitances and inductances, the electron transit time, and radio-frequency and radiation losses.

U.H.F. valves, such as acorn, doorknob, pencil, and lighthouse types, overcome these effects by reducing the physical size of the envelope and electrodes, spacing the electrodes close together, and by means of special lead arrangements.

CHAPTER SEVEN

TRANSISTORS AND SEMICONDUCTORS

In recent years the transistor—an entirely new type of electron device—has come into its own and bids to replace the bulky thermionic valve in most, if not all, applications. Transistors are far smaller than valves, have no filament and hence need no heating power, and may be operated in any position. They are mechanically rugged, have practically unlimited life, and can do some jobs better than valves, while catching up fast in other respects. In contrast to valves, which utilize the flow of free electrons through a vacuum or gas, the transistor relies for its operation on the movement of charge carriers through a solid substance, a semiconductor. Transistors are only one of the family of semiconductors; many other semiconductor applications are becoming increasingly popular and new ones are constantly being discovered.

You will remember from Chapter 1 that materials are classed as semiconductors if their electrical conductivity is intermediate between metallic conductors, which have a large number of free electrons available as charge carriers, and non-metallic insulators, which have practically no free electrons available to conduct current. There are many varieties of semiconductor, but the two most frequently used in electronics and transistor manufacture are germanium and silicon. Both elements have the same crystal structure and similar characteristics, so that the discussion that follows for germanium will also apply to silicon.

GERMANIUM CRYSTAL STRUCTURE

In Chapter 1 we described the structure of atoms and noted that only the outermost electron shell of an atom is of interest in electronics, since it contains the loosely held valence electrons, which are easily dislodged to become electric current carriers. Germanium has four valence electrons in its outer shell; for our purposes, the atom may be pictured as containing only these electrons and four protons in the nucleus to keep it electrically neutral. When germanium is in crystalline form its atoms assume the structure illustrated in Fig. 37. In this structure adjacent germanium atoms share their valence electrons in a strong bond, so that effectively four orbital electron pairs are associated with each nucleus. These electron pairs are termed covalent bonds

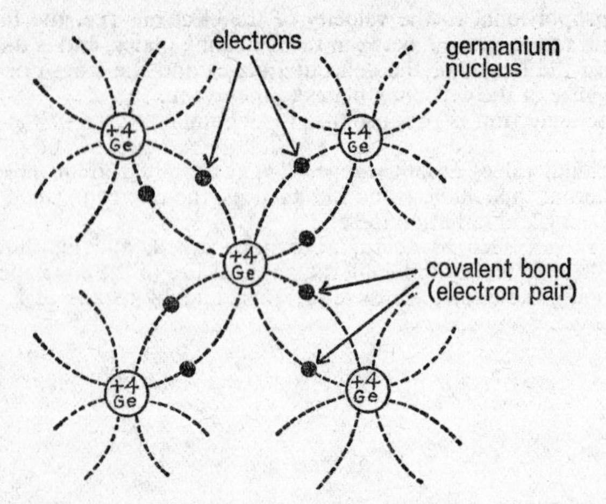

Fig. 37. Germanium crystal structure showing covalent bonds

and they are bound so strongly to each other and to the nucleus that no free electrons are available to conduct a current through the germanium. A pure germanium crystal, therefore, is practically a non-conductor of electricity. It is not completely non-conducting, since ordinary heat energy occasionally disrupts some of the covalent bonds, thus liberating free electrons as charge carriers.

If a small amount of an impurity is introduced into the germanium crystal, its current-conducting characteristics change radically. Thus, when atoms that have five electrons in their outer shell, such as antimony or arsenic, are introduced into the germanium atom (a procedure known as 'doping'), the fifth electron of the impurity atom does not find a place in the symmetrical covalent-bond structure and, hence, is free to roam around through the crystal.

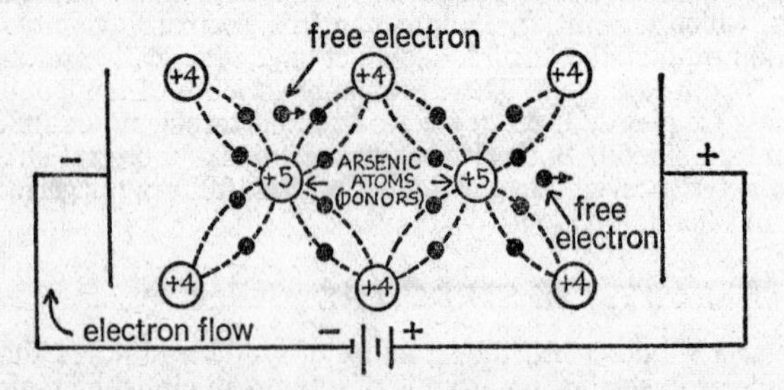

Fig. 38. Electron conduction through N-type germanium

These free electrons are then available as electric current carriers. By placing an electric field across the 'doped' germanium crystal, as shown in Fig. 38, the excess free electrons donated by the impurity atoms will travel towards the positive terminal of the voltage source. Relatively few impurity or 'donor' atoms within the germanium structure permit fairly substantial electron currents through the crystal when an electric field is applied. Germanium that has been doped by pentavalent donor atoms (i.e. five electrons in the outer shell) is known as *N-type* germanium, because current conduction is carried on with *negative* charge carriers, or electrons.

Consider now the situation when an impurity that has only three electrons in its outer shell, such as gallium or indium, is introduced into the pure germanium crystal. As shown in Fig. 39, the trivalent indium atoms take their place in the germanium structure, but one of the covalent bonds around each indium atom has an electron missing, or a hole in its place. Although the hole indicates the absence of an electron, it behaves like a real, positively charged particle when an electric field is applied across the crystal. Under the influence of the electric field, electrons within the crystal will tend to move towards the positive terminal of the voltage source and jump into the available holes of the indium atoms near the positive terminal. Since there are no free electrons available, the deficient indium atoms near the positive terminal 'steal' electrons from their neighbours to the left (in Fig. 39) by disrupting their covalent bonds.

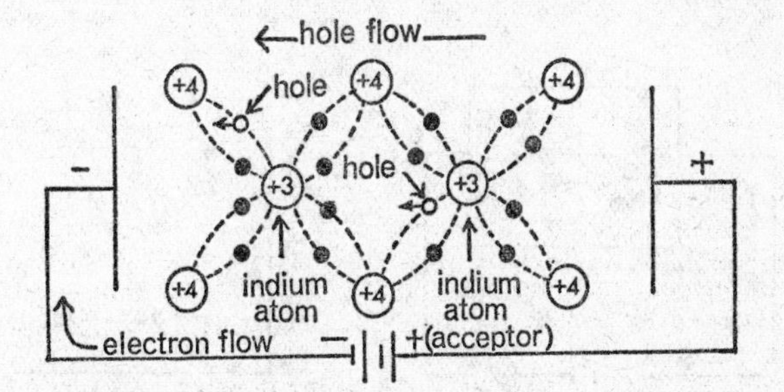

Fig. 39. Hole conduction through P-type germanium

This creates new holes in adjacent atoms to the left of those that have been filled. As electrons move to the right towards the positive terminal, the holes will move to the left towards the negative terminal, thus acting like mobile, positive particles. As the holes reach the negative terminal, electrons enter the crystal near the terminal and combine with the holes, thus cancelling them. At the same time, the loosely held electrons that filled the holes near the positive terminal, are attracted away from their atoms into the positive terminal. This, of course, creates new holes near the positive terminal, which again drift towards the negative terminal. Current conduction may thus be considered to occur by means of holes inside the crystal, and by means of electrons through the external connecting wires and battery.

An impurity that has three electrons in its outer shell (trivalent) is known as an acceptor atom, because it takes electrons away from surrounding germanium atoms. Germanium that has been doped with trivalent acceptor atoms is called *P-type* germanium, to specify that current conduction is carried on by holes, which are the equivalent of *positive* charges.

P-N JUNCTION DIODES

Conduction of electric current through P- or N-type germanium takes place equally well in either direction; hence, reversing the polarity of the battery in Figs. 38 and 39 will not affect the amount of current flow, although it reverses its direction.

Consider now what happens when P-type germanium is joined to N-type germanium and a voltage is applied across the junction, as illustrated in Fig. 40. In practice, such an abrupt P-N junction may be obtained in two ways. In the *grown* junction a single crystal is obtained from a melt which at first contains impurities of either the P- or N-type. In the middle of the growth process, impurities of the opposite kind are added to the melt, so that the remainder of the crystal abruptly grows into the opposite type.

In contrast, a *fused* P-N junction is obtained by pressing small 'dots' of indium (P-type) on a wafer of N-type germanium. After a few minutes of heat treatment, the indium fuses to the surface of the germanium and produces P-type germanium for a thin layer below the surface. A P-N junction is thus formed between this P-region and the remainder of the N-type germanium wafer. Both the grown and the fused P-N junctions are extensively used in junction diodes and transistors, each having specific characteristics and limitations.

With the P-type germanium biased positively, as illustrated in Fig. 40, the (positive) holes are repelled by the battery voltage towards the junction

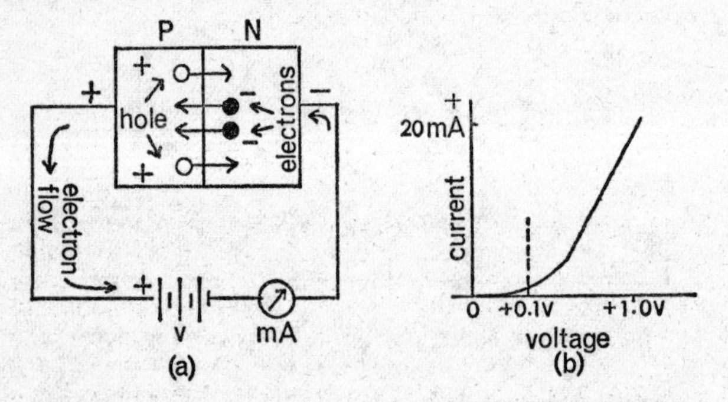

Fig. 40. Current flow across P-N junction with forward bias

between the P- and N-type material. Simultaneously, the electrons in the N-type germanium are repelled by the negative battery voltage towards the P-N junction. Although there is normally a potential barrier at the P-N junction that prevents electrons and holes from moving across and combining, under the influence of the electric field of the battery the holes move to the right across junction and the electrons move to the left. In the region of the P-N junction, therefore, electrons and holes meet and combine, thus ceasing to exist as mobile charge carriers. For each electron-hole combination that takes place near the junction, a covalent bond near the positive battery terminal breaks down, an electron is liberated and enters the positive terminal. This action creates a new hole which moves to the right towards the P-N junction.

At the opposite end, in the N-region near the negative terminal, more electrons arrive from the negative battery terminal and enter the N-region to replace the electrons lost by combination with holes near the junction. These electrons move towards the junction at the left, where they again combine with new holes arriving there. As a consequence, a relatively large current flows through the junction. The current through the external connecting wires and battery is carried by electrons, in the direction shown in Fig. 40a.

The battery connexion that permits current to flow across the P-N junction is known as forward bias. A minimum voltage of about 0·1 V is needed to overcome the potential barrier at the junction and permit any current to flow.

(See Fig. 40b.) The current then increases rapidly with increasing battery voltage and as little as 1 to 2 V permit currents of 20 to 100 mA.

If the battery voltage is reversed in polarity, as illustrated in Fig. 41a, an entirely different situation prevails. The holes are now attracted to the negative battery terminal and move away from the P-N junction, while the electrons also move away from the junction because of the attraction of the positive terminal. Since there are effectively no hole and electron carriers in the vicinity of the junction, current flow stops almost completely. A small reverse current of a few microamperes still flows across the junction, however, as illustrated by the voltage–current characteristic (Fig. 41b). This reverse current is due to thermally generated electron-hole pairs within both the P- and N-type materials. As mentioned before, some covalent bonds always break down

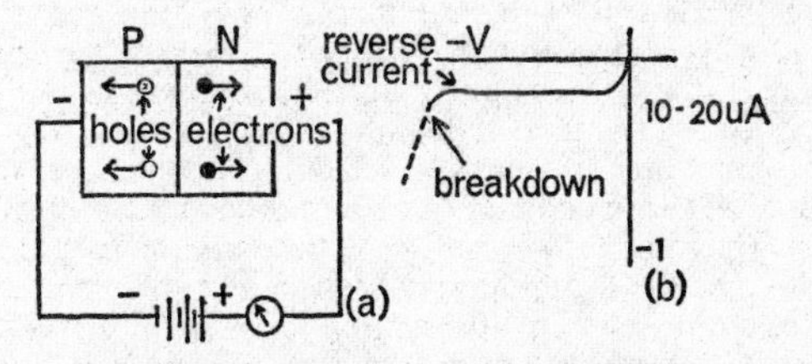

Fig. 41. P-N junction with reverse bias

because of the normal heat energy of the crystal molecules. Electrons liberated by this process in the P-material move right across the junction under the influence of the electric field, while holes generated in the N-material move to the left into the P-material. Thus a small electron–hole combination current is maintained by these so-called minority carriers. If the reverse bias is made very high, the covalent bonds near the junction break down, as indicated in Fig. 41b, and a large number of electron–hole pairs will be liberated; the reverse current then increases abruptly to a relatively large value.

The unilateral current conduction characteristic of a P-N junction is seen to be similar to that of the conventional diode valve discussed in Chapter 3. It was pointed out there that this characteristic permits diode valves to change alternating current into unidirectional current. Germanium and other semiconductor diodes (silicon, selenium, etc.) are, therefore, extensively used as rectifiers and detectors.

JUNCTION-TRIODE TRANSISTORS

Just as the triode valve followed on the heels of the vacuum diode, you might expect that the logical extension of the semiconductor diode junction would be a triode junction, consisting of two P-N junctions. This is indeed the case, and the modern P-N-P or N-P-N junction triode transistors are in many respects analogous to triode valves. A junction transistor can function as an amplifier or oscillator, as can a triode valve, but has the additional advantages of long life, small size, ruggedness, and absence of cathode heating power.

Fig. 42 illustrates a P-N-P junction, made up of a sandwich of two P-N germanium junction diodes, placed back to back. Although exaggerated in the illustration, the centre or N-type portion of the sandwich is extremely thin in comparison to the P-regions. The double junction may be either of the grown or fused crystal types, obtained in the manner discussed for junction diodes.

With the battery polarities as shown in Fig. 42, the P-regions are negative

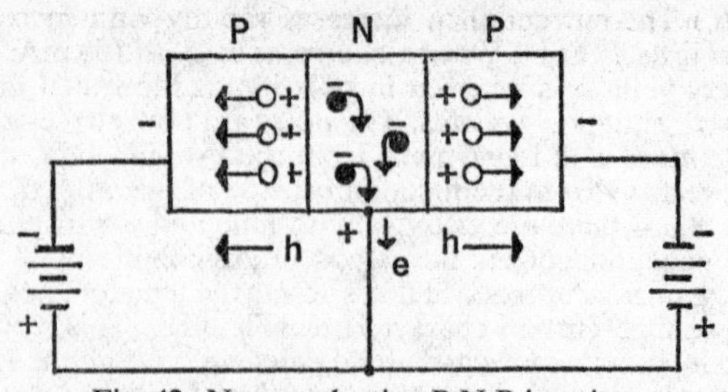

Fig. 42. Non-conducting P-N-P junction

with respect to the central N-region or conversely, the N-region (called the base) is positive with respect to the P-regions. The mobile electrons in the N-region, therefore, initially move away from both junctions in the direction of the positive connecting terminal. The holes in each of the P-regions also move away from the junctions and are attracted towards the negative terminals. After these initial displacements of holes and electrons the current flow stops.

Consider now the same P-N-P sandwich, but with the batteries connected as in Fig. 43. Note that the P-region at the left of Fig. 43 is biased positively, in the forward direction (see Fig. 40), while the P-region at the right is biased negatively, in the reverse direction (see Fig. 41). This is one of the basic operating connexions of a P-N-P junction transistor.

The holes in the left P-region, known as the *emitter*, are repelled by the positive battery terminal towards the left P-N or emitter junction. (The junction that is forward biased in a transistor is always termed emitter junction.) Under the influence of the electric field the holes overcome the barrier and cross the emitter junction into the N-type or *base* region. This region is very thin and only lightly 'doped' with impurity atoms, so that the majority of the

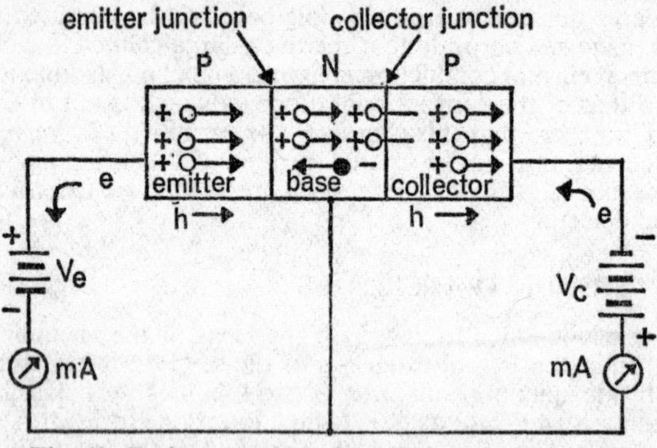

Fig. 43. Basic connexion of P-N-P junction transistor

holes are able to drift across the base without meeting electrons to combine with. A small number of holes (about 5%), however, are lost in this area because of recombination with electrons. The remainder penetrate through the almost porous base region and flow across the right-hand junction into the P-region or *collector*. (The junction with a reverse bias in a transistor is termed collector junction.) The negative collector voltage (V_c) aids in rapidly sweeping up the holes that pass into the collector region.

As each hole reaches the collector electrode, an electron is emitted from the negative battery terminal (V_c) and neutralizes the hole. For each hole that is lost by combination with an electron in the collector and base areas, a covalent bond near the emitter electrode breaks down and a liberated electron leaves the emitter electrode and enters the positive battery terminal (V_e). The new hole that is formed then moves immediately towards the emitter junction, and the process is repeated. It is evident, therefore, that a continuous supply of holes are injected into the emitter junction, which flow across the base region and collector junction, where they are gathered up by the negative collector electrode. Current conduction within the P-N-P transistor thus takes place by hole conduction from emitter to collector, while conduction in the external circuit is carried on by electrons. Furthermore, the collector current must be less than the emitter current by an amount proportional to the number of electron–hole combinations occurring in the base area.

The ratio of collector current to emitter current is known as alpha (symbolized α) and it is a measure of the possible current amplification in a transistor. From the definition, alpha cannot be higher than 1, but practical values of 0·95 to 0·99 are attained in commercial transistors.

Because of the reverse bias, no current can flow in the collector circuit unless current is introduced into the emitter. Since a small emitter voltage of about 0·1 V to 0·5 V permits the flow of an appreciable emitter current, the

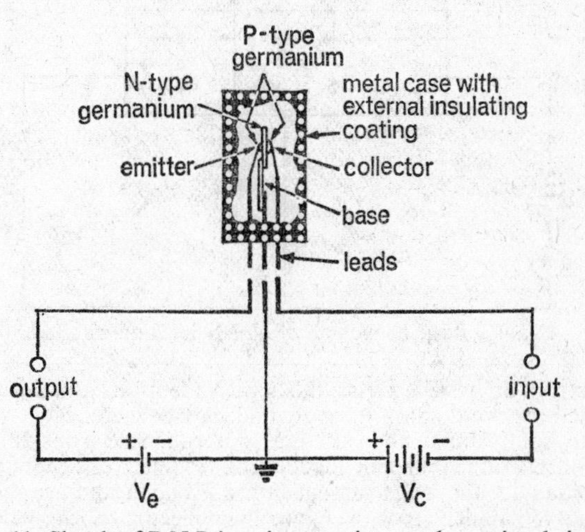

Fig. 44. Sketch of P-N-P junction transistor and associated circuit

input power to the emitter circuit is quite small. As we have seen, the collector current due to the diffusion of holes is almost as large as the emitter current. Moreover, the collector voltage (V_c) can be as high as 45 V, thus permitting relatively large output powers. A large amount of power in the collector circuit may thus be controlled by a small amount of power in the emitter circuit. The power gain in a transistor (power out/power in) thus may be quite high, reaching values in the order of 1,000.

A diagrammatic sketch, illustrating the structure of a P-N-P junction transistor and the associated input and output circuits, is shown in Fig. 44. The transistor shown is of the fused junction type. To operate as an amplifier, an a.c. signal must be introduced into the input circuit and a load resistance must be connected across the output.

N-P-N JUNCTION TRANSISTOR

An N-P-N type junction transistor is sketched in Fig. 45. Note that the positions of the N- and P-type germanium have been interchanged and the battery polarities have been reversed with respect to the P-N-P transistor. The emitter junction is still forward biased, however, in accordance with our previous definition, since electrons are repelled from the negative emitter battery terminal (V_e) towards the junction. Likewise, the collector junction has reverse bias because electrons are flowing away from the collector junction towards the positive collector battery terminal (V_c). The main difference between the P-N-P and N-P-N transistor, therefore, is that current conduction in the latter is carried by electrons, while the charge carriers in the former are holes.

The process of conduction within the N-P-N transistor is similar to that described before for the P-N-P type. With the emitter voltage, V_e, applied as

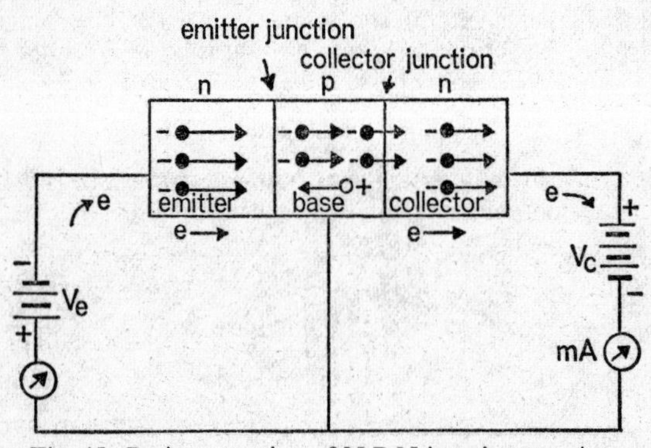

Fig. 45. Basic connexion of N-P-N junction transistor

shown, electrons are repelled from the negative terminal and injected into the emitter junction, overcoming the potential barrier there. Since the P-region (base) is only lightly doped and very thin, most of the electrons diffuse through the base and reach the collector junction. A few electrons (about 5%), however, combine with the holes present in the P-region and are lost as charge carriers. The remainder cross over into the collector region, where they are rapidly swept up by the positive collector voltage, V_c. For each electron flowing out of the collector and entering the positive battery terminal, an electron enters the emitter from the emitter-battery negative terminal. Electron conduction thus takes place continuously in the direction shown in Fig. 45.

The following conclusions about junction-transistor operation may be drawn from the above analysis:

1. The major charge carriers in the P-N-P junction transistor are holes.
2. The major charge carriers in the N-P-N junction transistor are electrons.
3. The collector current in either type of junction transistor is always less than the emitter current because of the recombination of holes and electrons occurring in the base area. The ratio of the collector to emitter current is known as the current gain or alpha (α).

TRANSISTOR CHARACTERISTIC CURVES

The performance of transistors may be determined from characteristic curves of their voltage and current relations, just as for valves. Fig. 46 illustrates the variation of the emitter current as a function of the emitter-to-base voltage (V_e) for a typical N-P-N junction triode transistor.

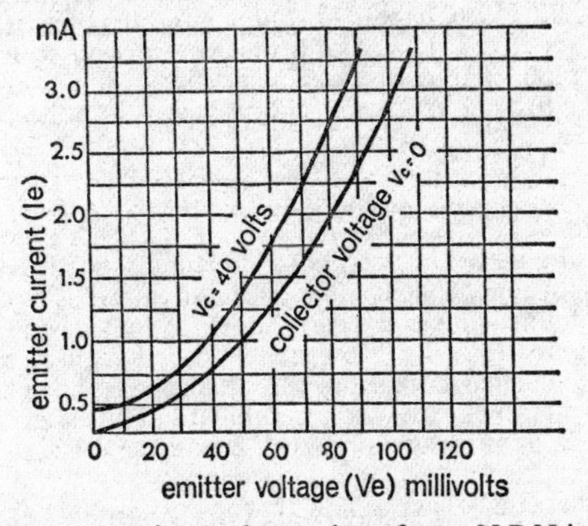

Fig. 46. Emitter current v. emitter-to-base voltage for an N-P-N junction transistor

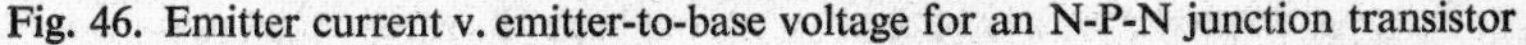

Note that the shape of the emitter-current/emitter-voltage (I_e–V_e) curves is somewhat similar to the characteristic anode-current/anode-voltage curve of a vacuum diode (Fig. 6). The emitter current increases rapidly with small increments in the emitter voltage and reaches a value of several milliamps with an emitter voltage no higher than about 0·1 V (100 millivolts). Moreover, the emitter current is almost independent of the collector-to-base voltage (V_c), although there is a small interaction, as shown by the two curves for V_c=0

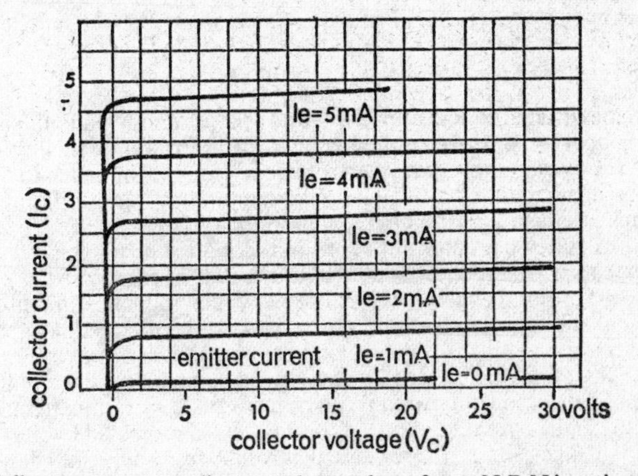

Fig. 47. Collector current v. collector-to-base voltage for an N-P-N junction transistor

and V_c=40 V. The main point to note is that a very small emitter voltage suffices to produce a large flow of emitter current. This also means that a small signal-voltage variation at the input of the transistor produces a large emitter-current variation, or equivalently, the input resistance to a small signal voltage impressed on the emitter is very low.

A family of collector-current/collector-voltage (I_c–V_c) characteristics for the same N-P-N junction transistor is shown in Fig. 47. The shape of the curves resemble very much the constant-current I_b–E_b curves of a pentode valve, shown in Fig. 19 (Chapter 5). The main difference between the valve curves and transistor characteristics is that current is used as the independent variable for plotting transistor characteristics.

Each curve in Fig. 47 represents the variation in collector current with changing collector-to-base voltage (V_c) for a constant value of the emitter current (I_e). Note that almost the entire variation in the collector current takes place at very low values of the collector voltage. When the collector voltage is raised above a value of about 1–2 V, it collects all the charge carriers that diffuse via the base to the collector junction, and the collector current becomes practically independent of the collector voltage. The collector current above this minimum voltage is nearly, but not quite, equal to the emitter current for each curve. You will remember that the collector current cannot equal the emitter current because of the small percentage of charge carriers lost in the base due to electron–hole combinations. As seen from the curves, the value of the current ratio, alpha, may be increased slightly by raising the collector voltage.

The I_c–V_c curves (Fig. 47) also show that a very large change in the collector voltage produces only a tiny change in the collector current, which means that the output resistance of the transistor (i.e. the ratio of voltage-change/current-change) is also very high. This sheds some light on the process of voltage and power amplification in a transistor, where the current gain (alpha) is necessarily less than one. We have already determined that a small signal voltage impressed in the low-resistance input (emitter) circuit of a transistor produces a relatively large emitter current. Almost the same amount of current will flow in the high-resistance output (collector) circuit of the transistor, where the voltage may be very high. Evidently, then, both the output voltage and output power can be quite large, compared to the tiny input voltage and power present at the emitter.

TRANSISTOR SYMBOLS AND CONNEXIONS

We have mentioned P-N-P and N-P-N transistors and found out something about their operation. In later chapters we will learn about their fascinating applications in various compact electronic circuits. But in order to recognize a transistor (in a circuit, of course) it is necessary to become familiar with some of the forms, connexions, and circuit symbols used by engineers.

Fig. 48 illustrates the common forms and circuit symbols used for N-P-N transistors (a) and P-N-P transistors (b). Note that in each case the emitter is distinguished from the collector by an arrow, which indicates the direction of 'conventional' current flow with forward bias. You will remember from basic electricity that conventional current flow is opposite in direction to electron flow. Checking back (in Fig. 45) for the N-P-N connexion, it is apparent that electrons flow into the emitter from the negative battery terminal; this means that conventional current flows out of the emitter, as indicated by the outgoing arrow in Fig. 48a. Similarly, we can determine from Fig. 43 that, for the P-N-P transistor, electrons flow out of the emitter towards the positive battery terminal. Consequently, conventional current flow is into the emitter, as indicated by the inward arrow in Fig. 48b.

When transistors are operated as amplifiers, three different basic circuit connexions are possible, as illustrated in Fig. 49. These are (a) common-base, emitter-input; (b) common-emitter, base-input; and (c) common-collector, base-input. Each of these circuit configurations has specific advantages and limitations.

Note that regardless of the circuit connexion the emitter is always biased in a forward direction, while the collector always has a reverse bias, in accordance with the basic connexions shown in Figs. 43 and 45. This necessitates a positive emitter bias and a negative collector bias for the P-N-P transistor, while opposite polarities are required for the N-P-N transistor. Except for this polarity reversal, the connexions for P-N-P and N-P-N transistors are identical.

Common-Base, Emitter-Input Connexion. The connexions shown in Fig. 49 can be understood by considering the emitter of the transistor roughly analogous to the cathode of a triode valve, the collector analogous to the anode of a triode, and the base analogous to the grid of a triode. The common-base connexion (Fig. 49a) is thus roughly analogous to an earthed-grid triode, which has a low input impedance and is useful as a high-frequency amplifier. We have already become familiar with the common-base connexion,

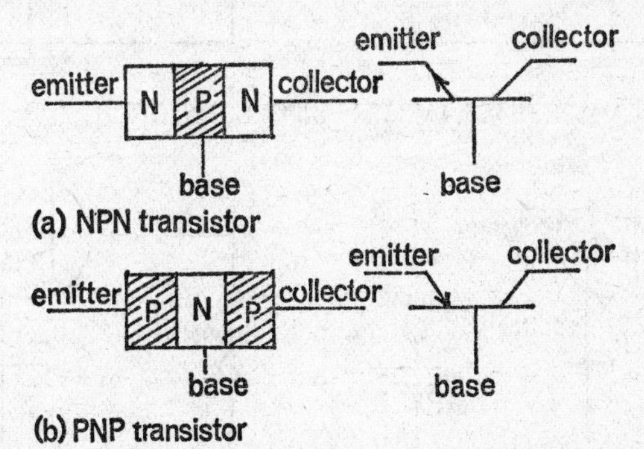

Fig. 48. Transistor forms and circuit symbols (a) N-P-N-type, (b) P-N-P-type

since it is convenient for illustrating transistor physics. As we found out, this circuit provides a very low input resistance, a high output resistance and a current gain (alpha) of less than one. Despite this low current gain, the common-base circuit provides respectable voltage and power amplification, as was mentioned earlier.

Common-Emitter, Base-Input Connexion. The common-emitter circuit (Fig. 49b) is roughly analogous to the conventional earthed-cathode triode valve circuit and is the most flexible and efficient of the three basic connexions. While its input resistance is somewhat higher and its output resistance is lower than that of the common-base connexion, the common-emitter connexion has the highest voltage and power gain of the three circuits. Like the equivalent valve circuit, the common-emitter connexion produces a phase reversal between the input and output signals. No such phase reversal occurs with the other two connexions.

Common-Collector, Base-Input Connexion. The common-collector connexion is the transistor equivalent of the earthed-anode triode valve (cathode-follower) circuit that will be discussed in Chapter 10. The circuit provides a relatively high input resistance, a low output resistance, and about the same current gain (alpha) as the common-emitter circuit. Its voltage gain, however, is always less than 1, as is the case in the equivalent cathode-follower circuit. As with the latter, the common-collector circuit is primarily used for impedance matching and as a buffer stage.

JUNCTION-TETRODE TRANSISTOR

The frequency response of conventional junction transistors is considerably limited because of the capacitance associated with the collector electrode and because of the finite time taken by the charge carriers to diffuse through the transistor base region. The collector capacitance acts as a shunt path for high

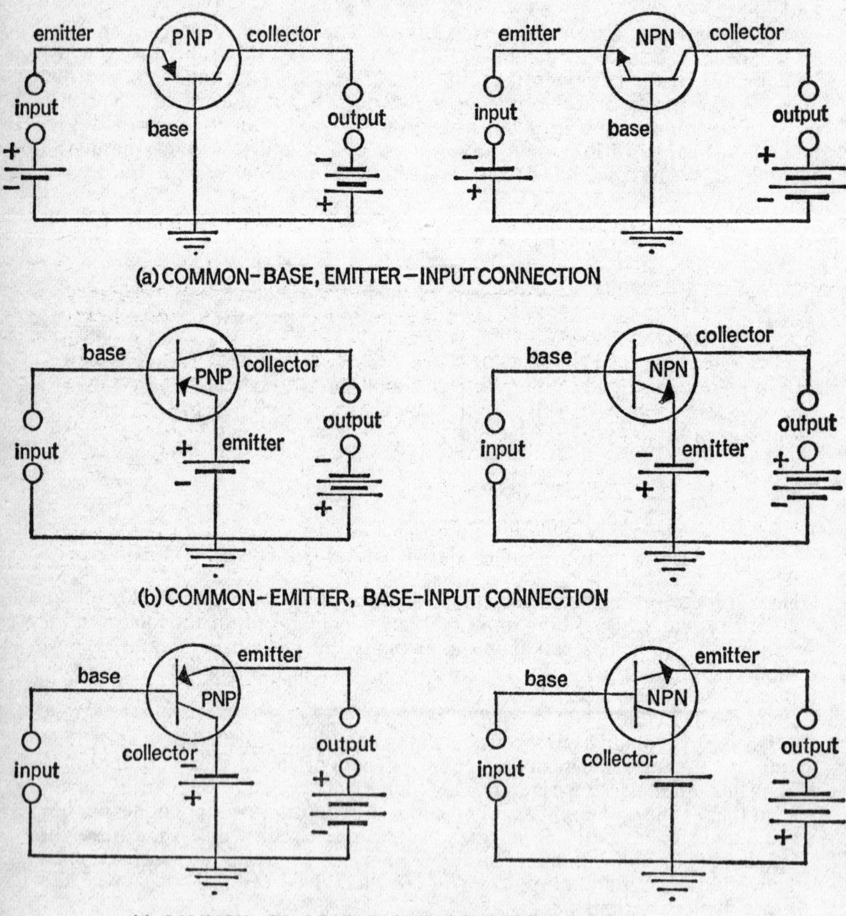

Fig. 49. Transistor amplifier circuit connexions

frequencies, just like the interelectrode capacitances in valves. The time taken by the charge carriers to arrive at the collector (transit time) places an upper frequency limit on the operation of the triode transistor. Obviously, if the polarity of the collector voltage reverses before the charge carriers have a chance to drift across to the collector, the current gain will fall off. The upper frequency limit for a transistor is termed the alpha-cut-off frequency and it is defined as the frequency at which the current gain falls to 70·7% of its low-frequency value.

A junction-tetrode transistor partially overcomes this unfavourable frequency-response characteristic of the conventional transistor triode. In the tetrode transistor, whose structural and circuit representations are illustrated in Fig. 50, a fourth electrode, b_2, is connected to the base region (P-layer) on

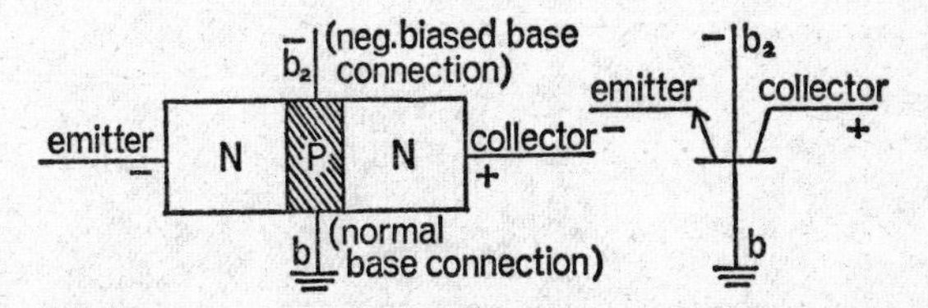

Fig. 50. Junction tetrode transistor (a) form, (b) circuit symbol

the opposite side of the conventional base connexion. When a negative bias of about -6 V is applied to this second base terminal, the emitter-to-collector current is forced by the field into a very small region near the vicinity of the regular base connexion, *b*. This produces the equivalent of a transistor with a very thin P-region and low base resistance. Both these factors reduce the transit time of charge carriers from emitter to collector, resulting in vastly improved response at high frequencies.

POINT-CONTACT TRANSISTORS

The point-contact transistor first produced in 1948 is older by about a year than the junction triode described above. It is related in some ways to the crystal detector used in the early days of radio, and at one time had the advantage over the junction type in high-frequency applications. This advantage has been wiped out by recent advances in junction design.

Fig. 51 illustrates the construction and approximate behaviour of a typical N-type point-contact transistor. In contrast to the early crystal detector, two sharply pointed tungsten electrodes (often referred to as 'cat's whiskers') make contact close together on the surface of a single germanium crystal pellet. A third electrode, the base, is soldered to the N-type germanium pellet. The two cat's whiskers are called emitter and collector, respectively, just as in the junction transistor. The entire assembly is enclosed in an impregnated plastic housing, sealed against the atmosphere.

Although the theory of the point-contact transistor is fairly complicated, you can gain a qualitative understanding of its operation by considering that the transistor must be 'formed' initially by passing a current through its electrodes during manufacture. This effectively produces a small P-region underneath each of the points, as is illustrated in Fig. 51b. With the two P-regions embedded within the N-type germanium base, a P-N-P type transistor results, which is similar to the P-N-P junction type. Because of the point contacts, however, the behaviour of the transistor differs in some respects from the corresponding junction type.

The circuit symbols and connexions for the point-contact transistor are the same as for the corresponding junction type, the emitter being biased in the forward direction and the collector in the reverse direction. The collector-current/collector-voltage characteristics of the point-contact transistor resemble those shown for the junction type (Fig. 47), except that there is considerable collector current even with zero emitter current and the curves slope

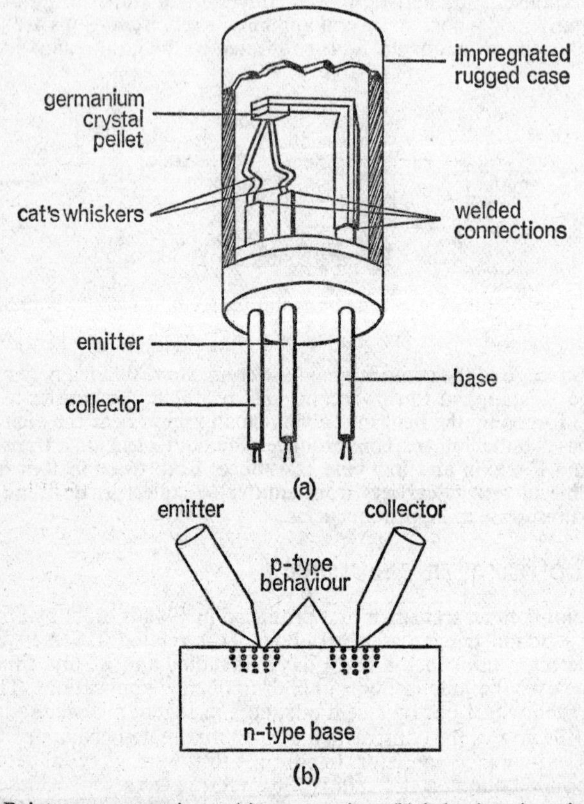

Fig. 51. Point-contact transistor: (a) construction; (b) behaviour in point region

upward, signifying that the collector current is not independent of the collector voltage, as is the case for the junction transistor. The increase in collector current with increasing collector voltage means that the collector (output) resistance of the point-contact transistor is substantially lower than that of the junction type, being of the order of 15,000 ohms. Moreover, the collector current can exceed the emitter current, in contrast to the junction transistor, and hence the current gain (alpha) is greater than 1. The emitter (input) resistance of the point-contact transistor, however, is higher (by about eight times) than that of the junction type.

Because of the small P-N areas of the point-contact transistor and its small internal collector capacitance, the alpha-cut-off frequency of the point-contact transistor tends to be much higher than that of the corresponding junction transistor. (The alpha-cut-off frequency is the point where the current gain drops to 70·7% of the low-frequency gain.) As a consequence, point-contact transistors can be easily made to amplify or oscillate up to 100 megahertz and above, which was far beyond the capabilities of early junction transistors. Recently developed junction transistors, such as the transistor tetrode and the 'mesa' transistor, have now made point-contact transistors obsolete, even at high frequencies.

UNIJUNCTION TRANSISTORS

As its name implies, the unijunction transistor has only one P-N junction, in contrast to the ordinary two-junction transistor. This single junction drastically modifies the electrical characteristics of the device and makes it suitable for a variety of oscillator applications.

Fig. 52 shows the construction and symbol of the unijunction transistor. Two contacts, base 1 (B1) and base 2 (B2), are made at opposite ends of a

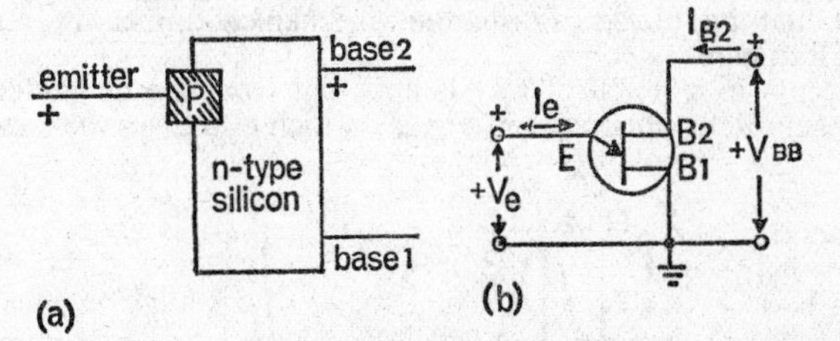

Fig. 52. Unijunction transistor: (a) construction; (b) symbol and principal voltages

small bar of N-type silicon. At the opposite end of the bar, close to base 2, is a single P-type rectifying junction, the emitter. A resistance of 5,000–10,000 ohms exists between the two bases (the interbase resistance, R_{BB}). In normal operation, base 1 is earthed and a positive voltage, V_{BB}, is applied at base 2. In the absence of emitter current, I_e, the N-type silicon bar acts simply as a voltage divider. Thus a certain fraction of the positive base-2 voltage, V_{BB}, appears at the emitter junction. As long as the emitter voltage, V_e, is less positive than this fraction, the junction is reverse-biased and only a small emitter leakage current flows. When the positive emitter voltage becomes greater than the fraction of the base voltage at the emitter, however, the junction will be forward-biased and an emitter current, I_e, flows. This current consists of holes, injected into the silicon bar, which move from the emitter to base 1. As the holes reach base 1, an equal number of electrons are attracted into the region between base 1 and the emitter. The net increase in electrons is equivalent to a reduction in the resistance between emitter and base 1. The upshot is that an increase in emitter current results in a decrease of resistance and in the emitter voltage, which is known as a negative-resistance characteristic. Such a negative-resistance characteristic is exploited in various oscillator and timing circuits.

TUNNEL DIODES

The most glamorous semiconductor-junction device that has come along in recent years is the tunnel diode, which promises to push semiconductor operating frequencies into the microwave range of several thousand megacycles. Invented by the Japanese scientist Leo Esaki, the tunnel diode—like the unijunction transistor—has a negative-resistance characteristic (i.e. a decrease of current with an increase in voltage), which makes it useful both as a high-frequency oscillator and as an amplifier. The tiny and reliable tunnel diode also enjoys great advantages in switching applications, such as are needed in electronic computers, since it can switch on and off in a few thousandths of a microsecond (10^{-9}s). Although the tunnel diode shares a negative-resistance characteristic with the unijunction transistor, its operating principles are completely different from the latter, as we shall see presently.

The increase in high-frequency and switching capability of the tunnel diode is achieved through a highly conductive, extremely narrow junction of P- and N-type germanium (or some other crystal, such as gallium-arsenide). Because of this extremely narrow junction, electrons are capable of 'tunnelling through' from one side of the junction to the other, though they have insufficient energy to surmount the 'potential barrier', or wall, always present at such a junction. You can conceive of this tunnelling effect in terms of a billiard ball that rolls over a hump in the table, though it has hardly been pushed at all, and does not (or should not) have the energy to do so. Neither common sense nor classical physics can explain this surprising situation, but it is explained by quantum physics as 'quantum-mechanical tunnelling', a matter we cannot go into here.

What occurs in a practical way is apparent from the characteristic of a typical tunnel diode, illustrated in Fig. 53, which also gives the characteristic

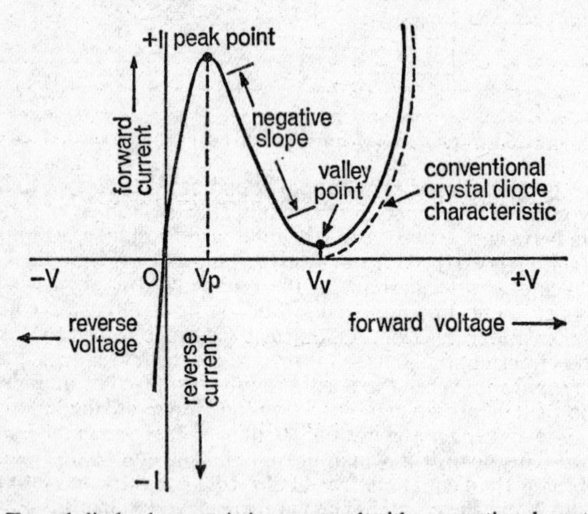

Fig. 53. Tunnel diode characteristic compared with conventional crystal diode

of a conventional crystal diode for comparison. When a reverse (negative) bias voltage is applied to the anode of a conventional crystal diode, it does not conduct, while a tunnel diode, in contrast, will conduct. At low values of an applied forward (positive) anode voltage, the tunnel diode passes a considerable current, which reaches a peak at a low voltage, V_p, when the conventional diode has not even begun to conduct. As the applied forward voltage is increased, the tunnel-diode current starts to decrease again and reaches a minimum at the valley point, for a voltage V_v. This decrease in tunnel-diode current with increasing voltage causes a negative slope, or negative-resistance characteristic, which is typical for all tunnel diodes. It is this negative-resistance characteristic that permits the tunnel diode to be used either as an amplifier, an oscillator, or a switching device (gate or 'flip-flop') in computers. When the forward anode voltage is further increased beyond the valley point, however, the tunnelling effect ceases and the current increases in a manner similar to that of a conventional diode.

ZENER DIODES

Semiconductor P-N junction diodes, usually made of germanium or silicon, have been used for many years as power rectifiers. They are capable of rectifying (i.e. converting to pulsating d.c.) relatively large alternating currents, but in comparison with valves have the disadvantage that their maximum safe

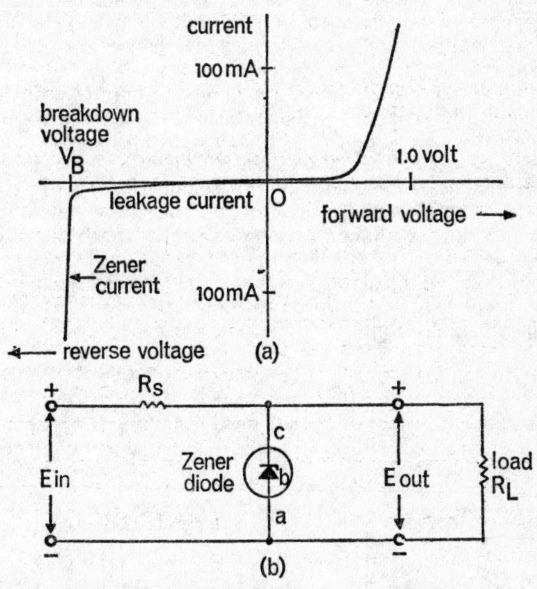

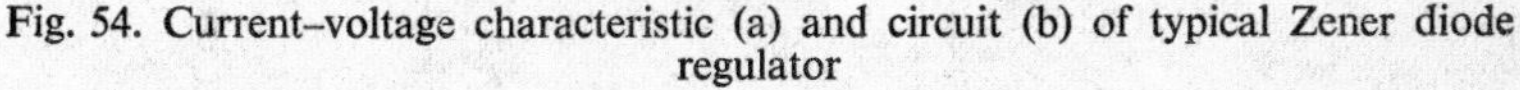

Fig. 54. Current–voltage characteristic (a) and circuit (b) of typical Zener diode regulator

inverse voltage (the voltage opposite to the direction of current flow) is relatively low. What happens if this safe inverse voltage is exceeded? This question is best answered by looking at the current–voltage characteristic of a typical silicon P-N junction diode (see Fig. 54a).

As is apparent from the characteristic, a substantial current flows through the diode when a forward voltage of approximately 1 V is applied. When a reverse voltage is applied, making the anode negative, conduction stops and the diode blocks the reverse current, like any other rectifier. As the reverse voltage is increased, a small but substantially constant leakage current begins to flow, which is characteristic of all semiconductor diodes. The small leakage current does not prevent the diode from functioning as a rectifier. If the reverse potential continues to be increased, however, beyond the safe inverse voltage, a critical voltage, called breakdown voltage, is eventually reached, where the reverse current increases sharply to a high value. The breakdown region is the 'knee' of the characteristic curve in Fig. 54a, and the sudden increase in current is known as avalanche or Zener current (after an explanatory theory by the scientist C. Zener). Conventional diodes and rectifiers never operate in the breakdown region, but the so-called Zener diode makes a virtue of it and operates at this very point.

The advantage of the strangely behaving Zener diode is that it makes a superb regulator at the output of a power supply. Consider the Zener regulator circuit shown in Fig. 54b. The input to the circuit is a direct voltage, E_{in}, whose voltage variations are to be regulated. (E_{in} may represent the output voltage of a power supply, such as described in Chapter 14.) The Zener diode is reverse-connected across the input voltage, so that its cathode is positive and its anode is negative. Since E_{in} is greater than the diode's breakdown voltage, it conducts and draws a relatively large Zener current through the series resistor R_S. A load resistance, R_L, is connected across the output (E_{out}). The total current through R_S, therefore, is the sum of the Zener diode and load currents.

If the input voltage, E_{in}, should increase, the current through both the Zener diode and the load, R_L, will increase. At the same time, however, the Zener diode resistance decreases and the current through the diode increases more than proportionately. As a result, a greater voltage drop will occur across the series resistor R_S and the output voltage (E_{out}) across the diode will be close to the original value. In this way, a Zener diode regulator can maintain the output voltage within a fraction of a volt, when the input voltage may vary over a range of several volts. Moreover, variations in the load resistance have a similar effect on the diode. As the load increases or decreases, the Zener shunt element will draw less or more current, respectively. The net result is a substantially constant output voltage across the Zener diode regulator.

THYRISTORS

The solid-state equivalent of the gas-filled switching tube, or thyratron, we have become acquainted with in Chapter 6 is the silicon controlled rectifier,

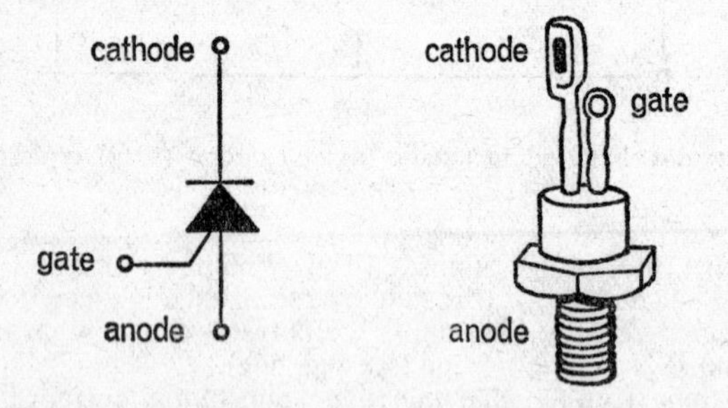

Fig. 55. Symbol and construction of silicon controlled rectifier (thyristor)

or thyristor. This extremely useful PNPN silicon device consists essentially of an alloyed P-N rectifier junction and a diffused P-N-P silicon pellet with a separate contact, known as a gate. The schematic symbol and external construction of a 16-ampere thyristor is shown in Fig. 55.

The addition of the gate drastically changes the characteristics of the rectifier. When the thyristor is operated with reverse voltage (anode negative and cathode positive), it blocks the flow of current until the breakdown or avalanche voltage is reached, as we have seen in the Zener diode (Fig. 54). When a positive voltage is applied to the anode, the thyristor also blocks the flow of forward current, in contrast to conventional rectifiers. Only when a certain critical value of the positive anode voltage, the forward breakover voltage (V_{BO}), is reached, the thyristor switches suddenly to a highly conductive state and the voltage across it drops to a low value (about 1 V). In this conducting state, the current through the device is limited only by the supply voltage and the resistance of the load. Once the thyristor is in the high-conduction state, current flow continues indefinitely until the circuit is interrupted for a brief moment, just like in the thyratron. In practice, the thyristor is operated with an anode voltage somewhat less than the forward breakover voltage and it is then 'turned on' by means of a small pulse (typically 1·5 V and 30 mA) applied to the gate. Once switched to the conducting state, the gate has no further influence over the current, until the anode current is completely interrupted for about 20 microseconds.

Since the thyristor permits control over powerful currents (30–100 A) by means of a small gate pulse, the device is very useful in many relay, switching, and control applications. Typical applications of silicon controlled rectifiers are in regulated power supplies, d.c.-to-a.c. invertors, radar modulators, servo systems, latching relays, electronic ignition systems, etc. The important advantage is, of course, that switching is extremely rapid and requires no moving parts.

SUMMARY

Atoms within a pure germanium (or silicon) crystal are strongly bound together by means of electron-sharing or covalent bonds. Each germanium atom completes its outer valence shell by combining its four electrons in electron pairs with those of adjacent germanium atoms.

A pure germanium crystal is practically an insulator.

When pure germanium is 'doped' with atoms that have five electrons in their outer shell (antimony or arsenic), the impurity or donor atoms make their excess electrons available as negative charge carriers.

When pure germanium is 'doped' with trivalent impurity atoms, such as indium or gallium, the impurity atoms (acceptors) borrow electrons from surrounding germanium atoms, leaving a deficiency of electrons, or holes, in their place. The holes act like mobile, positive charge carriers.

Germanium doped with pentavalent donor atoms is called N-type germanium; current conduction takes place through negative charge carriers, i.e. electrons.

Germanium doped with trivalent acceptor atoms is called P-type germanium; current conduction takes place through positive charge carriers, i.e. holes.

A P-N junction, consisting of wafers of P-type and N-type germanium, may be either grown or fused, depending on manufacturing technique.

When a P-N junction is biased in the forward direction, by applying a positive voltage to the P-region and negative voltage to the N-region, the holes and electrons are repelled towards the junction area and overcome the potential barrier there. Current conduction then takes place by means of electron-hole combinations in the vicinity of the junction.

With reverse bias applied to a P-N junction (P-region negative and N-region positive), holes and electrons are attracted away from the junction area, and current conduction stops except for a small reverse current. If the reverse bias is made very high, the junction breaks down, and a relatively large reverse current flows.

A junction triode transistor is a sandwich made up of two P-N junctions, either in P-N-P form or N-P-N form. The central region is called the base and the two outer layers are called the emitter and collector respectively.

The emitter junction of a transistor is always biased in a forward direction, while the collector junction is biased in reverse direction.

Current conduction in a P-N-P transistor takes place by hole conduction from emitter to collector, while current conduction in a N-P-N transistor is carried on by electrons as majority charge carriers.

The collector current in a junction transistor is less than the emitter current by an amount proportional to the number of electron–hole combinations occurring in the base area.

The ratio of collector-to-emitter current is known as current gain or alpha (α) and is always less than 1 (practical values 0·95 to 0·99).

Transistors may be connected in either of three basic (amplifier) circuits. These are: 1. common-base, emitter-input; 2. common-emitter, base-input; and 3. common-collector, base-input.

The common-base connexion is analogous to the earthed-grid triode valve and it provides a very low input resistance, a high output resistance, and a current gain of less than 1. There is no phase reversal.

The common-emitter connexion, the most flexible and efficient of the three basic connexions, is analogous to the earthed-cathode triode valve, and like the latter it reverses the phase of the output signal with respect to the input. Its input resistance is higher and its output resistance is lower than those of the common-base connexion, but it provides the highest voltage and power gains.

The common-collector connexion is the transistor equivalent of the earthed-anode triode (cathode-follower) and is used primarily as a buffer and for impedance matching. The connexion provides a high input resistance, a low output resistance and a voltage gain of less than 1.

A junction-tetrode transistor has an extra, negatively biased connexion to the base. This effectively extends the upper frequency limit (alpha-cut-off frequency) of transistor operation.

A point-contact transistor uses two 'cat's whisker' electrodes to form a type of P-N-P (or N-P-N) transistor. Its current gain (alpha) is greater than 1.

A unijunction transistor is a P-N junction with two bases, which has a negative-resistance characteristic that is controlled by the positive emitter voltage. This characteristic makes it useful in oscillator and timing circuits.

Tunnel diodes consist of extremely narrow P-N junctions that permit operation at microwave frequencies of several thousand megahertz or switching speeds approaching 10^{-9}s. Because of their negative-resistance characteristic, tunnel diodes are useful as oscillators and bistable switching devices, as well as high-frequency amplifiers.

The Zener diode consists of a reverse-biased silicon P-N junction, which is operated in the breakdown region. The large resulting reverse avalanche or Zener current is useful in voltage regulator applications.

The solid-state equivalent of the gas-filled thyratron is the thyristor (silicon controlled rectifier). It consists of a P-N rectifier junction and a P-N-P silicon pellet to which a contact (the gate) is made. When reverse-biased, the thyristor blocks the flow of current like conventional rectifiers. It also blocks current flow when forward-biased until the forward breakover voltage (V_{BO}) is reached, at which point it switches to a high-conduction state. The thyristor may be turned on by means of a small positive voltage pulse applied to the gate. Once switched on, a large current (30–100 A) flows and the gate loses control. Thyristors are useful whenever large currents must be controlled by means of a small trigger, such as in relay switching, and other control applications.

CHAPTER EIGHT

AUDIO AMPLIFIERS

We are now ready to apply the knowledge we gained in previous chapters about electrons, valves, transistors, and so on, to some practical matters. In the following chapters we shall consider a variety of circuits employing valves, or equivalently, transistors. Circuits are combinations of valves (or transistors) with other components, such as resistors, capacitors, and inductors, and form the basic building blocks of electronic systems: radio, radar, television, and so on. To understand the systems, you must be familiar with the circuits that make them up.

In this chapter we shall discuss amplifier circuits, or more specifically, audio amplifiers. An amplifier is a valve or transistor circuit, which builds up an a.c. signal applied to its input. It is called a voltage amplifier if the magnitude of the output voltage from the amplifier is considerably greater than that of the input voltage. As a matter of fact, the ratio of the output voltage to the input voltage is called the amplification or gain of the amplifier. There are also so-called power amplifiers. These are similar to voltage amplifiers, except that their main purpose is to supply a considerable amount of power (i.e. voltage times current) to the output or load circuit, although the a.c. input signal may not draw any grid current and, hence, the input power may be zero. A power amplifier may also build up the voltage to some extent, but this is of secondary interest. When a number of amplifiers are connected in series (called a casade), so that the output of one serves as the input to the next amplifier stage, the function of the early stages is usually to build up the voltage to a high level, while the last stage builds up the power to a level sufficient to operate a headset, loudspeaker, or similar output device. In the present chapter we shall talk about these early stages in an amplifier chain—the voltage amplifiers—and in Chapter 10 we shall describe the final stage, or power amplifier.

What about the 'audio' in audio amplifier? This term refers to sound or human hearing. As every hi-fi enthusiast knows, the range of human hearing extends from about 20 to 15,000 cycles per second, but varies considerably with age and individual. This means that we can hear air pressure changes or vibrations that vary in pitch from about 20 times per second to about 15,000 times per second. Pitch describes the subjective sensation of hearing an air pressure vibration, or sound. The pitch of a double bass is low, that of a flute or coloratura soprano is high. The more air pressure vibrations per second, the higher the pitch.

But neither valves nor transistors will amplify air pressure vibrations, or sound, directly. It is necessary, therefore, to first convert the air pressure vibrations into equivalent electrical vibrations by means of a microphone or similar acoustic-electrical transducer. This means that if you sing into a microphone a pure tone that has a pitch of, say, 1,000 vibrations per second, the microphone output voltage will be an electrical a.c. signal, or sinewave, that has a frequency of 1,000 cycles per second (c/s) or 1,000 hertz (Hz).

Audio amplifiers, therefore, amplify electrical a.c. signals that have a frequency range corresponding to the range of human hearing, or from about 20 to 15,000 c/s. In recent years, however, hi-fi 'purists' have greatly extended the frequency range of audio amplifiers, so that some commercial models are available with a range from about 5 to over 100,000 c/s. You may wonder at this, since nobody can hear these extreme frequencies; but as we shall see in Chapter 16 it really does make a difference.

BASIC VOLTAGE AMPLIFIER

In Chapter 4 we saw that a triode acts as an amplifier because the anode current is affected to a much greater degree by a change in the grid voltage than by a change in the anode voltage. We then defined the amplification factor (μ) as the ratio of a small change in the anode voltage to the small change in grid voltage required to produce the same change in the anode current (but in opposite direction, of course). We also placed variable voltages on the grid and anode of a triode (Fig. 9) and obtained a series of curves known as the static characteristics of a triode. From the latter, in turn, we were able to calculate the amplification factor (μ), the anode resistance (r_p) and the transconductance (g_m) of the valve.

GRID CIRCUIT

Now let us see what happens under dynamic conditions when an actual alternating signal voltage (e_g) is applied to the grid and a load resistor (R_L) is introduced into the anode circuit to extract useful power from the valve.

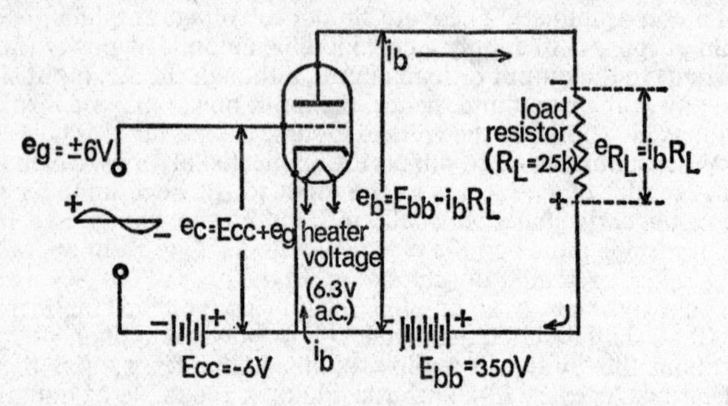

Fig. 56. Basic triode amplifier circuit and notation

Fig. 56 shows this basic triode amplifier circuit, together with the notation employed to designate the various voltages and currents.

In Fig. 56, the capital letters indicate steady or direct supply voltages or currents, while the lower-case letters indicate varying or alternating signal voltages and currents. Some typical circuit values have been assumed in order to make the example more illustrative. Thus, an alternating signal voltage (e_g, indicated by the sinewave) equal to 6 V peak amplitude has been introduced in the grid circuit of the triode, in series with a fixed bias voltage (E_{cc}) from a 6 V battery. You may well wonder why the bias battery is necessary. Assume for a moment, that it is absent, and the 6 V signal voltage is applied to the grid alone. The voltage on the grid will then vary between zero, +6 V, and −6 V, as the alternating signal goes through one cycle. Whenever the input signal goes positive and hits its +6 V peak value, a grid current will flow out from the grid of the valve. This grid current not only distorts the linearity of the characteristic curves, thus leading to distorted amplification, but it also requires a certain amount of power (i.e. grid voltage times grid current), which it extracts from the input signal. If the input signal is too weak to supply the power (as in the case of a tiny radio signal), further distortion takes place, with undesirable results.

For this reason, voltage amplifiers are always operated so as to consume no power in the grid circuit, particularly since the input signal is generally too weak to supply the required power. Moreover, it is the great advantage of valves that they are capable of being purely voltage-operated in the grid circuit, although power may be available from the anode circuit. To avoid consuming power in the grid circuit it is essential, therefore, that no grid current should flow during any portion of the cycle of the input signal. This, then, is the function of the bias voltage. The bias voltage in a voltage amplifier is always of sufficient magnitude and of such polarity as to maintain the grid negative (or at zero volts) during the positive peak of the alternating input signal. In our example (Fig. 56) the bias battery (another source could be used) has been connected to keep the grid at a potential of −6 V, in the absence of an input signal.

When an alternating signal voltage is now applied in series with the bias voltage, the total instantaneous voltage between grid and cathode (symbolized by e_c) is equal to the algebraic sum of the signal voltage and the steady grid bias. Thus,

$$e_c = E_{cc} + e_g$$

In our example, since $E_{cc} = -6$ V and the signal voltage (e_g) varies between +6 V and −6 V, the total instantaneous grid voltage, e_c, will vary between 0 and −12 V. Whenever the signal voltage reaches its positive peak of +6 V, the total grid voltage will go to zero $(+6-6=0)$, and whenever the signal reaches its negative peak of −6 V, the total grid voltage falls to −12 V $(-6-6=-12)$.

ANODE CIRCUIT

Let us now consider what happens in the anode circuit of the triode, while the grid-to-cathode voltage goes through its cycle from 0 to −12 V. Assuming that a voltage is applied to the heater of the tube (6·3 volts a.c., in this case), an anode current (i_b) will flow. This anode current will continue to flow as long as the signal voltage is not sufficiently large to drive the valve to anode-current cut-off. The bias, signal voltage and anode-supply voltage (E_{bb}) in our example (Fig. 56) have been so chosen that the valve does not reach anode-current cut-off, even for the most negative value of the grid voltage, which is −12 V. (You can verify this for yourself by consulting the characteristics shown in Fig. 10.) Anode current will, therefore, flow at all times during the input signal cycle. As a matter of fact, this is a necessary condition for all voltage amplifiers in order to obtain linear, distortionless amplification. You will remember that the characteristics become very distorted or non-linear near anode-current cut-off and, hence, the bias must be chosen to maintain the valve well above anode-current cut-off for the most negative value of the alternating input signal.

With anode current (i_b) flowing at all times, a voltage drop ($i_b \times R_L$) is produced across load resistor R_L in the anode circuit (25,000 ohms, in this case). The voltage drop across this fixed load resistor depends on the value of the anode current, and this current, in turn, is controlled by the grid voltage.

Because of the amplifying action of the valve, a small change in the signal or grid voltage produces a large change in the anode current and, hence, in the voltage across the load resistor. (The voltage across the load resistor is $e_{RL}=i_bR_L$.) By making the load resistance sufficiently high, a large voltage drop is produced across it, resulting in high voltage amplification. Amplification, or gain, is the ratio of the output voltage across the load to the input signal voltage. A large load voltage and, hence, high amplification, may be obtained even with quite small anode currents, since there is no theoretical limit for the size of the load resistor. The output power, therefore, may be quite small.

Since the total anode current, i_b, never becomes negative (or even reaches zero), it is not alternating current. Actually, it consists of two components, one direct and one varying, like the grid voltage (e_c). The d.c. component of the anode current, identified as I_b, is the normal anode current that flows with the anode voltage and bias applied, but with the alternating input signal voltage (e_g) absent. The a.c. component of the anode current (identified as i_p), on the other hand, is the variation in the total anode current caused by the input signal at the grid. The total instantaneous anode current i_b, thus, is the sum of the d.c. (I_b) and a.c. (i_p) components.

Note that in Fig. 56 the anode current (i_b) flows out of the anode, through the load resistor, and then towards the positive terminal of the anode-supply battery (E_{cc}). Since electrons flow from negative to positive, the top of the load resistor is more negative than the bottom or battery-connected end. The load voltage, thus, is in opposition to the anode-supply voltage, and since it is connected in series with it, subtracts from it. The greater the anode current, the greater is the load voltage (i_bR_L) and, hence, the less anode-supply voltage is left over to reach the anode of the valve. The total instantaneous voltage on the anode (identified by e_b), therefore, is the difference between the anode-supply voltage, E_{bb}, and the voltage across the load, e_{RL}. Expressed mathematically,

$$\text{total instantaneous anode voltage } e_b=E_{bb}-i_bR_L$$

From the above expression, it is clear that the total anode voltage (at any instant) becomes smaller as the anode current becomes larger. Furthermore, since the anode current increases directly with the input signal voltage, the total anode voltage decreases with increasing grid or signal voltage. Thus the anode voltage (e_b) is a maximum when the grid voltage (e_c) is a minimum, and vice versa. The grid and anode voltages, therefore, are in *phase opposition* or, equivalently, the valve is said to produce a 180° phase reversal of the anode voltage with respect to the grid voltage. This is a very important fact to keep in mind. The amplification is, of course, not affected by this phase reversal.

GRAPHICAL ANALYSIS

The picture we have just given of the voltage and current relations in a basic triode amplifier circuit can be considerably simplified by plotting the input and output waveforms against the characteristics of the triode, as shown in Fig. 57.

The anode-current/grid-voltage characteristics are shown in the upper left-hand portion of Fig. 57. These are not the static characteristics with which we have become familiar, but the dynamic transfer characteristic, so called because it has been obtained under actual operating conditions for a specific anode-supply voltage (350 V) and a specific (25,000 ohm) anode-load resistor. A new characteristic must be obtained each time the load resistor or the supply voltage is changed. We have assumed an operating point (*N*) on the dynamic transfer characteristic of -6 V grid voltage, equal to the value of the bias battery (E_{cc}) in Fig. 56. The operating point determines the quiescent values of the grid voltage (bias) and anode current in the absence of a signal voltage.

Now let us apply a 6-V alternating input signal to the grid and see what happens at the output. The input signal sinewave *A-B-C-D-E-F-G-H-I-J-K* is plotted against time at the lower left of Fig. 57. The waveform of the output current may be obtained by drawing lines from the points of the input waveform to their intersections with the dynamic transfer characteristic at points *N*, *L*, *M*, *O*, *Q* and then extending the lines from the intersections to the right to plot the corresponding points *A′-B′-C′-D′-E′-F′-G′-H′-I′-J′-K′* of the output-current waveform. By choosing the same time scale for both input and

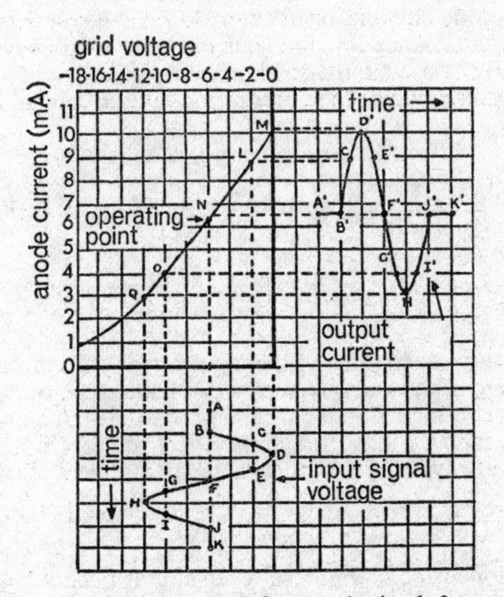

Fig. 57. Typical input and output waveforms obtained from dynamic transfer characteristic of a triode for 25,000-ohm load

output waveforms, the two waveforms may be directly compared. It is seen from Fig. 57 that the output current is a reasonably pure sinewave and not distorted in comparison with the input signal. This excellent amplifier characteristic has been achieved by choosing the operating point (*N*) so that the largest 'swing' of the input signal (from 0 to -12 V on the grid) is accommodated within the linear portion of the dynamic transfer characteristic. If the operating point had been chosen near the curved portion of the characteristic, say at point *Q*, the output-current waveform would have been considerably distorted with respect to a pure sinewave. Note, however, that the bottom portion of the dynamic transfer characteristic (Fig. 57) is much less curved than the corresponding portion of the static anode-current/grid-voltage characteristic (Fig. 10) and, hence, operation is more linear (less distorted) even for large signals extending into the lower portion of the characteristic. This 'straightening out' of the characteristic curve is achieved solely by the insertion of the load resistor in the anode circuit.

AMPLIFICATION

Besides the graphic portrayal of the anode-current waveform, Fig. 57 also permits us to compute the amount of amplification or voltage gain of the triode, when operated in the circuit and with the voltages shown in Fig. 56. You will remember that the voltage gain is the ratio of the output voltage across the load to the input signal voltage on the grid. Since we are dealing with a ratio, a simple way of obtaining the gain is to compare the peak value of the alternating output voltage with the peak value of the signal voltage. The peak output voltage is the product of the peak a.c. anode current (i_p) and the load resistor (R_L). To find the peak value of the a.c. anode current, we must determine the total change in the anode current from its quiescent or no-signal value to its full value at maximum input signal. From Fig. 57 we obtain a quiescent value (at point N) of about 6·4 mA and a full-signal value at point D' of the anode current, or at point M on the characteristic, of about 10·1 mA. Hence, the change in anode current is $10{\cdot}1-6{\cdot}4=3{\cdot}7$ mA, which is the peak amplitude of the alternating current.

Knowing the peak value of the current, the output voltage is

$$e_{out}=i_p\times R_L=0{\cdot}0037\text{ (amps)}\times 25{,}000\text{ (ohms)}=92{\cdot}5\text{ V}$$

The peak value of the input signal we have arbitrarily fixed at 6 V and, hence, the

$$\text{voltage gain}=\frac{e_{out}}{e_g}=\frac{92{\cdot}5}{6}=15{\cdot}4$$

If we had chosen any other value of the input signal and computed the output voltage for this signal, the voltage gain would have come out about the same. Note also that the voltage amplification is somewhat less than the amplification factor (μ), which we have previously determined (in Chapter 4) to be about 20. This is always true for any practical amplifier.

PHASE RELATIONS

We have mentioned the 180° phase reversal that occurs between the input signal and the anode voltage. It is of interest to compare the phase relations between all major input and output waveforms in the basic triode amplifier circuit of Fig. 56. These phase relations are illustrated in Fig. 58 for our previous example of -6 V bias, 350 V anode-supply voltage and 25,000 ohms load resistance. Although this specific example has been chosen, the phase relations shown in Fig. 58 hold true regardless of the particular values of the voltages and currents in the amplifier. The broken vertical lines passing through the four waveforms make it possible to compare corresponding waveforms at the same instant in time.

Curve A represents the instantaneous signal voltage (e_g), which is seen to vary between $+6$ V and -6 V. Curve B shows the variation in the total grid voltage (e_c) with the signal. When the signal is absent, or zero, the grid voltage is equal to the bias, or -6 V. When a signal is present, the total grid voltage rises to zero volts at maximum positive ($+6$ V) signal swing, and it drops to -12 V for the maximum negative ($-$ 6 V) signal swing. Curve C shows the variation in the instantaneous value of the total anode current (i_b) with signal. As noted before, the anode current has a quiescent value (I_{bo}) of 6·4 mA for zero signal voltage, and varies between 3 mA and 10·1 mA for the full (± 6 V) signal swing. Since the anode current increases with increasing grid voltage, waveforms *A*, *B*, and *C* are all in phase (i.e. they reach their positive and negative peaks at the same time).

Finally, curve *D* illustrates the variation in the instantaneous value of the total anode voltage (e_b) with the signal voltage. Note that the quiescent value of the anode voltage (E_{bo}) with the signal zero or absent, is 190 V. You can verify that this must be the correct value by subtracting the voltage drop across the load resistor (with the signal absent) from the anode-supply voltage.

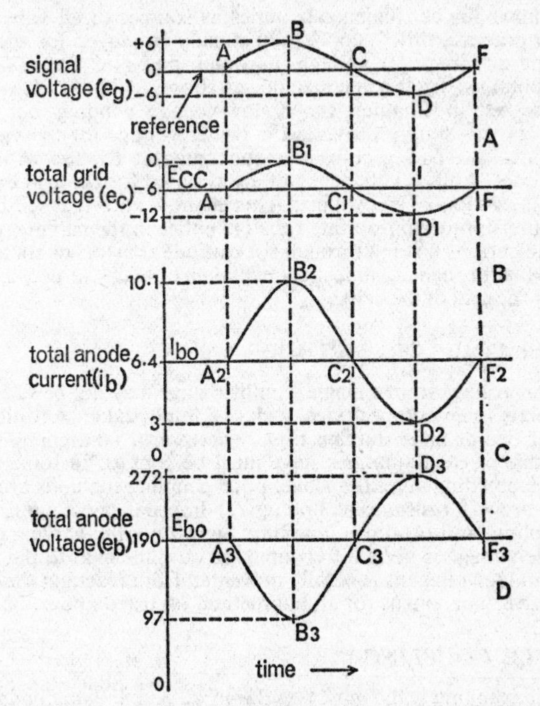

Fig. 58. Phase relations in triode amplifier circuit of Fig. 56

The drop across the load resistor (E_{RL}) is the product of the quiescent (no-signal) value of the anode current (I_{bo}) and the load resistance (R_L), or $I_{bo} \times R_L = 0{\cdot}0064 \times 25{,}000 = 160$ V. The quiescent value of the anode voltage, therefore, is

$$E_{bo} = E_{bb} - E_{RL} = 350 - 160 = 190 \text{ V}$$

Now when the anode current rises to its maximum value of 10·1 mA at full signal (and develops an output voltage across the load of about 92·5 V), the anode voltage is seen to fall in direct proportion to the rise in anode current. It reaches a minimum value of 97 V when the input signal, grid voltage, and anode current reach their maximum values. The drop in total anode voltage from 190 V to 97 V (i.e. 93 V) is seen to be just about equal to the peak value of the load voltage (92·5 V). (Theoretically, it should be exactly equal, of course.) Since the anode voltage reaches its negative peak when the input signal reaches its positive peak, and vice versa, the two waveforms are one half-cycle, or 180°, out of phase with respect to each other. This is the reason why a valve is said to introduce a 180° phase reversal between input and output voltage. Summing up the relations shown in Fig. 58, we see that the anode current is in phase with the grid voltage, but the anode voltage is 180° out of phase with the grid voltage.

SUPPLY VOLTAGE SOURCES

Fig. 56 shows for convenience batteries as sources of all valve operating voltages. In practice, other supplies are usually employed for the operating voltages. The anode-supply voltage (E_{bb}) and fixed grid bias (E_{cc}) are most frequently obtained from a rectifier power supply, with which we shall become acquainted in Chapter 14. A low-voltage winding on the power transformer of this supply provides the heater voltage for the valves of the amplifier, since alternating current is quite suitable for use with indirectly heated cathodes. A fixed voltage is not always used for the grid bias, but the valve sometimes supplies its own bias. This method, known as self-bias, makes use of the insertion of appropriate resistors either in the cathode or grid circuits. Anode current flowing through the cathode resistor, or sometimes grid current flowing through a grid resistor, will then develop the necessary voltage drop to bias the grid of the valve.

AMPLIFIER COUPLING METHODS

The output voltage from a single amplifier stage may not be sufficient to be applied directly to an output device, such as a loudspeaker. Additional amplification over two or three stages is usually necessary. To accomplish this, the output voltage of each amplifier stage must be coupled in some way to the grid of the succeeding amplifier valve. Four coupling methods are in general use. These are (1) resistance-coupling, (2) impedance-coupling, (3) transformer-coupling, and (4) direct coupling. Since only the varying component of the anode voltage is needed to couple the output signal to the next stage, the steady anode voltage is generally prevented from reaching the grid of the following valve. This is true for all but method (4) listed above.

RESISTANCE COUPLING

Resistance coupling is the most popular of all coupling methods because it is cheap and provides excellent audio fidelity over a wide frequency band. Typical resistance-coupled triode and pentode amplifier circuits are illustrated in (a) and (b), respectively, of Fig. 59. The voltage supplies for the heaters of the valves are not shown, but it is understood that they are there.

Note that the first stage of the triode circuit (Fig. 59a) is the same as the basic amplifier circuit of Fig. 56, except that the bias battery (E_{cc}) has been replaced by a self-bias resistor, R_k, in the cathode circuit. The cathode-resistor is shunted by a so-called by-pass capacitor, C_k. You can easily see that this device enables the valve to develop its own bias. When anode current flows from the negative terminal of the anode-supply battery (E_{bb}), through resistor R_k, and from cathode to anode of the valve, the bottom end of bias resistor R_k becomes negative and the top (or cathode) end becomes positive, since electrons flow from negative to positive. Since the bottom end of R_k is connected to the control grid in series with signal voltage e_g, the grid becomes negative with respect to the cathode, thus furnishing the necessary bias voltage. This method of biasing is to some extent self-regulating. If the steady anode current should increase for some reason (such as a fluctuation in the supply voltage), the negative bias developed across R_k will also increase, which, in turn, tends to reduce the anode current again. If the steady anode current should decrease, on the other hand, the bias voltage will also decrease, which will tend to increase the anode current again. Self-bias, thus, assures stable amplifier operation.

Although we have shown, for convenience, a battery for the anode-supply voltage E_{bb}, this voltage is usually obtained from the positive output of a rectifier power supply. The alternating component of the first valve's output voltage, e_{RL}, is coupled to the grid of the following stage through a coupling

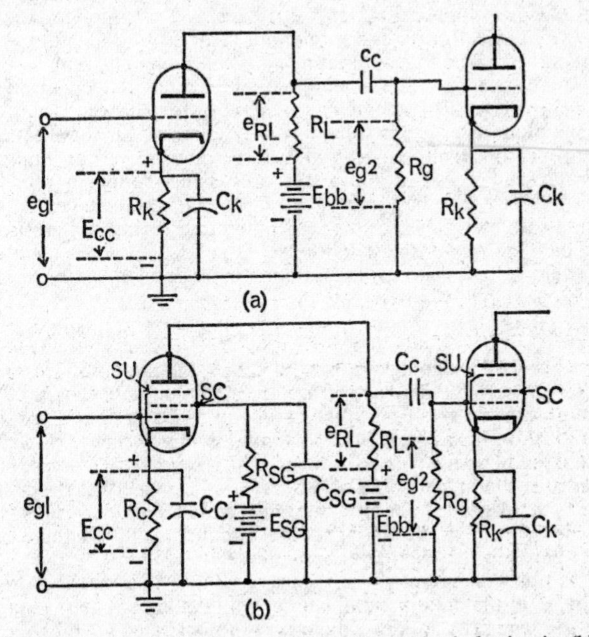

Fig. 59. Resistance-coupled amplifier with self-bias: (a) triode circuit; (b) pentode circuit

capacitor, C_c. The coupling capacitor prevents the steady anode voltage of the first valve from reaching the grid of the second valve and, thus, overloading it. (Remember that a positive grid voltage will result in grid current and distortion.) The input signal to the second stage, e_{g2}, is developed across a grid resistor, R_g, which also acts as a grid return for the bias voltage developed across R_k in the cathode circuit.

You may wonder about the function of the by-pass capacitor C_k across bias resistor R_k. Think for a moment what would happen if it was not there. The current flowing through R_k would then develop an output signal across it, as happens across load resistor R_L. This output signal is in phase opposition to the grid input signal, and since the voltage across R_k is in series with the input signal, e_g, the two signals would tend to partially cancel themselves out, a process known as degeneration. The total input signal to the grid of the first stage would then be considerably reduced, and so would the available voltage amplification. By placing capacitor C_k across the bias resistor, no alternating signal voltage is developed across it, since the capacitor is large enough to present practically a short circuit to alternating currents. The alternating voltage across R_k is therefore by-passed to earth and cannot cancel the input signal at the grid. Care must be taken, however, that capacitor C_k is sufficiently large (i.e. has a low reactance in relation to the bias resistance) to offer a short circuit to the lowest frequency present in the output signal. A value of about 25 microfarads is usually needed. Sometimes the capacitor is purposely omitted to obtain degeneration of the input signal, which also has some advantages, as we shall see later (p. 115).

Pentode Circuit. The pentode circuit shown in Fig. 59b is essentially the same as the triode amplifier, except for the method of supplying the voltages to the additional electrodes. The suppressor grid (Su) of the pentode is generally operated at cathode voltage and, hence, is simply connected to the cathode. The screen grid (Sc), however, requires a positive operating voltage, which is generally less than that of the anode. For convenience, a separate screen-supply battery, E_{SG}, is shown in the figure, which supplies the screen with the proper positive operating voltage through screen dropping resistor R_{SG}. In practice, the screen and anode are usually supplied from the same source, either battery or power supply, and the correct voltages are obtained by choosing suitable values for the screen and anode dropping resistors, R_{SG} and R_L. The screen by-pass capacitor, C_{SG}, serves to by-pass the alternating voltage developed across R_{SG}, in the same manner as by-pass capacitor C_k.

DESIGN CONSIDERATIONS AND FREQUENCY RESPONSE

One good way of designing a resistance-coupled audio amplifier is to choose a valve likely to give the desired voltage amplification, and look up the values of all circuit components and voltages in the manufacturer's data manual. The theoretical design and prediction of amplifier performance is a somewhat complicated mathematical procedure into which we cannot here go, but it may be of interest to point out some of the factors that affect the design and performance of a resistance-coupled amplifier stage. These considerations are equally true for the triode and pentode circuits, although there are some differences between the two in performance.

The easiest part is determining the bias resistor R_k. From the valve characteristics (given in the data manual) we know the quiescent anode current that flows for the available anode voltage and the desired grid bias voltage. By Ohm's law, the bias resistance R_k is then simply the quotient of the grid bias divided by the anode current ($R_k = E_{cc}/I_b$, at the operating point). The by-pass capacitor C_k should have a value of 20 to 25 microfarads to assure effective by-passing at the lowest audible frequencies, as we have mentioned before.

The choice of the anode-load resistor R_L is somewhat more complex. Three main factors determine it—namely, the available anode-supply voltage (E_{bb}), the desired voltage gain and the frequency response. These factors conflict with each other to some extent and a reasonable compromise must be made in each case. As we have seen before, the output voltage, and hence the amplification (gain), increase in direct proportion with the value of the load resistor R_L. We have also seen that the voltage drop across the load resistor subtracts from the available anode-supply voltage. If the anode-load resistor is too large, the voltage available at the anode (E_b) becomes too low for proper operation of the circuit. It then becomes necessary either to reduce the value of the load resistor or to increase the anode-supply voltage. Since the latter is generally fixed, the maximum value of the anode-load resistor is also determined.

It is not immediately apparent why the value of the load resistor should have any effect on the frequency response of the amplifier (i.e. the relative amplification throughout the audio frequency range). We know that inductors and capacitors are frequency-sensitive, but resistors are not. But there is more to the circuit of the audio amplifier than we have shown in Fig. 59. Remember the triode's interelectrode capacitances (Chapter 4): grid-to-cathode capacitance (C_{gk}) and grid-to-anode capacitance (C_{gp}) are directly across the input of the stage, while anode-to-cathode capacitance (C_{pk}) is in parallel with the output of each stage. The operation of the amplifier makes the effect of these capacitances appear even larger than they actually are. In addition, there are various stray and distributed capacitances, due to wire leads and components, which shunt the input and output of the stage.

Imagine now the output capacitance of the first valve, plus the input capacitance of the second valve, plus various stray and distributed capacitances, all connected effectively in parallel with the anode-load resistor R_L, as well as with the grid resistor R_g. Capacitors in parallel act like the sum of the individual capacitors and, hence, one fairly substantial capacitance (50 to 100 micromicrofarads) shunts the anode-load and grid resistors. Since the reactance (opposition to the flow of alternating current) of a capacitor decreases as the frequency goes up, the shunting effect of these various capacitances will make itself felt at the higher audio frequencies and will effectively cut off all frequencies above a certain limiting frequency. The frequency at which this cut-off happens depends on the anode-load resistor R_L and to some extent also on the grid resistor R_g. If the anode-load resistor has a relatively low value, most of the alternating output signal will be developed across it and relatively little will flow through the parallel capacitance. The frequency at which capacitive shunting becomes excessive will then be high and possibly beyond the audible range. If we attempt to make the load resistor too large, however, we will force a substantial part of the output signal to flow through the relatively low shunt reactance of the capacitor and this shunting effect will increase with frequency (since the capacitor's reactance decreases with frequency). The cut-off frequency beyond which the amplifier gain becomes insufficient will then be relatively low.

The desired frequency response of the amplifier is perhaps the most important consideration in limiting the values of the load resistor and grid resistor. Here again a compromise must be made. If we desire large voltage amplification, large values of anode-load and grid resistors are necessary (about

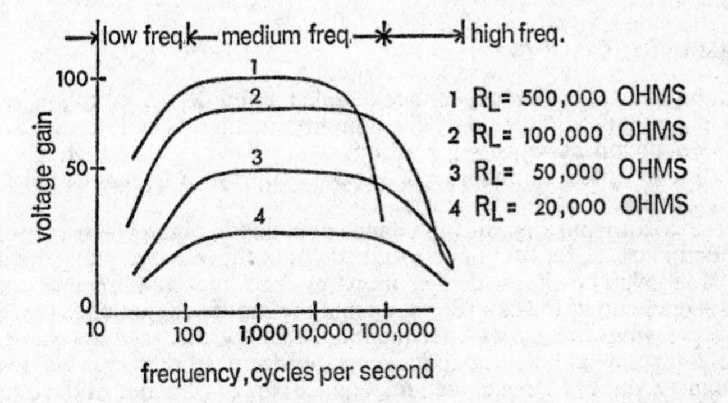

Fig. 60. Frequency response of resistance-coupled amplifier for various values of the load resistance R_L

500,000 ohms in practice), but then the frequency response at the higher audio frequencies will be defective. On the other hand, if we desire a 'flat' frequency response curve over the audible range (see Fig. 60), the anode-load and grid resistor must be made small at the expense of the available voltage amplification. The grid resistor (R_g), however, must always be sufficiently large not to shunt the anode-load resistor appreciably. Fig. 60 shows some typical frequency response curves of a resistance-coupled audio amplifier for various values of the load resistance.

Note in Fig. 60 that the voltage gain of the resistance-coupled amplifier also drops off at low frequencies, below approximately 100 c/s. Again capacitors are to blame. This time it is primarily the effect of the coupling capacitor C_c, but also to some extent the combined effect of the cathode- and screen-by-passing capacitors, C_k and C_{SG} (if present). We have already seen that the by-pass capacitors must have a value sufficiently large so as not to cause any degeneration and resultant loss of gain at the lowest audio frequency we wish to hear. The effect of the coupling capacitor C_c is the converse of the valve's input and output capacitances, since the coupling capacitor is in series with the input voltage to the second stage. This being so, a part of the signal voltage of the second stage is wasted across the coupling capacitor, while the remainder is developed across grid resistor R_g. Since the reactance of the capacitor decreases with frequency, the voltage developed across it (which is proportional to the reactance) is almost always negligible in comparison to that across R_g at medium frequencies, above about 200 c/s. At low frequencies, however, the reactance of C_c may become appreciable and cause a falling off in voltage gain. The value of the coupling capacitor must, therefore, be made large enough so that it has a negligible reactance at the lowest audio frequency to be reproduced. This generally requires values of about 0·01 to 0·1 μF.

This leaves only the design of the screen resistor R_{SG} in a pentode circuit. It is simply made sufficiently large so that the voltage drop developed across it, when subtracted from the screen-supply voltage (E_{SG}), will leave the correct voltage on the screen. For example, if a direct-supply voltage of 300 V is available for both anode and screen, and the screen voltage should be 100 V for proper operation at 0·5 mA screen current, a voltage drop of 200 V (i.e. 300 − 100) is required to be developed across screen resistor R_{SG}. The value of the screen resistor, thus, is 200 V/0·0005 A, or 400,000 ohms. Since it shunts a large resistance, the value of the screen by-pass capacitor C_{SG} can be relatively small, and is generally of the order of 0·1 to 0·5 μF.

IMPEDANCE COUPLING

Fig. 61 illustrates an impedance-coupled amplifier, a coupling method which is rarely used. Since only the coupling method is of interest, batteries have been shown as sources for all operating voltages, although a self-bias resistor for the grid bias and a rectifier power supply for the anode voltage are used in practice.

As is evident from Fig. 61, impedance coupling is obtained by substituting an inductor coil, L, for the anode-load resistor in the resistance-coupled amplifier of Fig. 59. The circuit derives its name from the impedance of the coil, which is made up of the coil resistance and the inductive reactance ($2\pi fL$), the latter increasing directly with frequency. Since the d.c. resistance of the inductor coil is low and the inductive reactance (opposition to a.c.) is relatively high, a large value of the a.c. load impedance can be obtained at the output of the first stage without an excessive voltage drop, as occurs with resistance-coupling. Because of the low voltage drop, the valve can be operated at higher anode voltages than a resistance-coupled stage and, hence, a greater voltage amplification can be obtained. The price paid for this increase in gain is a non-uniform frequency response. Since the inductive reactance of the coil (and hence the load impedance) goes up with frequency, both the output voltage (e_L) and amplification will rise with increasing frequencies. The frequency

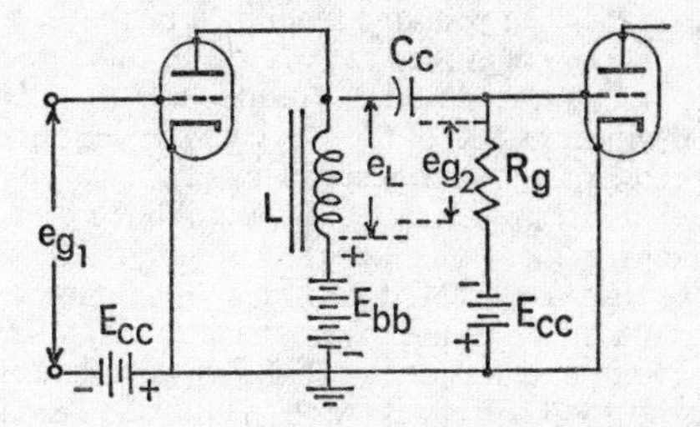

Fig. 61. Impedance-coupled triode amplifier

response of an impedance-coupled stage, therefore, shows a slowly rising amplification with increasing frequencies; or equivalently, a marked falling off in response at lower frequencies. At very high frequencies the amplification begins to drop off because of the distributed capacitances of the valve and of the coil, as in the case of the resistance-coupled circuit. In other respects the impedance-coupled circuit is the same as the resistance-coupled type.

TRANSFORMER COUPLING

The transformer-coupling method, which used to be more popular than it now is, is shown in Fig. 62.

A so-called interstage transformer is used to couple the two amplifier valves with the primary winding connected in the anode circuit of the first valve and the secondary winding to the grid circuit of the following valve.

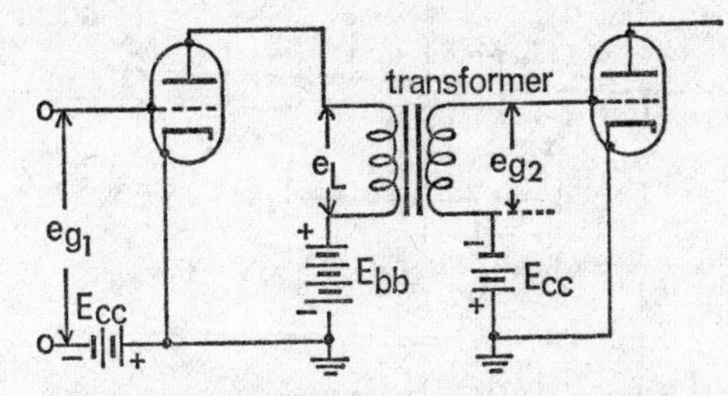

Fig. 62. Transformer-coupled triode amplifier

Since there is no direct connexion between the two windings, the anode voltage of the first valve is isolated from the grid circuit of the second. As in impedance coupling, the primary reactance of the transformer has a high inductive reactance and a low d.c. resistance, and thus wastes little of the d.c. supply voltage. This permits operation at high anode voltages, resulting in large amplification. In addition, transformer coupling has the advantage that the output voltage (e_L) of the first stage may be stepped up by the ratio of the secondary turns to the primary turns. This ratio (symbolized N) may in practice be about 3 : 1. If the primary reactance of the transformer is large in order to obtain a large output voltage (e_L), the amplification of a transformer-coupled stage may approach the product of the valve's amplification factor and the secondary-to-primary turns ratio (i.e. $A \simeq \mu \times N$).

The frequency response of a transformer-coupled circuit is generally rather poor. At low frequencies the primary reactance of the transformer (which is proportional to frequency) begins to fall off, resulting in decreased gain below a certain limiting frequency. At high frequencies the distributed shunting capacitances of the windings and of the valve go into resonance with the transformer secondary reactance, which results in a pronounced peak or rise in the frequency response at the resonant frequency. Above and below this resonant frequency, the response falls off rapidly. Nevertheless, in a properly designed, expensive transformer, the various factors can be so balanced as to achieve a fairly uniform response over the audio-frequency range. But a transformer that achieves a frequency response comparable to a resistance-coupled stage may cost from 10 to 20 times as much as the inexpensive resistance-capacitance coupling network. A further disadvantage of the transformer is its generally large size and weight, and the requirement for proper shielding to prevent interference pickup from stray magnetic fields.

Because of these considerations, transformer coupling is generally reserved as a means of impedance-matching one stage to the next (the primary and secondary impedances can be adjusted to match the output and input of any two stages) and for phase inversion to drive a push-ball power amplifier, which we shall become acquainted with in Chapter 10. Transformer coupling is also employed in tuned r.f. and i.f. amplifiers (see p. 145).

DIRECT COUPLING

In a direct-coupled amplifier the anode of one valve is connected directly to the grid of the next without any intervening capacitor, transformer, or other coupling device. (See Fig. 63.) As a result, both the d.c. anode voltage

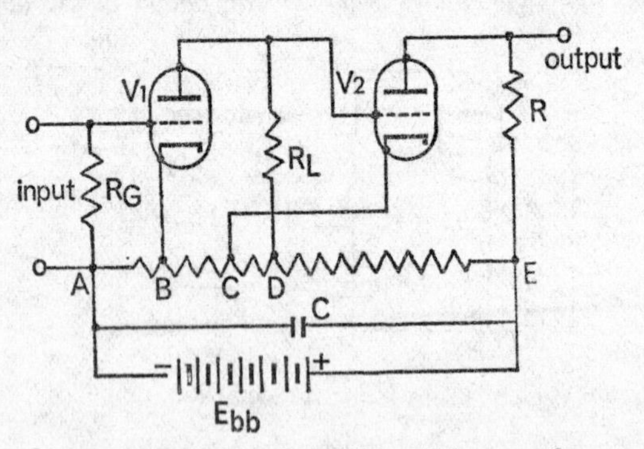

Fig. 63. Direct-coupled Loftin–White amplifier

and the alternating component of the first valve's output signal pass on to the grid of the next stage. As you can imagine, this causes some complications. Since the anode of the first valve must have a positive voltage with respect to its cathode, and the grid of the next valve must have a negative voltage with respect to its cathode, a special voltage divider must be used to obtain the voltages for proper circuit operation. Fig. 63 shows a simple direct-coupled amplifier, called the Loftin–White circuit, in which this neat trick is successfully accomplished.

The proper voltage distribution is obtained through the voltage divider A-B-C-D-E, which is connected across the direct voltage supply (E_{bb}). The most positive point (E) of the supply is connected to the anode of the second valve through anode-load resistor R. The anode of the first valve obtains its d.c. operating voltage from a somewhat less positive point (D) on the divider through anode-load resistor R_L. The output voltage of the first valve is directly coupled to the grid of the second valve so that the latter is at anode-voltage potential. However, by connecting the cathode of the second valve (V2) to a potential more positive than the anode of V1 and the grid of V2 (at point C of the divider), the grid of V2 is actually negative with respect to its cathode. Although it appears that the cathode of V2 is connected to a point (C) less positive than the anode of V1, this is not actually the case, since the anode voltage of the first valve is diminished by the large voltage drop across R_L. Finally, the cathode of V1 is connected to a point (B) on the voltage divider that is more positive than the grid of V1, which is connected to the negative terminal of the supply at point A. This assures the necessary grid bias for the first stage.

Although the voltage-divider network looks deceptively simple, it is quite difficult to design and adjust, since the various currents flowing through it must be taken into account. When the valve voltages are adjusted properly, the circuit serves as a distortionless amplifier with a uniform frequency response over a wide range of frequencies, from zero to considerably above the audio range. The direct-coupled amplifier is especially useful for amplifying very slow variations in the input voltage, since the impedance of the coupling elements does not vary with frequency. Furthermore, since its response is practically instantaneous (because of the absence of time constants), the direct-coupled amplifier is also useful for amplifying pulse signals.

TRANSISTOR AMPLIFIERS

Fig. 64a shows a two-stage resistance-coupled amplifier and Fig. 64b shows a transformer-coupled circuit, which correspond to the circuits shown in Figs. 59 and 62, respectively. The outstanding advantages of these circuits are, of course, that no heater voltage supply is required and that a small (6 to 12 V), d.c. source takes care of all transistor current needs, thus making battery operation quite economical.

The two-stage resistance-coupled amplifier (Fig. 64a) makes use of two P-N-P-type transistors in a common-emitter connexion, which is quite similar to the conventional triode-valve amplifier circuit. The main difference between this circuit and the resistance-coupled valve amplifier shown in Fig. 59a is the method used to obtain correct biasing for the emitter and collector electrodes. You will recall from Chapter 7 (Figs. 43 and 49) that in a P-N-P transistor the emitter junction must be forward (positively) biased, while the collector junction is reverse (negatively) biased. In the circuit of Fig. 64a the necessary bias voltage and polarity is obtained from a single voltage source in a rather interesting manner. Typical values are shown to illustrate the operation. A voltage divider, consisting of resistors R_1 and R_{g1} (in stage 1), is connected across the battery, and the junction of the two resistors is connected to the base of the transistor. Because of the ratio of the two resistors (10 kΩ to 100 kΩ, or 1 : 10), the base is only about 1·1 V negative with respect to earth and the emitter, or conversely, the emitter is about 1·1 V positive with respect to the base and, thus, is furnished with the required forward bias. The current in the emitter

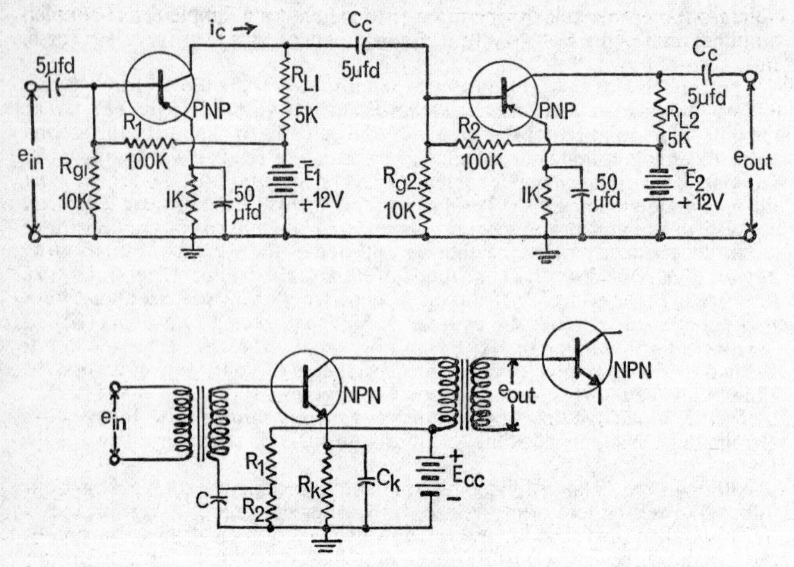

Fig. 64. Transistor amplifiers: (a) two-stage resistance-coupled circuit; (b) transformer-coupled circuit

circuit is then essentially the bias voltage divided by the 1 kΩ emitter resistor, or about 1·1/1000 = 1·1 mA. To prevent degeneration of the input signal, the emitter resistor is by-passed with a large (50 μF) capacitor.

The required negative collector voltage is obtained through load resistor R_{L1}, which is connected directly to the negative terminal of the battery. The output voltage developed across R_{L1} is coupled through the 5 μF capacitor (C_c) to the base of the following transistor. The input voltage to this transistor is developed across resistor R_{g2}, which also serves as part of the bias voltage divider for stage 2. In all other respects stage 2 is identical to stage 1. The final output voltage (e_{out}) from load resistor R_{L2} is coupled through C_c to an output device, such as a loudspeaker or headphones. The voltage amplification obtained from this circuit is approximately 5,000.

Fig. 64b shows one stage of a transformer-coupled transistor amplifier. Here N-P-N transistors have been used in a common-emitter connexion, which requires a positive collector bias and a negative emitter potential, as you will recall from Chapter 7. Except for this polarity reversal, biasing is obtained in the same way as for the circuit shown in Fig. 64a. The input signal from the transformer secondary is returned to earth through by-pass capacitor *C*. In other respects the circuit corresponds to transformer-coupled triode amplifier shown in Fig. 62. Note, however, that in both these transistor circuits the values of the by-pass and coupling capacitors are rather large, compared to those used with valve amplifiers. This is necessary so that the capacitive reactances at the lowest audio frequencies of interest will be low compared with the relatively small associated resistors and transistor impedances.

SUMMARY

In a voltage amplifier, the amplification or gain is the ratio of the output voltage across the load resistor to the input voltage on the grid.

Audio amplifiers amplify alternating electrical signals that have a frequency range corresponding to the range of human hearing, from about 20 to 15,000 c/s.

Voltage amplifiers are always operated so as not to consume any power in the grid circuit. To avoid grid current and the resulting power consumption and distortion, a negative bias voltage equal to or greater than the positive peak of the input signal is applied to the grid. The total grid voltage, thus, is equal to the difference between the input signal and the bias. For distortionless amplification, the bias is chosen to maintain the flow of anode current during the entire input-signal cycle.

The amount of voltage amplification depends primarily on the amplification factor and the value of the anode-load resistor.

The total instantaneous anode voltage is the difference between the direct anode-supply voltage and the alternating signal voltage across the load.

The input signal at the grid and the output signal or anode voltage are 180° out of phase (i.e. in phase opposition). The anode current, however, is in phase with the grid voltage.

The operating point on the dynamic transfer characteristic determines the quiescent values of the grid bias, anode voltage and anode current, in the absence of an input signal voltage. The operating point must be chosen to accommodate the largest 'swing' of the input signal within the linear portion of the characteristic.

The operating voltages for the amplifier are not usually secured from batteries. The anode voltage is generally provided from a rectifier power supply, while a low-voltage winding on the power transformer furnishes heater power. Either fixed bias, from a tap on the output of the power supply, or self-bias is employed. Self-bias is obtained by inserting a resistor in the cathode circuit of the valve, whose value equals the desired bias divided by the value of the anode current that flows in the absence of a signal.

Amplifier stages are coupled together by (1) resistance-coupling, (2) impedance-coupling, (3) transformer-coupling, and (4) direct-coupling.

Resistance-coupling is most popular because it is economical and provides excellent fidelity over the audio-frequency range.

The by-pass capacitors must be sufficiently large to have a low reactance compared with the associated resistances at the lowest audio frequency of interest.

The high-frequency response of a resistance-coupled amplifier is limited by the valve's input and output (interelectrode) capacitances and by the stray wiring and distributed capacitances.

The low-frequency response of a resistance-coupled amplifier is limited by the coupling capacitance and also by the value of the by-pass capacitors.

The choice of the anode-load and grid resistors in a resistance-coupled amplifier is a compromise between large voltage gain (for large resistors) and flat high-frequency response (for small resistors).

Impedance coupling and transformer coupling permit an increase in gain at the price of uniform frequency response. Transformer coupling produces a resonant peak in the medium-frequency range. The frequency response falls off above and below this peak. Transformer coupling is used primarily for impedance-matching the output of one stage to the input of the next and to provide phase inversion for a push-pull power amplifier. It is also used in tuned r.f. and i.f. amplifiers.

In a direct-coupled amplifier the anode of one valve is connected directly to the grid of the next without intervening capacitors, transformers, or other coupling devices. Consequently, both the anode voltage and the alternating component of the output voltage pass on to the grid of the next stage.

Direct-coupled amplifiers are useful for amplifying slow variations in the input voltage and for amplifying pulse signals.

CHAPTER NINE

WIDEBAND (VIDEO) AMPLIFIERS

We have already mentioned (p. 74) that some audio amplifiers reproduce frequencies far beyond the audible range. This extended response becomes of great importance in the so-called wideband or video amplifiers. In order to merit the description 'wideband' an amplifier must be capable of responding to frequencies from a few hertz (cycles per second) to several million hertz (see Fig. 65). The term video amplifier originated from the fact that wideband

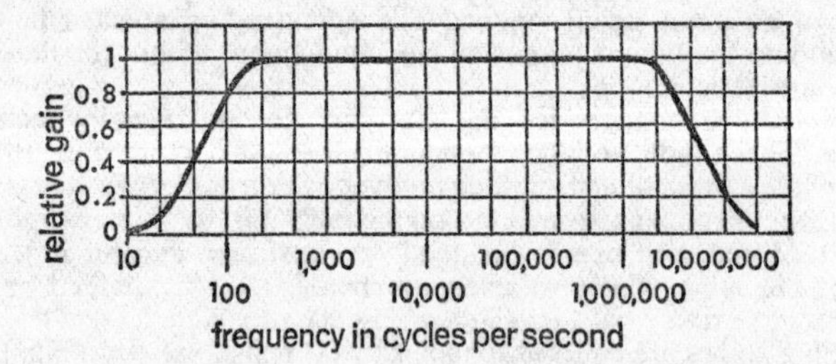

Fig. 65. Frequency response of typical video amplifier

amplifiers were first developed to handle the picture (video) signals of television systems, which require an almost 'flat' frequency response from about 60 cycles per second to 4 megacycles per second. At the present time, however, video amplifiers are used in a variety of applications besides television, such as radar systems and other devices employing pulse-type signals.

VIDEO-AMPLIFIER PULSE RESPONSE

So far we have been dealing exclusively with sinewaves, which are gradually changing alternating currents or voltages. Each cycle consists of a positive and a negative alternation, and the frequency of the voltage or current refers to how many cycles are completed each second. In contrast, a pulse is an abruptly changing voltage or current, which may or may not repeat itself. The simplest non-repetitive pulse is the step voltage (or current) shown in Fig. 66a. Such a step voltage may be obtained, for example, by connecting a voltmeter across a battery through a switch and then suddenly closing the switch. The voltmeter would read zero up to the time when the switch is closed, whereupon the voltage would suddenly rise to its maximum value and stay there. But even the most sensitive voltmeter would take a certain amount of time to indicate the maximum value of the battery voltage, while the ideal voltage step takes no time at all to reach its maximum. This sudden rise to maximum amplitude

is characteristic of all square or rectangular-shaped pulses. In practice it is somewhat easier to obtain a square or rectangular pulse waveform, as illustrated in Fig. 66b (at left), than a step voltage. By applying such a square (or rectangular) waveform to the input of an amplifier and repeating it a number of times per second, the response of the amplifier to the pulse waveform may be conveniently studied.

The pulses shown in Fig. 66 are, clearly, not sinewaves. The French mathematician Fourier demonstrated long before the advent of video amplifiers, that an abruptly changing non-sinusoidal waveform may be made up mathematically of a series of sinewaves of varying amplitudes and harmonically related frequencies. As a matter of fact, an infinite number of sinewave frequencies are needed to make up or reproduce the abruptly changing pulses shown in Fig. 66. If only a few sinewave frequencies are available to compose the pulses, either mathematically or in an amplifier, the pulse waveforms will be distorted in various ways and the distortion will become more serious the fewer frequencies are available. Hence, the greater the frequency range of the amplifier (called bandwidth), the smaller is the distortion of pulse waveforms.

Now it is a very laborious procedure to determine the frequency response of an amplifier by plotting it step by step. In contrast, it is very simple to apply square or rectangular pulses to the input of the amplifier and to display the resulting output on a cathode-ray oscilloscope. Since the relative distortion of the pulses, as reproduced by the amplifier, tells a great deal about its frequency response, the second method is much preferred. Moreover, since most video amplifiers are used for the amplification of complex pulse waveforms (such as in radar and television) rather than sinewaves, the square-wave method of amplifier testing and design directly gives the amplifier response to abruptly changing waveforms (pulses).

Fig. 66b shows some of the things that can happen when an ideal square wave (good approximations of which are easily obtained) is applied to the input of an amplifier. The distorted waveform present in the output of the average amplifier is shown at right of Fig. 66b. The principal changes introduced by the amplifier are: (1) the amplitude of the pulse rises and falls at a finite rate, rather than instantaneously; (2) the amplitude of the reproduced pulse initially overshoots the correct value, and when falling, it will undershoot the minimum value; (3) the top of the reproduced square wave falls off or sags with time, instead of being flat. Any or all of these defects may be present simultaneously in the amplifier output.

The rate at which the output voltage rises when an input pulse is suddenly applied, called the rise time, represents the amplifier's capacity to reproduce sudden changes in amplitude of the applied signal. The initial response of an amplifier to the sudden rise of a pulse is termed the transient response and, as we have seen, it is related to the sinewave frequency range or bandwidth of the amplifier. The rise time of a pulse reproduced by the amplifier is usually measured by the time it takes the output voltage to rise from 10% to 90% of its maximum value (not considering the overshoot). As thus defined, the rise time is inversely proportional to the bandwidth of the amplifier and it is approximately given, for zero or small overshoots, by the relation

$$\text{rise time (seconds)} = \frac{0{\cdot}35}{\text{bandwidth (c/s)}}$$

The bandwidth in the relation just stated is defined as the highest frequency, in cycles per second, for which the overall response of the amplifier does not fall below 70·7% of its maximum response at medium frequencies. Of course, the above relation can be worked the other way; the bandwidth in c/s can be computed by measuring the 10–90% rise time of a square-wave or rectangular pulse applied to the amplifier and dividing the result into 0·35 (i.e. bandwidth = 0·35/rise time).

The amount of overshoot present in the amplifier output depends on the sharpness with which the high-frequency response of the amplifier falls off and also on the amount of phase shift introduced by the various capacitive and inductive elements. A response curve that falls off only moderately at the high-frequency end, together with small phase shift, produces only a small or zero overshoot. When overshoot is present, however, undershoot is also there.

Finally, the amount of falling off or sag, of the top of the reproduced pulse, depends on the phase shift and frequency response of the amplifier at the low-frequency end. If the low-frequency response is good, sag will be small, resulting in the desired flat top.

HIGH-FREQUENCY COMPENSATION

The resistance-coupled amplifier described in the last chapter represents a fair attempt at a wideband amplifier, provided the anode-load resistor (R_L) of each stage is made small enough to achieve the desired wide frequency response. You will recall that by sacrificing voltage gain and using low anode-load and grid-coupling resistors, a frequency response can be achieved that is uniform to frequencies far beyong the audio response range. The amplification at medium frequencies of such an uncompensated resistance-coupled amplifier is approximately equal to the product of the valve transconductance (in mho) and the anode-load resistance. Using, for example, a miniature valve, which has a transconductance of 5,100 micromho, and making the anode-load resistor (R_L) of the amplifier equal to 10,000 ohms, we obtain for the maximum mid-frequency amplification $= g_m R_L = 0{\cdot}0051 \times 10{,}000 = 51$.

Such an arrangement has a rise time in microseconds equal to approximately 2·2 times the product of the anode-load resistance and the combined input and output capacitance (in μF) of the valve. The manufacturer lists an input capacitance of 4 μμF and an output capacitance of 2·8 μμF, making a total (C_{Tot}) of 6·8 μμF.

Thus, for our present example,

the rise time $= 2{\cdot}2 R_L C_{Tot} = 2{\cdot}2 \times 10{,}000 \times (6{\cdot}8 \times 10^{-6}) = 0{\cdot}15$ microsecond.

Consequently, from our previous relation, the

$$\text{bandwidth} = \frac{0{\cdot}35}{\text{rise time (sec)}} = \frac{0{\cdot}35}{0{\cdot}15 \times 10^{-6}} = 2{\cdot}33 \times 10^{6} \text{ c/s}$$

or 2·33 megahertz, which is rather good for a resistance-coupled amplifier.

While an uncompensated resistance-coupled amplifier provides a fair bandwidth and rise time without overshoot, its characteristics fall short of the requirements of television and pulse systems. As we have seen, the resistance-coupled amplifier is beset with deficiencies at both ends of the frequency band. At high frequencies, the valve's combined input and output capacitances together with stray capacitances shunt the anode-load resistor, thus limiting the amplification at these frequencies. At low frequencies, the reactance of the coupling capacitor becomes large compared to the grid input resistor, which again limits the gain and leads to a falling off in response at low frequencies. Equivalently, the resistance-coupled amplifier reproduces pulses with a relatively long rise time and with a pronounced sag at the top of the waveform.

SHUNT COMPENSATION

It is possible to improve the high-frequency response of a resistance-coupled amplifier markedly by adding a small inductance L_1 in series with the anode-load resistor R_L. (See Fig. 67a.) This arrangement is known as shunt

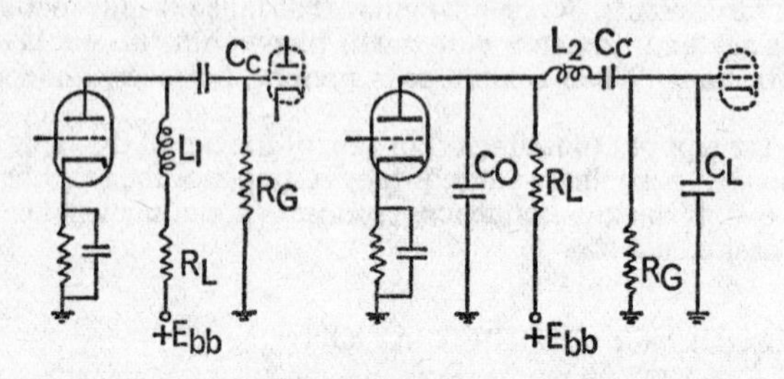

(a) shunt compensation (b) series compensation

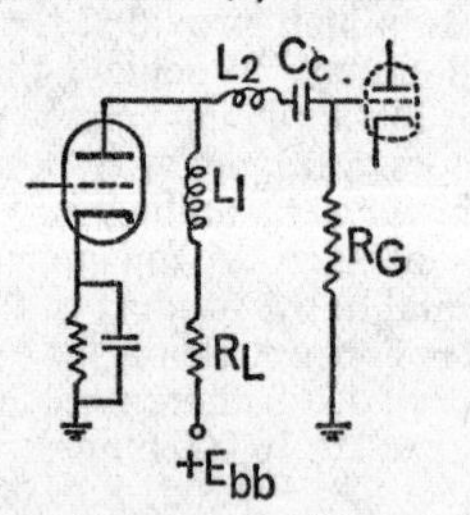

(c) shunt–series compensation

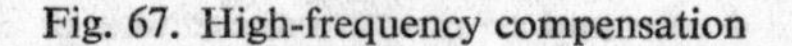

Fig. 67. High-frequency compensation

compensation or shunt peaking, because the increased reactance of the shunt coil L_1 at high frequencies serves to peak the response at these frequencies to compensate for the falling off in response due to the total shunt capacitance (C_{Tot}). Shunt peaking improves the high-frequency response and the rise time, provided the bandwidth demands are not too great and there are only a few stages of amplification. The amount of peaking introduced by the coil depends on a quality or 'Q' factor (see p. 141), which is defined as

$$\text{shunt-peaking } `Q' = 2\pi f L_1/R_L = 6{\cdot}28 f L_1/R_L$$

where L_1 is the shunt-peaking inductance, R_L is the anode-load resistance and f is the frequency at which the response of the uncompensated amplifier (i.e. without the coil) has dropped to 70·7% of the mid-frequency gain.

Fig. 68 shows a universal response curve for a shunt-peaked amplifier, which graphically illustrates the dependence of the frequency response on the 'Q' factor, as defined above. The vertical axis shows the relative voltage amplification, while the horizontal axis gives the ratio of the actual frequency to the frequency at which the amplification is down to 70·7% of the mid-frequency value, when no compensation is used.

For comparison, the case of the uncompensated resistance-coupled amplifier has been dashed in. By definition, its response is down to 70·7% of the mid-frequency value, when the frequency ratio is 1. The response at that frequency is strikingly lifted as the 'Q' of the shunt-peaking coil in a compensated amplifier is increased, so that for a Q of 0·55, the response is even greater than at the mid-frequency. This large amount of peaking is, however, not desirable, since it, too, represents a distortion of the frequency response and, also, because it leads to large overshoots in the reproduction of pulse waveforms.

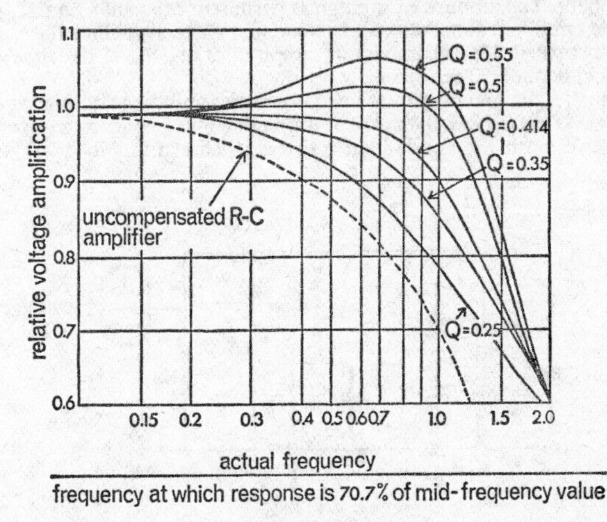

Fig. 68. Relative frequency response of shunt-peaking amplifier for various values of quality factor 'Q'

The curve for $Q=0{\cdot}5$ gives nearly constant amplification up to the compensated frequency, while the curve for $Q=0{\cdot}414$ is of maximum flatness, having no peak at all. In the curves for Q equal to 0·35 and 0·25, the response drops off moderately at high frequencies, but these are characterized by small or zero overshoot in the pulse rise and excellent phase-shift characteristics. Even for $Q=0{\cdot}25$, the speed of rise is 1·4 times as great as that of the uncompensated amplifier, while for $Q=0{\cdot}414$, it is 1·7 times as great.

SERIES COMPENSATION

Another method of boosting the high-frequency response of a resistance-coupled amplifier, shown in Fig. 67b, is known as series compensation or series peaking. In this case a small inductance coil, L_2, is connected in series with the coupling capacitor (C_c). At high frequencies the coil L_2 resonates with the input capacitance (C_i) of the next stage, thus causing an increased voltage across C_i. Since both C_i and grid resistor R_g are across the input of the next valve, the input voltage and, hence, the gain are boosted.

Fig. 67c shows a shunt–series compensation circuit that combines the high-frequency peaking of shunt coil L_1 with the resonant effect of series-peaking coil L_2. The high-frequency response of an amplifier can be further improved by so-called four-terminal compensating networks, which use additional elements besides series and shunt-peaking coils.

LOW-FREQUENCY COMPENSATION

We have seen that the low-frequency response of an uncompensated resistance-coupled amplifier falls off because of the increasing reactance of the coupling, grid and screen by-pass capacitors at low frequencies. These low-frequency deficiencies make themselves felt in the inability of the amplifier to reproduce the flat top of a square or rectangular wave without a pronounced sag (Fig. 66b). The amount of sag that is permissible depends on the application of the amplifier. For example, in television video amplifiers the sag is not allowed to exceed 5% (of maximum amplitude) at 50 c/s, the time of one picture field being 1/50 second.

Fortunately, the loss of gain at low frequencies and the resulting sag can be compensated for fairly easily by the addition of a low-frequency compensating filter in series with the anode-load resistor, as shown in Fig. 69. The filter,

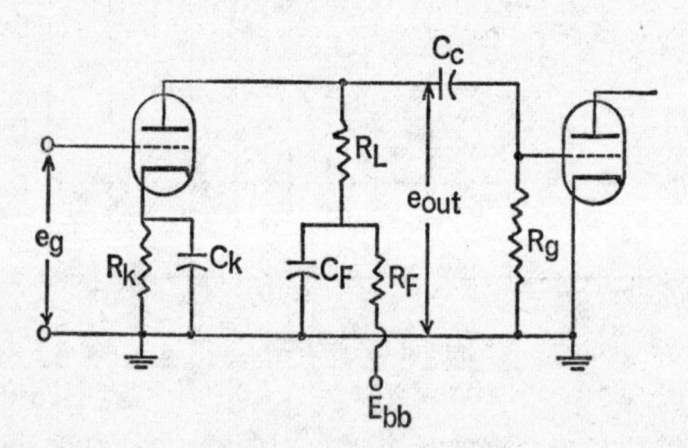

Fig. 69. Low-frequency compensation in resistance-coupled amplifier

consisting of R_F and C_F, has a double effect. First, it introduces a phase shift in the anode circuit that compensates for the phase shift in the coupling circuit R_gC_c. Second, by increasing the anode-load impedance at low frequencies because of the increased reactance of C_F, the low-frequency voltage gain of the amplifier is maintained practically constant. This, of course, eliminates sag. The actual proportioning of the filter elements, R_F and C_F, is usually a compromise between the allowable voltage drop across R_F and the filter values needed for perfect compensation.

SUMMARY

An amplifier having a relatively uniform frequency response from a few hertz to several megahertz is called a wideband or video amplifier: video amplifiers are used in television and pulse-type systems.

A pulse is a non-sinusoidal, abruptly changing voltage or current waveform, which may be distorted by an amplifier in various ways. The response of an amplifier to a pulse or square wave is indicative of the amplifier's transient response and its total frequency range (bandwidth).

The principal modifications of a square wave, applied to the input of an amplifier, are (1) the finite rise time (measured from 10 to 90% of maximum amplitude), (2) an initial overshoot and final undershoot of the correct square-wave amplitude, and (3) a sag or falling off of the top of the reproduced square wave.

The rise time in seconds for a reproduced pulse is inversely proportional to the bandwidth of the amplifier, and for small overshoots is approximately given by 0·35/bandwidth, where the bandwidth is measured by the highest frequency for which the response does not fall below 70·7% of the mid-frequency value.

The amount of overshoot and undershoot present depends on the sharpness of the high-frequency fall-off in amplifier response.

The amount of sag present depends on the phase shift and low-frequency response of the amplifier.

The high-frequency response of a resistance-coupled amplifier may be compensated for by adding a small inductance coil either in series with the anode-load resistor (shunt compensation), or in series with the coupling capacitor (series compensation), or in series with both.

The low-frequency response of an amplifier may be compensated for by adding a suitable resistor–capacitor filter in series with the anode-load resistor.

CHAPTER TEN

POWER AMPLIFIERS, DISTORTION, AND FEEDBACK

We have not yet talked about the process of obtaining useful power from an amplifier to operate a loudspeaker or other output device. The distinction between voltage and power amplification is somewhat artificial, since useful power (i.e. the product of voltage and current) is always developed in a load resistance through which current is flowing and across which a voltage drop exists. The distinction between the two types is really one of degree: it is a question of how much voltage and how much power. If you have a tiny radio signal, which may have a magnitude of a ten-millionths of a volt, you could never hope to obtain sufficient amplification from one valve to operate a loudspeaker directly; and there are other problems you would run into. However, if you magnified the amplitude (or voltage) of the signal first by a number of voltage-amplifier valves, you could apply the strengthened signal to the input of a final power amplifier stage, specially designed to provide large power output to a loudspeaker. Not only are there amplifiers designed to provide ample voltage amplification in the early stages and adequate output power in the final stage, but the valves themselves are specially designed for each job, voltage amplifiers having large amplification factors and a relatively high anode resistance, while power-amplifier valves provide large anode currents through relatively low anode resistances and with small amplification.

In the present chapter we shall explore some of the means and circuits used to secure large power amplification. We shall also have a closer look at the various distortions a signal may undergo when passing through an amplifier, and a powerful method of combating this distortion—negative feedback.

AMPLIFIER CLASSES

Besides being classified as to frequency (audio, video, r.f., etc.) and type (voltage or power), amplifiers are also frequently classified according to their mode of operation as either Class A, Class AB, Class B, or Class C. This classification depends on the portion of the input signal cycle during which anode current is expected to flow, and it is especially applicable to power amplifiers.

In a *Class A amplifier* the signal voltage and grid bias are so adjusted that anode current flows at all times, throughout the full cycle of the applied signal voltage. This type of amplifier is characterized by excellent fidelity and low distortion, relatively low power output for a given valve type and low (20–35%) efficiency (i.e. the a.c. power output for a given d.c. input power).

A *Class B amplifier* is adjusted so that the grid bias is approximately equal to the cut-off value of the anode current, when no input signal is applied. Thus, when a signal voltage is applied, the anode current in one valve of a Class B amplifier (usually two are used) flows only for about one-half of each cycle. Class B amplifiers have somewhat higher distortion than Class A, and provide intermediate power output and efficiency (50–60%).

A *Class AB amplifier* is designed to operate with an input signal voltage and grid bias that permits the anode current to flow for considerably more than half, but less than the entire cycle of the input voltage. Since these amplifiers automatically operate as Class A at low signal levels, they have the advantage of low distortion for small signals and medium power output and efficiency at high signal levels.

In a *Class C amplifier* the grid bias is adjusted to be greater than the value required for anode-current cut-off. No anode current flows, therefore, in the absence of a signal, and the anode current in each Class-B valve flows for less than half of each cycle of the input voltage, when a signal is applied. Class C amplifiers are characterized by relatively high distortion, high power output and excellent efficiency (70–75%). Because of their inherent distortion, Class C amplifiers are never used as audio amplifiers, but they are primarily employed in the final stages of radio transmitters. Audio amplifiers use almost exclusively Class A or Class AB amplification, if high-fidelity reproduction is desired, while Class B amplifiers are occasionally used in car radios.

You may sometimes find the suffix '1' or '2' added to the amplifier class; this denotes the presence or absence of grid current flow during the input cycle. Thus, a Class AB_1 amplifier does not draw any grid current during any part of the cycle, while the grid of a Class AB_2 is driven positive during a part of the input signal cycle, so that grid current flows during this portion of the cycle. The use of Class A_2 or Class AB_2 is frowned upon by designers of hi-fi audio amplifiers because of the difficulty of avoiding distortion in the presence of grid current. If a small amount of grid current can be tolerated, Class AB_2 operation with large input signals is an economical means of raising the available output power from a valve. Amplifier classifications are not rigid, but depend to some degree on the user. While a specific output stage of an amplifier may have been designed to operate strictly as Class A, you may easily be able to overload it and change its classification by operating it from a large input signal source. When turning up the volume for such a large input signal, the power stage may at first operate as Class AB_1 with increased power output and hardly noticeable distortion, but if you turn the gain up high enough, it will eventually operate as Class AB_2, with high output and high distortion.

CLASS A POWER AMPLIFIERS

Let us now study the operation of Class A power amplifiers in more detail, since this is the type most frequently used in well-designed audio amplifiers. The discussion that follows also applies to some extent to Class AB and Class

B amplifiers, since it is usually only necessary to modify the grid bias and anode voltage slightly to change from Class A to Class AB or B.

The output of a power amplifier stage in an audio amplifier generally supplies power to the voice coil of a loudspeaker, designed to convert the electrical audio signal variations into air pressure vibrations, or sound. In this application, sufficient power to move the coil of the loudspeaker through relatively large displacements is more important than high voltage amplification. Consequently, the valves used for this final stage are designed to sacrifice voltage amplification for high power handling capacity. As we have seen, this generally means high anode currents and high voltages, a high transconductance, a low amplification factor and a relatively low anode resistance. Either triodes, pentodes, or beam-power valves can be used for Class A power amplifiers. The triodes have the least amount of distortion, but also the smallest power output for a given signal, and the lowest efficiency. Pentodes have higher power output with a small input signal and also somewhat more distortion. Finally, beam-power valves have still higher power output for a given signal, and increased efficiency than either type, with about the same distortion as pentodes. Although hi-fi purists favour triodes, pentodes and especially beam-power valves are quite popular in high-quality audio power amplifiers, because they require less voltage amplification and give more output than triodes. By using beam-power valves in a push-pull circuit and applying negative feedback, the distortion can be practically eliminated, as we shall see later (p. 108).

Fig. 70 shows a typical triode power output stage, which is transformer-coupled to the speech coil of a loudspeaker, represented by the load impedance

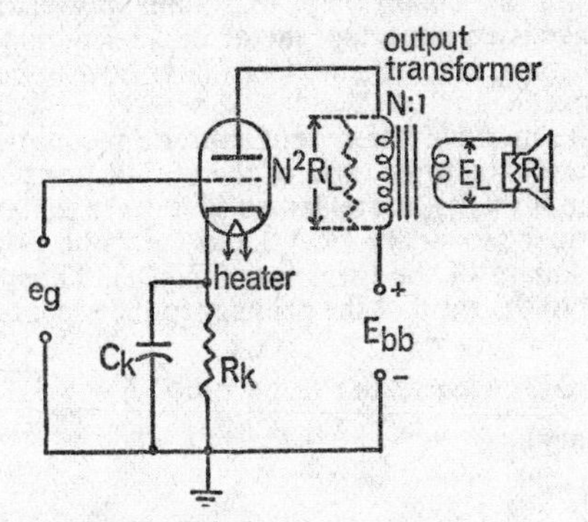

Fig. 70. Triode power-output stage

R_L. A pentode or beam-power valve circuit would be similar, except that provisions would have to be made to supply the screen with a d.c. operating voltage through a screen resistor.

Transformer coupling is always used in the output stage of valve amplifiers, rather than resistance coupling, to match the anode resistance of the power amplifier valve to the speech-coil impedance of the loudspeaker. The speech coil of a loudspeaker consists of a few turns of wire, which have a d.c. resistance of a few ohms and an a.c. impedance (opposition to the flow of alternating current) between 4 and 16 ohms, depending on the speaker. Since the anode resistance of the power amplifier may be anywhere from about 1,000 ohms to over 100,000 ohms, very little power could be developed across the tiny impedance of the speech coil if it were directly inserted into the output circuit of the valve. Theoretically, maximum power would be transmitted to the loudspeaker if the speech-coil impedance could somehow be made equal to the valve's anode resistance. For various reasons, primarily excessive distortion, this is not practicable for pentodes. However, the speech coil impedance must be made comparable in value to the anode resistance if useful power is to be extracted from the valve. This impedance matching is easily achieved by the output transformer, which makes the speech-coil impedance 'look' as if it were multiplied by the square of the turns ratio (N^2). The transformer turns ratio is thus chosen to achieve the desired impedance matching between the anode resistance of the output valve and the speech-coil impedance.

The performance of a power-amplifier stage, such as that shown in Fig. 70, is usually determined graphically from the valve's characteristic curves. As an example, let us assume that a loudspeaker with an 8-ohm (impedance) coil is to be operated Class A from a triode power amplifier in the circuit of Fig. 70. The available anode-supply voltage (E_{bb}) is 250 V, and it is desired to determine the required signal input voltage to the grid and the available power output, when the valve is operated under the conditions recommended by the manufacturer for Class A amplification. From the manufacturer's data manual, we find that this power triode has an anode resistance of 800 ohms, an amplification factor of 4·2 and a transconductance of 5,250 μmho. This combination of low amplification factor and low anode resistance with high transconductance is typical for power triodes, which must produce high output currents. For an anode voltage of 250 V and Class A operation, the manufacturer recommends a negative grid bias of −43·5 V and a load resistance of 2,500 ohms to obtain low distortion at reasonable power output. The choice of the load resistance is an important factor in determining the power output as well as the distortion, and we shall learn later how to compute both for a given load resistance.

Accepting for the moment the manufacturer's recommendations, we can determine immediately the turns ratio of the output transformer to make the 8-ohm loudspeaker impedance appear as if it were 2,500 ohms. The load across the transformer secondary (R_L) 'looks into' the primary as if it were multiplied by the square of the turns ratio (N^2R_L). Thus, the square of the turns ratio is equal to the ratio of the primary to the secondary impedance, or

$N^2 = \frac{2{,}500 \text{ ohms}}{8 \text{ ohms}} = 313$. Hence, the turns ratio, $N = \sqrt{313} = 17{\cdot}7$ (stepdown primary-to-secondary).

LOADLINE

Fig. 71 is a reproduction of the triode characteristics, on which we have placed a 2,500-ohm loadline to determine the dynamic performance of the valve for the chosen operating conditions. To draw this loadline, we start out with the chosen operating point 'O', for an anode voltage of 250 V and a grid bias of −43·5 V. The quiescent or d.c. anode current (I_o) at this point is seen to be 60 mA. We have talked rather loosely about an anode voltage (E_b) of 250 V at the operating point, although the entire available anode-supply voltage (E_{bb}) is only 250 V. With a d.c. anode current of 60 mA, we should expect some d.c. voltage drop across the load resistance and, hence, the voltage at the anode should be the difference between the anode-supply voltage and this voltage drop across the load, as in the case of resistance coupling. As a matter of fact, there is a small voltage drop across the 'winding resistance' of the transformer primary, but this is only a few volts and may be neglected. The 2,500-ohm load impedance that is 'reflected' into the primary from the 8-ohm loudspeaker is only present for the *alternating* component of the anode current (i.e. the output signal), while the resistance to *d.c.* is only that of the primary transformer winding. Thus, we see that practically the entire supply voltage appears at the anode of the valve and only a negligible amount is lost in the primary winding of the transformer. This is the reason that we may use the terms 'anode-supply voltage' and 'anode voltage' synonymously for transformer coupling.

As we shall see, the loadline is a very handy device for graphically portraying the distribution of the anode-supply voltage across the load and across the internal (anode) resistance of the valve. By tracing the path of the input signal along the loadline, we obtain the valve's performance in a nutshell. To construct the loadline, we have to keep in mind that the load resistance represents

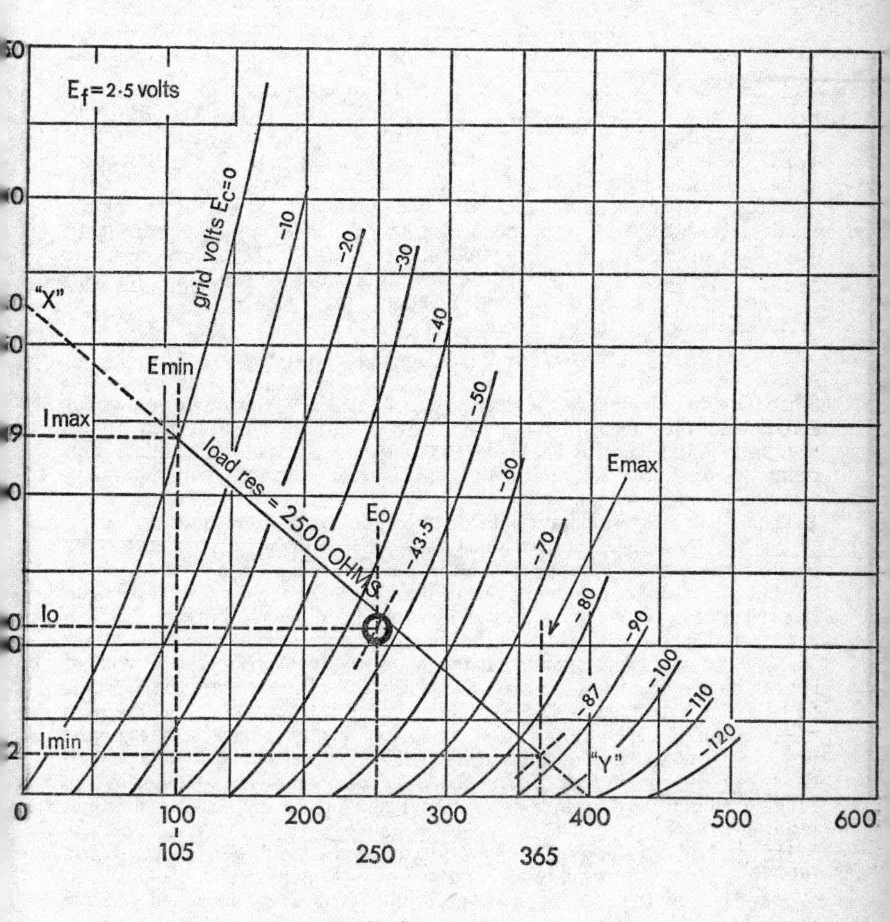

Fig. 71. Triode characteristics for 2,500-ohm load: vertical axis gives the anode current (I_b) in milliamps

the ratio of the a.c. load voltage (E_L) to the changing or a.c. component of the anode current. But we have seen before (in the discussion of resistance coupling), that the load voltage is exactly equal to the change in anode voltage, since the two together make up the anode-supply voltage. Thus, we must construct a line whose inverse slope or ratio of anode-voltage-change to anode-current-change is everywhere equal to the load resistance (2,500 ohms, in this case). Since the operating point must be on the loadline, we need only one other point to fix the line. The simplest procedure is to determine the intersection of the loadline with either the anode-current or anode-voltage axis. Let us find the intersection with the anode-current (I_b) axis first. At the axis, the anode voltage is zero and the change in the anode current equals the change in anode voltage (with respect to the operating point) divided by the load resistance. Hence,

$$\text{change in anode current} = \frac{\triangle E_b}{R_L} = \frac{250}{2{,}500} = 0{\cdot}1 \text{ amp or } 100 \text{ mA.}$$

As the anode voltage goes to zero, therefore, the anode current increases by 100 mA (from 60 mA at the operating point to 160 mA at the intersection with the anode-current axis. We can thus mark off point 'X' on the anode-current axis at 160 mA. Point 'X' together with the operating point 'O' determines, of course, the loadline completely, but for exercise let us now determine the intersection of the loadline with the anode-voltage axis. At that point the anode current must be zero and, hence, the *change* in anode current (with respect to the operating point) is 60 mA or 0·06 A. The change in anode voltage to produce this current must, therefore, equal the product of the load resistance times the change in anode current ($\triangle E_b = \triangle I_b R_L$).

Hence, the change in anode voltage

$$= \triangle I_b \times R_L = 0{\cdot}06 \times 2{,}500 = 150 \text{ V.}$$

Since the anode voltage at the operating point is 250 V, the anode voltage for zero current must be 250+150, or 400 V. We therefore mark off point 'Y' on the anode-voltage axis, at 400 V. We can now draw a 2,500-ohm loadline from point 'X' to point 'Y', which will pass, of course, through the operating point 'O'.

The actual path of operation does not extend throughout the entire length of the loadline, however, since we do not want to draw grid current on the positive swing of the input signal; nor do we wish to drive the valve to anode-current cut-off on the negative swing of the signal. We therefore limit the positive swing of the input signal arbitrarily to +43·5 V, which is just sufficient to overcome the −43·5 V grid bias and, thus, drive the grid to zero grid voltage ($E_c = 0$). An a.c. input signal that has a positive peak of +43·5 V will, of course, have a negative peak of −43·5 V and thus drive the grid of the triode to −87 V grid voltage (−43·5−43·5=−87) on the negative swing. The path of operation on the loadline, consequently, extends to zero grid voltage on the positive swing of the input signal and to −87 V on the grid during the peak of the negative swing of the a.c. input signal, as is shown by the solid line in Fig. 71. To obtain this grid swing, an a.c. input signal of 43·5 V peak amplitude is required.

Having laid out the path of operation on the loadline, we can now determine some useful information from it. Of special interest are the anode voltages and currents at the extremes of the input-signal swing, since these extreme values permit us to compute the power output and distortion of the amplifier. Thus, at the extreme positive swing of the a.c. input signal, when the total grid voltage is zero ($E_e = 0$), we find from Fig. 71 that the maximum anode current

(I_{max}) is about 119 mA, while the anode voltage has dropped to its minimum value (E_{min}) of about 105 V. This means, of course, that the voltage across the 2,500-ohm load (i.e. the primary of the output transformer) has reached its negative peak equal to the difference between the anode-supply voltage and the anode voltage, or $250 - 105 = 145$ V. The load voltage is negative because of the 180° phase reversal occurring within the valve.

Similarly, at the extreme negative swing of the a.c. input signal, when the total grid voltage is -87 V, we obtained from Fig. 71 a minimum anode current (I_{min}) of about 12 mA and a maximum instantaneous anode voltage (E_{max}) of about 365 V. Again we see that the anode voltage is 180° out of phase with the input signal, the latter reaching its negative peak when the anode voltage is at its positive peak, and vice versa. The voltage across the transformer primary (load) during the maximum negative swing of the input signal now reaches its positive peak, equal to $365 - 250$ or 115 V. Since the negative peak of the load voltage is 145 V, while the positive peak is only 115 V, the output-voltage waveform is clearly not quite as symmetrical as a sine-wave. This leads us to suspect strongly the presence of some distortion in the signal output, a suspicion which we shall confirm a little later on.

From the operating point 'O' in Fig. 71 we can determine the d.c. input power to the valve, called the anode dissipation, which is an important factor in determining the life of the valve. Multiplying the (quiescent) values of the d.c. anode voltage and the d.c. anode current at 'O' we obtain an anode dissipation of $250 \times 0{\cdot}06 = 15$ W. This happens to coincide with the maximum rating listed in the valve manual.

A.C. POWER OUTPUT

Although the a.c. power developed across a resistor is simply the product of the r.m.s. (root-mean-square) values of the current and the voltage, the output power in the loadline construction of Fig. 71 is more easily determined by using the maximum and minimum current and voltage values at the extremes of the input signal swing. The power output of a Class A amplifier is given by the following approximate relation:

$$\text{Power output} = \frac{(I_{max} - I_{min}) \times (E_{max} - E_{min})}{8}$$

Substituting the values obtained from Fig. 71, after converting milliamps to amperes, we obtain for this particular triode:

$$\text{Power output} = \frac{(0{\cdot}119 - 0{\cdot}012) \times (365 - 105)}{8} = 3{\cdot}48 \text{ W.}$$

Although this power, when applied to a loudspeaker, is sufficient to pierce the eardrums of any sensitive listener we shall see a little later (p. 106) that the power output may easily be more than doubled by employing two triodes in a push-pull circuit.

DISTORTION

We all seem to have a pretty good idea of what distortion is, since we all have listened to the tinny, squeaky and unnatural sounds of miniature radios; the noisy, crackling and tortured ones emanating from an overloaded portable radio; and the boomy and screechy ones being projected from some so-called high-fidelity gramophones. Being aware of this intuitive knowledge of distortion of the average radio listener, we have glibly talked about distortion being produced in amplifiers by 'non-linear' operation. But in order really to know what we are talking about we have to analyse the phenomenon of distortion somewhat more closely and classify it according to the various forms it may take.

Ideally, if some waveshape that is an exact electrical representation of a sound passes through an audio amplifier completely unmodified, except as to amplitude, there is no distortion and the amplified output from a speaker will sound exactly like the original (provided, of course, that the speaker does not introduce any distortion). In practice, a reproduced sound never sounds like the original, even with an ideal amplifier and speaker. For one thing, a speaker is not a musical instrument and can never reproduce the spatial sound distribution of a symphony orchestra, or even a grand piano. Another reason is that sound is usually reproduced monaurally, that is, by a single channel from one microphone through one amplifier and one speaker, while we listen binaurally with two ears. The time difference with which sound strikes each ear of the listener produces a spatial realism in depth, which permits us to tell approximately where each sound is coming from. This effect cannot be reproduced by monaural, or single-channel reproduction, but has been simulated quite successfully by the so-called stereophonic or binaural reproduction devices.

Leaving aside these refinements, let us define the main modifications, or distortions, of an input waveform caused by a deficient voltage or power amplifier. These are frequency distortion, phase or time-delay distortion, and amplitude or non-linear distortion.

Frequency distortion. We have already met frequency distortion: it is the limitation in the frequency range or bandwidth reproduced by an amplifier, caused by various coupling elements and unavoidable shunt capacitances. Frequency distortion exists whenever the full audio range from about 20 to about 15,000 cycles per second is not completely reproduced. If the high-frequency response is deficient, you will not be able to hear the 'overtones' of some high-pitched instruments, such as the violin, flute, or horn, and they will all sound very much alike. The sound from the speaker will then appear very boomy or bassy. On the other hand, if the low-frequency response of the amplifier is deficient, you may not hear at all the gorgeous sound of the large (30–50 c/s) organ pipes, and the bass drum may sound no different from the kettle drum. In general, the sound will then be tinny and squeaky.

Phase or time-delay distortion. We have become acquainted with phase or time-delay distortion during the discussion of the pulse response of wideband amplifiers. The various capacitive and inductive coupling elements of an amplifier cause phaseshift because the current in an inductance lags behind the applied voltage (in time or phase), while the current in a capacitor leads (in time or phase) the applied voltage. This in itself would not cause distortion if the resultant time delay of the output with respect to the input would be the same for all frequencies. Unfortunately, this is usually not the case, and different frequencies are delayed by different amounts, resulting in output waveforms that differ in appearance from the input waveform. The primary effect of phase or time-delay distortion is the production of overshoots and undershoots in a pulse waveform, such as was illustrated in Fig. 66b. Although this is important in the exact reproduction of pulse waveforms in a video amplifier, it does not appear to have much audible effect in an audio amplifier. However, the phase-shift characteristics and frequency response of an amplifier together determine the transient response, i.e. the response to sharply rising and suddenly changing waveforms. Combined frequency and time-delay distortion may thus lead to poor transient response in an audio amplifier, which, in turn, may wash out the sharp 'attack' of a piano or percussion instrument.

Amplitude or non-linear distortion. Amplitude or non-linear distortion is perhaps the most serious form of distortion occurring in an amplifier and it is usually this type we have in mind when we say that a sound is 'terribly distorted'. As the term implies, amplitude distortion causes a modification in the relative amplitudes of a signal. It does not mean that the amplitudes of the two half-cycles of a sinewave cannot be enlarged, as indeed they must be in an amplifier, but it means that they are unequally amplified with respect to each other and the input waveform. Thus, amplitude distortion exists if the top or bottom of a sinewave is 'clipped off' or reduced in amplitude, as shown in

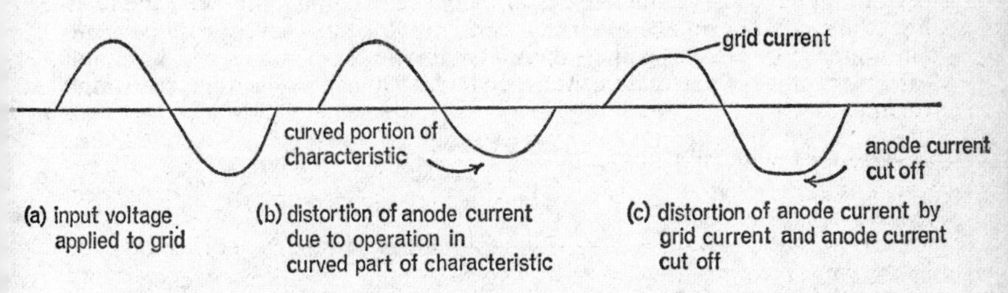

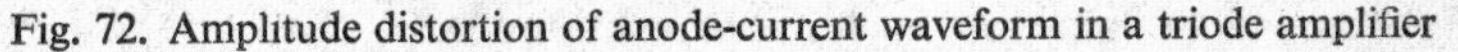

Fig. 72. Amplitude distortion of anode-current waveform in a triode amplifier

Fig. 72. The pure sine-waveform of the applied input voltage is shown in Fig. 72a. If the amplifier is operated Class A, in the linear portion of the characteristic, the waveform of the anode current and the output voltage should look exactly as in (a), except that it will be larger, of course. However, the characteristics of amplifier valves are never completely linear and, hence, some 'non-linear' distortion is always present. Fig. 72b shows the slightly flattened anode-current waveform of an amplifier that is driven into the curved bottom portion of the characteristic during the negative swing of the input signal. Finally, Fig. 72c shows the severe distortion of the anode-current waveform when the grid of the amplifier draws grid current during the positive swing of the applied signal and the anode current is cut off during the negative swing of the input signal. The result is a waveform that is clipped at the peak of each half-cycle and looks more like a square wave than a sinewave. The distortion of the anode-voltage waveform (output signal) is, of course, the same in each case, except that it occurs reversed in phase because of the 180° phase reversal between input and output.

Fig. 72 illustrates graphically that non-linear operation of an amplifier results in first slightly flattening the sinewave amplitude and finally 'clipping' or 'squaring' it with severe distortion. The waveform of Fig. 72c is similar to the pulse waveforms illustrated in Chapter 9, and like those it can be thought to be made up of a fundamental sinewave frequency (equal to the frequency of the input sinewave, shown in Fig. 72a) and a number of multiples or harmonics of that frequency. Thus, the waveform of Fig. 72c, for example, may consist of a fundamental 1,000 c/s sinewave, equal to the input voltage (Fig. 72a), plus a second harmonic of 2,000 c/s that is perhaps 20% in amplitude compared to the fundamental, plus a third harmonic of 3,000 c/s, which is perhaps 10% of the fundamental amplitude, plus a fourth harmonic of 4,000 c/s, which may be 5% of the fundamental amplitude, and so on. The more severe the distortion, the greater is the number (and the amplitudes) of the harmonic frequencies present in the output signal. Because of the production of harmonic frequencies not present in the input signal, this form of distortion is sometimes called harmonic distortion. The presence of harmonic frequencies might by itself not be so unpleasant, if it were not for the fact that these new frequencies interact with each other, the high-frequency components being modulated by the low frequencies, and the resulting intermodulation distortion gives rise to sounds that are guaranteed to make you leave the room quickly.

As an example, let us determine the amplitude distortion of the Class A triode amplifier whose characteristics are shown in Fig. 71. Although it is possible to calculate the amplitude of all the harmonics present in the amplifier output from the characteristics of the valve in conjunction with the loadline, the second-harmonic distortion is usually the most severe, and hence its amplitude is of greatest importance. The amplitude of the second harmonic (as a percentage of the fundamental) can be determined by the following simple formula:

$$\text{2nd harmonic distortion } \% = \frac{\dfrac{I_{max}+I_{min}}{2} - I_o}{I_{max}-I_{min}} \times 100$$

Substituting the values previously obtained from the loadline,

$$\text{2nd harmonic distortion } \% = \frac{\dfrac{0{\cdot}119+0{\cdot}012}{2} - 0{\cdot}06}{0{\cdot}119-0{\cdot}012} \times 100 = 5{\cdot}1\%$$

A second-harmonic distortion of 5·1% (of the fundamental amplitude) in an audio amplifier is generally considered quite tolerable for ordinary music and speech reproduction, although it is far too noticeable to be accepted as high-fidelity reproduction. The latter usually requires the total harmonic distortion (of all harmonics together) to be less than 1%. Further reduction of distortion may be accomplished by operating strictly Class A, limiting the amplitude of the input signal and, consequently, the power output, by using a push-pull circuit and by the application of negative feedback. Let us consider push-pull operation next.

PUSH-PULL POWER AMPLIFIER

One way of obtaining greater power output from an amplifier is to connect two identical valves in parallel, anode-to-anode, grid-to-grid, and cathode-to-cathode. This will assure roughly twice the power output of a single valve, but the total distortion will also go up. A far better way is to connect two identical tubes in such a manner that their grids are excited by equal but 180° out-of-phase input signals and their outputs are combined in a centre-tapped output transformer. The resulting push-pull amplifier circuit is shown in Fig. 73 for two beam-power tetrodes, although triodes or pentodes can be used equally well. The outstanding advantage of this circuit is that it cancels out all even-harmonic distortion (i.e. 2nd, 4th, 6th harmonic, etc.) and thus permits more power output per valve for a given permissible distortion. As a consequence, the circuit will give more than twice the power output of a single valve, with considerably less distortion. For example, a single beam-power tetrode operated at 250 V anode voltage under optimum conditions, will give a power output of about 6·5 W with 10% total harmonic distortion, while the push-pull circuit shown in Fig. 73, operated Class A under similar conditions, will give a power output of about 17·5 W with only 2% total harmonic distortion.

Although for simplicity batteries are shown in Fig. 73 as sources of bias and

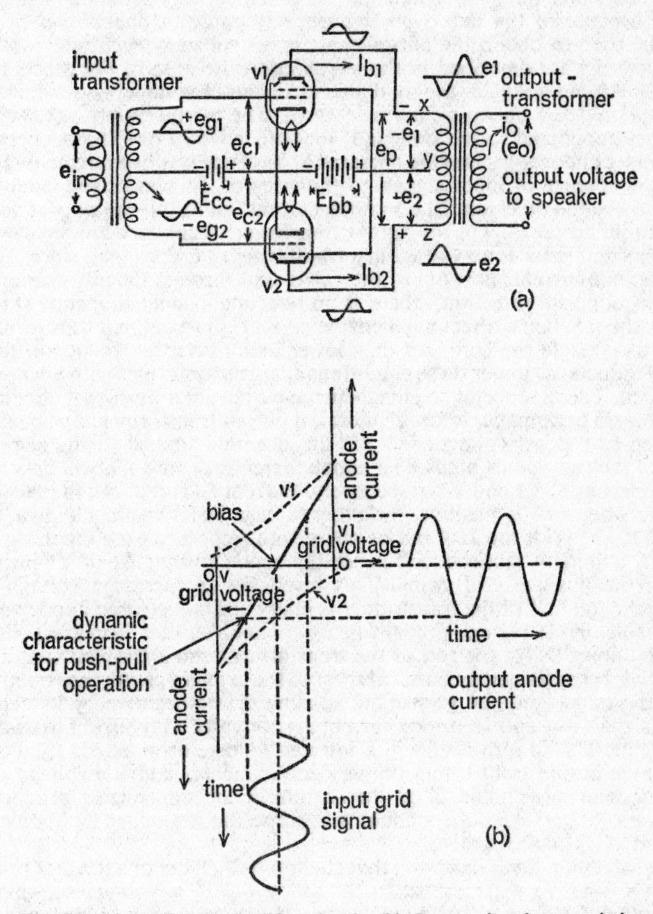

Fig. 73. (a) Push-pull amplifier circuit, (b) Composite characteristic

anode voltage, in practice the anode and screen potentials are derived from a rectifier power supply, while grid bias is obtained from a cathode-biasing resistor that is inserted into the common cathode circuit of the two valves. Cathode biasing has the advantage that no by-pass capacitor is required, since in a push-pull circuit the alternating signal currents do not flow through the common cathode return or the anode supply.

The symbol notations in Fig. 73 are the same as previously used, except that the numerical subscripts refer to valve 1 (V1) or valve 2 (V2), respectively. As mentioned, the circuit requires two input signal voltages, e_{g1} and e_{g2}, which are equal in amplitude, but 180° out of phase with respect to each other. Two out-of-phase voltages are easily secured by means of a centre-tapped input transformer, as shown, since whenever the top of the transformer secondary is positive with respect to the centre tap ($+e_{g1}$), the bottom is negative with respect to centre tap ($-e_{g2}$), and vice versa. When resistance coupling is preferred because of the improved frequency response, a phase-inverter stage must be used to obtain the out-of-phase input voltages. Such phase-inverter arrangements are described in the next section. Because of the necessity for two equal input voltages, the total alternating input voltage, $e_{g1}+e_{g2}$, is twice that required for operating a single valve. The anode-supply voltage, E_{bb}, provides anode and screen potentials for both valves. For Class A operation, the anodes and screens may be operated at the same positive direct voltage.

In the absence of input signals to the grids of the two valves, each valve draws the same quiescent (d.c.) anode current, and both the output voltage and output current (i_o) in the output transformer secondary are zero, since a direct current induces no voltage in a transformer. Furthermore, since the two direct anode currents flow out of the valves and through the transformer primary in opposite directions, there is no resulting magnetizing current in the transformer. When a direct magnetizing current is present in a transformer, it tends to saturate the core and thus lower its inductance. To obtain the required inductance under these conditions a large transformer with a large core is needed. The absence of the magnetizing current in a push-pull circuit permits the use of a smaller, more economical output transformer.

When two signal voltages, e_{g1} and e_{g2}, are now applied to the respective grids of the two valves, sinusoidal anode currents, I_{b1} and I_{b2}, will flow in the anode circuits of V1 and V2, respectively. Current I_{b1} is 180° out of phase with I_{b2} (i.e. when one is positive, the other is negative) because the two input signals to the grids are 180° out of phase with respect to each other. Assume that e_{g1} is initially positive and hence the anode current I_{b1} of V1 increases during the positive grid swing. This results in an increased voltage drop across the top half of the transformer primary and a decreased anode voltage of V1 (the anode voltage drops, when the anode current increases). Consequently, point 'X' at the top of the transformer primary, which is at anode potential, becomes negative with respect to the primary centre tap (point 'Y').

At the same time, the grid of V2 is being driven negative by its negative input signal, e_{g2}, and its anode current I_{b2} decreases with respect to its quiescent value. This in turn results in a lowered voltage drop across the bottom half of the output transformer primary and in increased anode voltage at V2. As a consequence, point 'Z' at the bottom of the transformer primary becomes positive by the same amount with respect to the centre tap (point 'Y'), as point 'X' is made negative.

The foregoing analysis shows that the top half of the transformer primary becomes negative with respect to the centre tap by a voltage e_1, while the bottom half becomes positive with respect to the centre tap by an equal voltage, e_2. These two voltages are added in series across the transformer primary, resulting in the total primary voltage, e_p, which is exactly twice the amplitude of e_1 or e_2. The total alternating primary voltage induces an output voltage (e_o) in the transformer secondary and an output current (i_o) flows through the speaker. A half-cycle later all the input and output voltage polarities reverse, of course, and the primary voltage e_p is of opposite phase, though still equal to the sum of e_1 plus e_2.

The power output and distortion analysis of the push-pull circuit is somewhat complex, but the basic effect of push-pull operation is clearly shown by the combined dynamic transfer characteristic (Fig. 73b). This composite push-pull characteristic is obtained by adding together the individual transfer characteristics of V1 and V2, and it is seen to be completely straight. Because of the resulting linear Class A operation, the anode-current waveform in the output is practically undistorted.

PHASE INVERTERS

If resistance coupling to the input of a push-pull amplifier is desired, a phase inverter must be used to supply the two 180° out-of-phase grid input voltages. Fig. 74 illustrates two popular circuits out of a great variety of existing phase inverters. The circuit shown in (a) requires only one valve to provide the two out-of-phase voltages and thus is quite economical.

Phase inversion in the circuit of Fig. 74a is obtained by splitting the total anode-load resistance and, hence, the output voltage equally between the

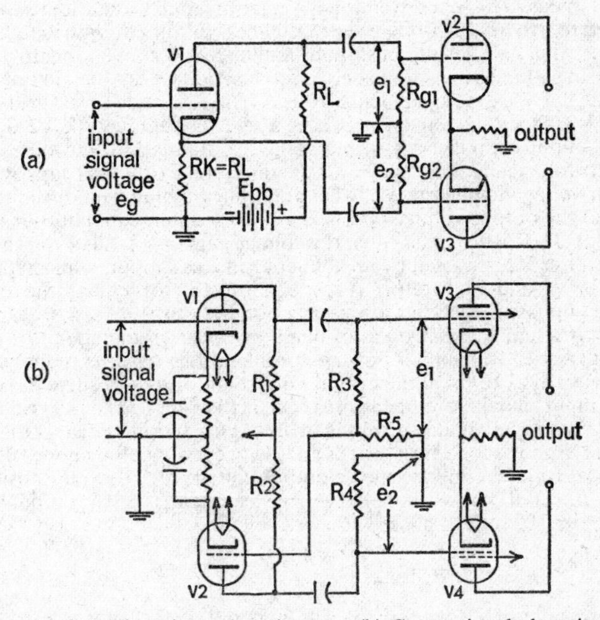

Fig. 74. (a) Cathode-resistor phase inverter, (b) Conventional phase inverter

anode and cathode circuits of V1. Since R_L and R_k are equal in value and the anode current flows through both of them, voltages of equal amplitude are developed across the two resistors. But because of the phase reversal taking place within the valve, the output signal taken off the anode-load resistor R_L is 180° out of phase with the input signal, while the output signal taken off the cathode-load resistor R_k is in phase with the input signal. The two output signals, consequently, are 180° out of phase with respect to each other. You can verify for yourself that, for a positive input signal, the anode voltage of V1 will drop, thus making the anode-connected end of R_L negative, while the cathode-connected end of R_k simultaneously becomes more positive (since electrons flow from negative to positive).

The two out-of-phase output voltages of V1 are coupled through d.c.-blocking capacitors and grid resistors R_{g1} and R_{g2} to push-pull amplifier valves V2 and V3, where they develop out-of-phase grid input voltages e_1 and e_2. The push-pull amplifier is biased by a common cathode resistor. The output circuit of the push-pull stage is not shown.

As in the case of an un-bypassed cathode-biasing resistor, a degenerative feedback action takes place across R_k, which is common to both the anode and grid circuits of V1. In effect, the output voltage developed across R_k opposes the signal input voltage to the grid of V1 and so prevents the valve from realizing its amplification. Each of the output voltages, e_1 and e_2, is therefore slightly smaller than the input signal voltage, although the total output voltage (e_1+e_2) is slightly less than twice the value of the input voltage. Since phase inversion is the desired aim, the lack of amplification does not matter. Besides, the cost of V1 and the associated coupling components is still less than the expense of a high-quality push-pull input transformer.

Fig. 74b shows the conventional phase-inverter circuit that is usually employed to drive a push-pull power amplifier. In this circuit triode V1 drives the upper valve, V3, of the push-pull amplifier as a conventional resistance-coupled stage. Phase inversion is provided by triode V2. A portion of the output voltage of V1, developed across resistor R5, is applied to the grid of phase inverter V2. Since the valve introduces a 180° phase reversal, V2 drives the lower push-pull amplifier V4 with a voltage e_2 that is 180° out of phase with respect to voltage e_1 applied to V3. If the ratio of R5 to R3 is properly chosen, the input voltage applied to V2, when multiplied by the gain of the stage will just equal the output voltage e_1 of stage V1. To attain equal output voltages, the ratio (R3 + R5)/R5 must equal the voltage gain of V2. Since the gain of the phase inverter V2 is wasted by applying only a small input voltage, the amplification of V1 and V2 together is just equal to that of a single valve. Again, the price of phase inversion is the sacrifice of one valve.

You can visualize the process of phase inversion in the circuit of Fig. 74b somewhat better from the following consideration. Assume the input signal voltage is such as to drive the output voltage of V1 in a positive direction. A positive input signal, developed across R5, is then applied to the grid of phase inverter V2 and its anode current increases. This action increases the voltage drop across anode-load resistor R2 and, hence, swings the anode of V2 in a negative direction. Thus, when the output voltage of V1 swings positive, the output voltage of V2 swings negative, or vice versa. The two output voltages are therefore 180° out of phase with each other.

FEEDBACK AMPLIFIERS

The principle of feedback is probably as old as the invention of the first machines, but it has only dawned upon us rather recently that feedback is not only marvellously useful, but applicable to almost everything. Technically, feedback simply means transferring a portion of the energy from the output of some device back to its input. By making the input thus dependent on the output of the device, very fine control of any process becomes possible. In effect, this amounts to spying on the end result (output) of some process and controlling the input to the process in accordance with the effectiveness of the output. Such an arrangement, in which the input depends on feedback from the output, is known as a closed-loop control system and it is the most effective control system possible. (See Fig. 75.) The governor of a steam engine, which regulates the flow of steam in accordance with the speed of the output drive wheel and thus controls the speed, is an early example of such a closed-loop feedback system. The thermostat, which controls the fuel flow to an oil burner or the operation of an electric radiator in accordance with the desired room temperature, is another excellent example.

Although the applications of feedback are myriad, it has only been about 20 years that feedback has come into use in connexion with electronic amplifiers. Here it has been found eminently useful in reducing distortion caused within amplifiers and making amplifier operation more stable in respect to variations in gain due to mains-voltage changes, valve differences, ageing, etc. To understand how this is possible, we must first of all distinguish between two basic types of feedback: regenerative or positive feedback and degenerative or negative feedback. When the feedback energy (voltage or current) is in phase with the applied signal and thus aids it, regenerative or positive feedback takes place. Degenerative or negative feedback is the term used, when the feedback energy is out of phase with the applied signal and thus opposes it. Regenerative feedback increases the gain (and also the distortion) of an amplifier, and if sufficiently large, leads to oscillations, as we shall see in the chapter on oscillators. In contrast, degenerative feedback decreases the gain, as well as the distortion of an amplifier, and for the latter reason, we shall be primarily interested in degenerative (negative) feedback.

The basic principle underlying the operation of a feedback amplifier can be understood from the schematic diagram shown in Fig. 75. Here the amplifier

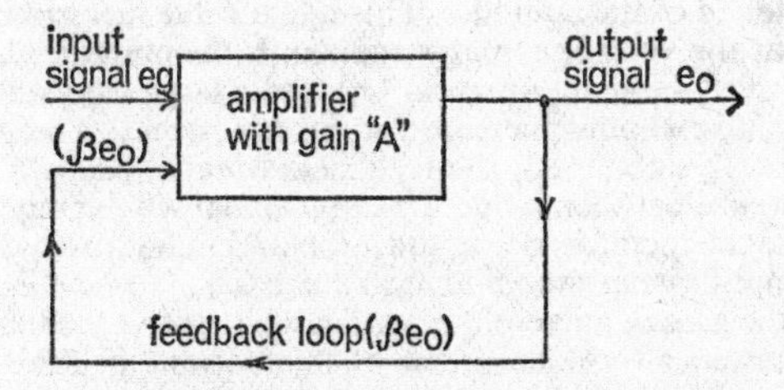

Fig. 75. Schematic diagram of feedback amplifier

gain in the absence of feedback is A. Feedback is then applied by feeding a portion β of the output voltage e_o back to the amplifier input. The actual input to the amplifier, thus, consists of the sum of the signal voltage, e_g, plus the feedback voltage, βe_o, or a total of $(e_g+\beta e_o)$. This total input voltage, amplified by the gain A of the amplifier, must be equal to the output voltage. Thus we may write

$$(e_g+\beta e_o)\times A=e_o$$

But since the voltage gain of an amplifier, by definition, is the ratio of the output voltage to the input voltage (e_o/e_g), we can solve the above expression for this ratio and, thus, obtain the gain in the presence of feedback:

$$\frac{e_o}{e_g}=\text{voltage gain with feedback}=\frac{A}{1-\beta A}$$

where A is the gain of the amplifier without feedback.

We have obtained the above expression by assuming that the portion fed back, β, is positive and, hence, positive feedback or regeneration takes place.

The quantity βA represents the amplitude of the feedback voltage (sometimes called feedback factor) superimposed upon e_g. The larger this feedback factor, the smaller is the denominator of the above expression and, hence, the larger is the gain with feedback. Unfortunately, increasing the gain by positive feedback also increases the distortion and noise in the same proportion and, hence, positive feedback is rarely used except in oscillators (Chapter 11).

For negative feedback, the fraction β and hence βA becomes negative (i.e. $-\beta A$) and the

$$\text{voltage gain with negative feedback} = \frac{A}{1+\beta A}$$

The greater the feedback factor βA, the smaller is the feedback gain of the amplifier, but also the smaller is the distortion introduced by the amplifier. When the feedback factor βA is made large compared to 1, as is the case for large amounts of feedback, the denominator of the above expression reduces simply to βA, and the previous expression becomes

$$\text{voltage gain with large negative feedback} = \frac{1}{\beta}$$

The gain of the amplifier in this case is quite small, but—as is apparent—depends only on the feedback fraction β and is thus substantially independent of the actual gain A of the amplifier. This remarkable fact can be explained by considering that the voltage actually applied to the amplifier is the difference between a relatively large input signal (e_g) and a large feedback voltage (βe_o). If the amplification should increase for some reason (such as mains-voltage fluctuations or valve changes), the feedback voltage (βe_o) will be larger and hence the difference between it and the input signal will be much smaller. This tends to oppose the increase in amplification and maintain it stable. The same is true if the amplification should tend to decrease.

As a result, the gain of an amplifier with large negative feedback is extremely stable and is practically independent of fluctuations in the supply voltage, ageing of valves, or differences between valves caused by replacements. Moreover, since the gain depends only on the feedback fraction β, the variation in gain with frequency (i.e. the frequency response) is entirely controlled by the nature of β. If the feedback fraction is obtained through a resistive network, as is usually the case, the feedback, and hence the gain, does not vary with frequency as long as the feedback is negative. The frequency response of an amplifier may thus be considerably improved by using large amounts of feedback. This improvement is not absolute, because phase shifts caused within the amplifier by various coupling elements, tend to change the sign of the feedback from negative to positive at the extremes of the audio band, thus nullifying the degenerative feedback action.

By designing the feedback loop with capacitive and inducive elements, rather than with resistors, the feedback may be made to vary in any desired manner with frequency and, thus, any required frequency response may be achieved. Feedback circuits are, therefore, very popular to boost the bass and treble response of an audio amplifier. The recording characteristics of various gramophone records may be compensated for easily by a properly designed feedback loop.

REDUCTION OF DISTORTION AND NOISE

Hum, noise, and distortion introduced within an amplifier are reduced with negative feedback by the same factor $(1+\beta A)$ as the gain. Since the loss in gain can always be compensated for by increasing the amplitude of the input signal, the net effect of negative feedback is a reduction in distortion, noise, and hum. To understand how this reduction takes place, consider Fig. 76.

Let us see first what happens without the use of feedback. Assume that when an input signal e_g is applied to the grid of an amplifier valve, the anode current i_p is afflicted with a pronounced irregularity during its positive half-cycle owing to highly non-linear operation. (For the purposes of illustration we have exaggerated the irregularity beyond anything likely to happen in practice.) The anode voltage, e_p, and the output signal will be, of course, reversed in phase and thus have the irregularity in its negative half-cycle.

Now suppose negative feedback is applied to the amplifier. The voltage fed back to the grid of the amplifier, e_{gf}, has the same waveform and phase as the anode or output voltage, but is smaller in amplitude. Since the original input

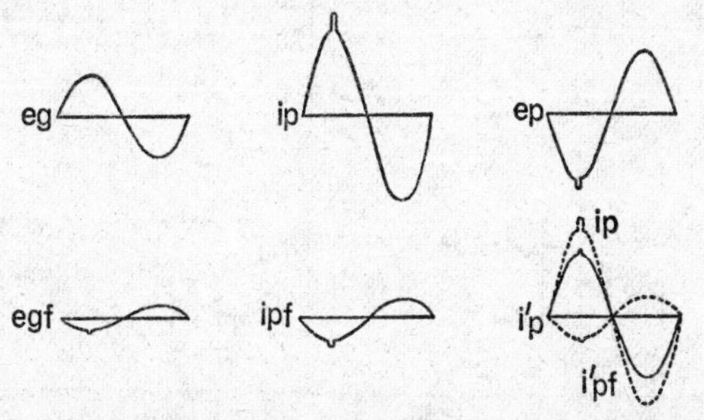

Fig. 76. Reduction in distortion due to negative feedback

signal, e_g, and the feedback signal, e_{gf}, are opposite in phase, it is evident that the two input signals oppose each other and produce degeneration. The feedback voltage (e_{gf}) produces a component of the anode current, i_{pf}, that is opposed in phase to the original anode current, i_p. Since the irregularity appearing in the feedback component of the anode current (i_{pf}) is out of phase with the irregularity in the original anode current, i_p, the two oppositely phased irregularities will tend to cancel each other. The greater the feedback component, the more cancellation will take place. The final result is the algebraic addition of the two anode-current components i_p and i_{pf} (shown dotted in Fig. 76) to produce the resultant anode current, i_p'. Although smaller in amplitude than the original anode current, the irregularity in the resultant anode current has also been substantially reduced.

The same reduction that is attained with negative feedback for non-linear distortion also takes place for extraneous noises and power-supply hum introduced within the amplifier. As a consequence, less hum filtering of the power supply is required with a feedback amplifier than would otherwise be necessary.

OUTPUT DISTORTION AND LOUDSPEAKER DAMPING

When applied to the power output stage of an audio amplifier, negative feedback has two important beneficial effects: it reduces the distortion caused by the variation in the loudspeaker impedance and it produces speaker damping by reducing the apparent anode resistance of the output valve. The impedance of the speech coil of a loudspeaker is not constant at all audio frequencies (as might be implied by the numerical impedance rating, such as an '8-ohm' speech coil), but it varies considerably over the audio range. This causes a corresponding variation in the anode-load impedance 'seen' by the valve,

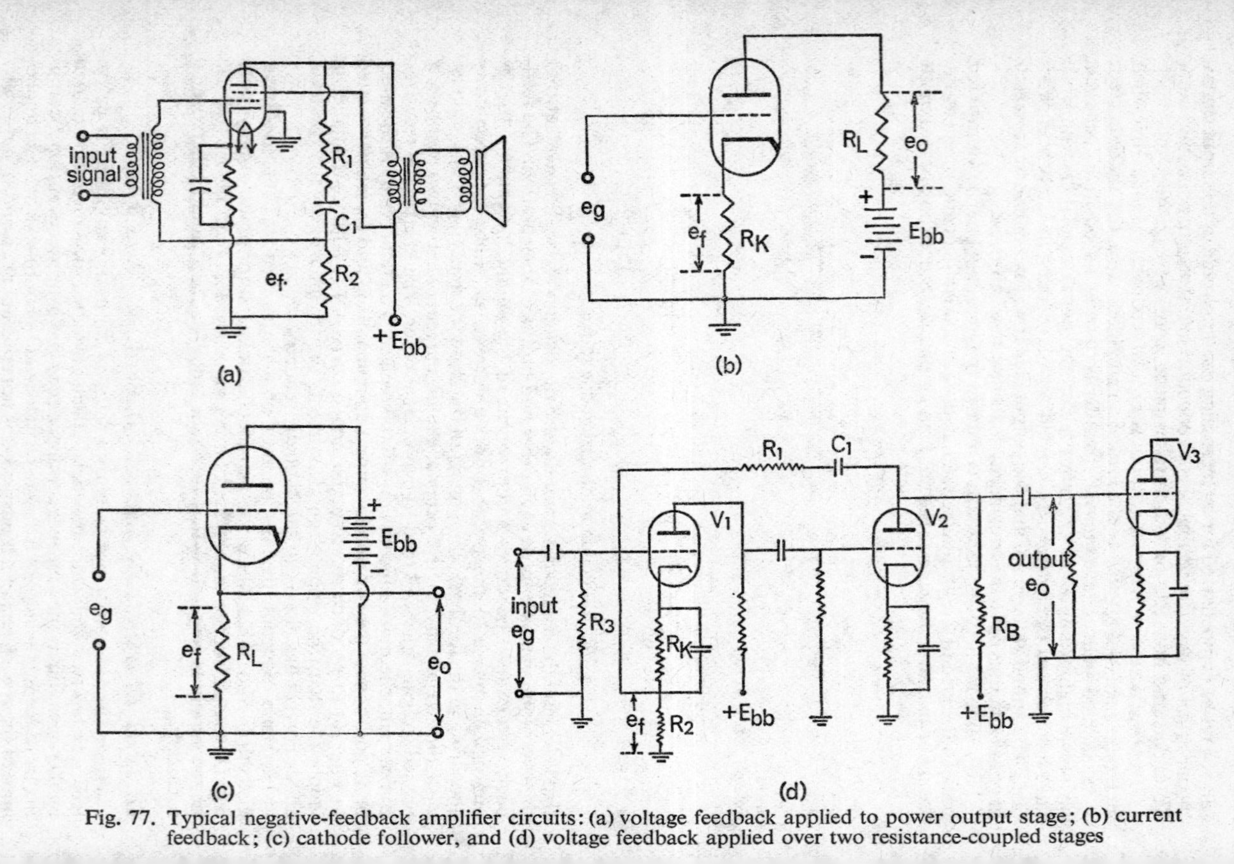

Fig. 77. Typical negative-feedback amplifier circuits: (a) voltage feedback applied to power output stage; (b) current feedback; (c) cathode follower, and (d) voltage feedback applied over two resistance-coupled stages

which leads to distortion in the output. This distortion is especially serious for pentodes, which have an inherently high anode resistance. Negative feedback lowers the anode resistance of the valve and, thus, the distortion produced by the varying speech-coil impedance.

The high anode resistance of pentodes has another deleterious effect on loudspeaker reproduction. The source impedance 'looking back' from the speech coil into the pentode anode circuit is very high. This causes oscillations of the speech coil (with resultant distortion) when sharp transient waveforms are applied to the loudspeaker, an effect known as hangover. The cure for this is to damp the loudspeaker by lowering the source impedance of the output valve. Again negative feedback does the trick by making the high anode-resistance characteristics of a pentode look like those of a low anode-resistance triode. Triode output valves do not require negative feedback, since they have already a sufficiently low source impedance. Negative feedback applied to a power output valve is shown in Fig. 77a.

PRACTICAL DEGENERATIVE-FEEDBACK CIRCUITS

Fig. 77 illustrates a few typical amplifier circuits employing negative or degenerative feedback.

The application of *voltage feedback* to a tetrode power amplifier stage is shown in Fig. 77a. This type of feedback is termed voltage feedback because it is derived from the output voltage. An alternating-voltage divider, consisting of R_1, R_2, and C_1, is connected across the output (anode circuit) of the tetrode. A fraction of the alternating output voltage ($e_f = \beta e_o$), developed across R_2, is applied to the grid of the valve through the secondary of the input transformer. Capacitor C_1 prevents the direct component of the anode voltage from reaching the grid. Since the feedback voltage is developed across the output of the valve, it is automatically 180° out of phase with the input voltage and, hence, is degenerative. The feedback voltage (e_f) in this case is approximately equal to the output voltage multiplied by the fraction $R_2/(R_1 + R_2)$. (This fraction is the 'β' we have previously referred to.) We have already mentioned the beneficial effects of negative feedback in a power output stage, which consist of a reduction in the distortion caused by the varying loudspeaker impedance and the prevention of 'hangover' effects by increased loudspeaker damping. Both these improvements are attained by a reduction in the source impedance (i.e. effective anode resistance) of the power output valve. Feedback is especially popular in beam-power valves because full power output can be attained with a comparatively small input driving signal. As was explained before, the reduction in gain caused by feedback must be compensated for by increasing the grid input voltage. In triodes and some pentodes the increase in driving voltage required to produce full power output may be inconveniently large, requiring extra stages of amplification.

Current feedback. The circuit of Fig. 77b is the familiar cathode-biased resistance-coupled amplifier, with the by-pass capacitor left off. We have already discussed the degenerative action caused by the feedback voltage e_f (across R_k) being out of phase with the input signal e_g. This type of feedback decreases the gain and distortion of the amplifier, but it increases the anode resistance of the valve. It is known as current feedback because the feedback voltage e_f is proportional to the anode current of the valve, rather than the output voltage. Current feedback improves the high-frequency response of an amplifier, but makes it worse at low frequencies. Also, by increasing the source impedance of the stage, loudspeaker hangover effects in a power amplifier are accentuated.

Cathode follower. The circuit of Fig. 77c is known as a cathode follower because the output voltage, taken off cathode resistor R_L, is in phase with and follows the input signal. The input voltage is applied between grid and earth, and the output is taken off between cathode and earth. You can easily verify that the output is in phase with the input by considering that an increasing positive input signal (e_g) will increase the anode current and, hence, produce an increased voltage drop, e_f, across cathode-load resistor R_L. Since electrons flow from negative to positive, the cathode end of resistor R_L becomes more positive than before and the rise in output voltage is in phase with the positive-going input signal. Similarly, when the input signal goes negative, the cathode end of R_L becomes less positive (or more negative) by the amount of the drop across R_L, and the output signal is again in phase with the input.

Since cathode-load resistor R_L is common to both the output and grid circuits, the output voltage e_f across R_L is also in series with the input signal (e_g) and thus causes degenerative current feedback. It is current feedback because the amount of the feedback voltage depends on the anode current, and it is degenerative because a positive-going cathode for a positive input signal will make the grid negative with respect to the cathode and thus oppose the positive input signal. Because of the large amount of degeneration, the voltage amplification of the cathode follower is always less than unity.

Although the gain of the cathode follower is less than 1, the circuit is capable of producing a large output voltage at large anode currents and, hence, considerable power gain. Furthermore, the effect of degenerative current feedback is to increase the input impedance to the grid of the valve, while at the same time lowering the anode resistance and output impedance by a factor $\mu+1$. This permits the use of the cathode follower as an impedance matching device (similar to a transformer) for coupling a high-impedance source to a low-impedance output, such as a transmission line. When used as a power amplifier, the cathode follower produces extremely low distortion and excellent loudspeaker damping because of the very low source impedance of the circuit. Because of the high input impedance, cathode followers are also frequently used as isolation devices to produce low loading of a circuit, as might be required in a valve voltmeter or an oscilloscope.

Multiple-stage feedback. Feedback may be applied over several stages and is sometimes used from the secondary of an output transformer back to an early voltage-amplifier stage. When this is done the benefits of feedback accrue to all the stages included in the feedback loop, but the correct design is somewhat difficult due to excessive phase-shifts. Fig. 77d shows voltage feedback being applied over two resistance-coupled stages. The feedback voltage is derived from an output voltage divider, C_1-R_1-R_2, in the same manner as shown in Fig. 77a. However, because of the double phase reversal occurring over two stages, the feedback voltage (e_f) is in phase with the input signal (e_g) and, hence, cannot be injected into the grid circuit. By injecting the feedback voltage into the cathode circuit, in this case, it is opposed in phase to the input signal, as required.

SUMMARY

In a Class A amplifier the signal voltage and grid bias are adjusted so that the anode current flows throughout the entire cycle of the applied input voltage. Class A amplifiers have excellent fidelity, relatively low distortion and low power output.

A Class B amplifier is adjusted so that the grid bias is equal to the anode-current cut-off value and, hence, anode current flows only for one-half of each input voltage cycle. A Class B amplifier has higher distortion than Class A, intermediate power output and efficiency.

A Class AB amplifier permits anode-current flow for more than half but less than the entire input voltage cycle. Class AB amplifiers have low distortion at low signal levels and medium power output and efficiency at high signal levels.

In a Class C amplifier the grid bias is adjusted to be greater than the anode-current cut-off value, so that anode current in each valve flows for less than half of each input cycle. Class C amplifiers have high power output and efficiency together with high distortion, and for the latter reason, are not used as audio amplifiers.

Transformer coupling is always used in a power output amplifier to match the anode resistance of the valve to the speech coil of a loudspeaker. The impedance ratio equals the square of the turns ratio.

The loadline is a graphic presentation of the distribution of the anode-supply voltage across the load and the internal resistance of the valve. It permits the power output and distortion of the valve to be determined.

The three main types of distortion are frequency distortion, phase or time-delay distortion, and amplitude or non-linear distortion.

Frequency distortion is the limitation in the audio bandwidth caused by the amplifier coupling elements.

Time-delay distortion is the unequal delay of different audio frequencies caused by phase shifts in the amplifier coupling elements. Frequency distortion in conjunction with time-delay distortion produces poor transient response in an amplifier (i.e. distorts pulses).

Amplitude distortion of the output waveform of an amplifier is caused by non-linear operation, such as anode-current cut-off and grid current flow. Curved characteristics also cause amplitude distortion. Amplitude distortion results in the production of harmonic frequencies not present in the input signal. These cause distorted sound.

In a push-pull amplifier the grids are excited by oppositely phased input signals and the output is combined in a centre-tapped output transformer. The out-of-phase exciting voltages are supplied either by a centre-tapped input transformer or by a phase inverter. Push-pull operation cancels out even-harmonic distortion and produces more than twice the power output of a single valve.

In a feedback amplifier a portion of the output voltage (or current) is fed back to the input. If the feedback voltage is in phase with the applied signal and aids it, regenerative or positive feedback takes place. If the feedback voltage is opposed to the input signal (i.e. out of phase), degenerative, negative, or inverse feedback is said to occur. Regenerative feedback increases the voltage gain and may lead to oscillations. Degenerative (negative) feedback decreases the amplifier gain, but also decreases distortion, noise, and power-supply hum within the feedback loop. Large negative feedback makes an amplifier almost independent of voltage fluctuations and ageing of valves. It also improves the frequency response.

Voltage feedback is proportional to the output voltage, while current feedback is proportional to the anode current of the valve.

CHAPTER ELEVEN

OSCILLATORS

Rapid to-and-fro motion of electrons in a conductor results in radiation of electromagnetic (radio) waves—the basis of all radio communication. It is the purpose of electric oscillators to generate these rapidly alternating electron currents from a direct-current supply. When Heinrich Hertz devised the first radio transmitter in 1887, he utilized an electric spark in a tuned circuit to produce very rapid electron oscillations and, consequently, radio waves. Higher powers than possible with spark transmitters were obtained in turn by the Poulsen electric-arc oscillators and the Fessenden–Alexanderson high-frequency electric generators. Although these devices eventually attained good efficiency, they were noisy, often had poor frequency stability, and usually produced several radio frequencies, in addition to the desired one.

The invention of the triode valve in 1907 finally made possible the development of powerful transmitting oscillators in their present form, with excellent frequency stability, silent operation, and ease in changing frequencies. Because of this combination of merits, valve oscillators are the most widely used generators of radio frequencies, and they will be the only ones discussed in this chapter. It is, however, important to understand that oscillations are not confined to valves, but that they occur in a wide variety of electrical and mechanical devices. A pendulum, an ordinary watch, a swing, windscreen wipers, etc., are all excellent examples of mechanical oscillators. You can visualize the basic principles underlying all oscillating systems most clearly by considering such a mechanical oscillator.

MECHANICAL OSCILLATOR

Fig. 78a shows a simple spring pendulum, consisting of a weight suspended from one end of a coiled spring. If the weight is pulled downward from its resting position and then released, as in (b), it will swing back beyond its original position to that shown in (c). It will then continue to oscillate up

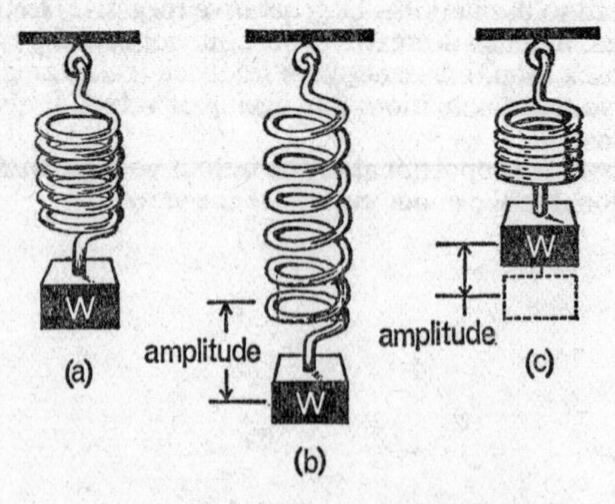

Fig. 78. Simple mechanical oscillator

and down until it gradually comes to rest. It stops when all the energy initially imparted to it is dissipated in heat because of the friction in the spring and bearings. One oscillation or cycle is completed when the weight has moved from its initial position (a), down to position (b), then up to position (c), and finally back to its original position (a). The frequency (f) is the number of oscillations or cycles completed in one second. The period is the time required to complete one oscillation, and is equal to $1/f$, the reciprocal of the frequency. It takes exactly the same time for each cycle to be completed, regardless of whether the weight swings through large or through small distances.

The maximum distance the weight moves from its original point, either up or down, is called the amplitude. The amplitude of the spring pendulum becomes progressively smaller as the oscillations die down, and reaches zero when the pendulum comes to a stop. It is possible to obtain an accurate record of the action by attaching a fine ink-fed brush to the weight and drawing a paper tape horizontally past the swinging pendulum, in a slow and even

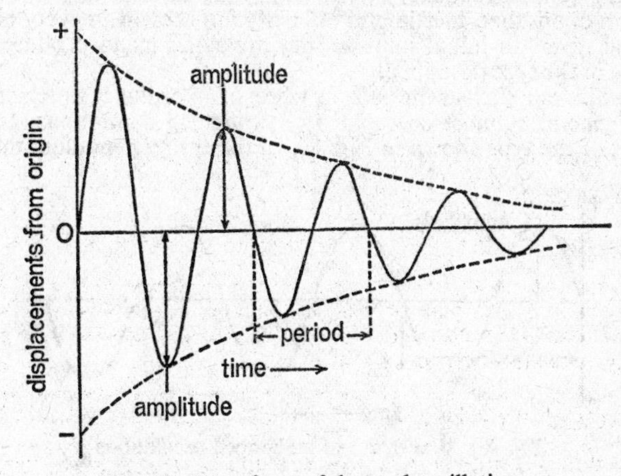

Fig. 79. Waveform of damped oscillations

motion. The result, shown in Fig. 79, is a graph of the displacements of the weight from its original position (the axis) against time. Since the oscillations eventually die down, the waveform of Fig. 79 is called a damped oscillation.

An analysis of the action described above discloses that two elements are required for oscillations to occur, the spring and the weight. Neither a weightless spring nor the weight alone can produce an oscillation. When the spring is extended by pulling the weight down, *potential energy* is stored in it in the form of tension. When the weight is let go, this tension or energy is released. The spring pulls back the weight and assumes its original slack position. During the motion, however, the potential energy of the spring has been transformed into the *kinetic energy* of motion of the weight. Because of its inertia or flywheel effect, the weight resists any sudden change in its motion. It does not stop, therefore, when the spring is slack again, but on the contrary continues to compress the spring until all its kinetic energy is again stored as potential energy in the compressed spring. Now the spring releases its tension again in the form of kinetic energy of motion of the weight, and the process is repeated all over again. The action would continue indefinitely if it were not for the fact that some energy is lost in the form of heat because of friction in the spring and bearings. As a result, a little less energy is stored in the spring during each cycle, until the oscillations finally die out.

If the weight of the spring pendulum is increased, the oscillations will take place more slowly, and if the weight is decreased, the oscillations occur at a more rapid rate. In other words, the frequency of oscillations is inversely proportional to the weight. If, on the other hand, the spring is shortened, the oscillations take place at a more rapid rate, and if it is elongated, the oscillations will be slower. Consequently, the frequency is also inversely related to the length of the spring. Instead of lengthening the spring, a more elastic or compliant spring could be substituted, and instead of shortening it, a stiffer spring could be used. Thus, the frequency can also be said to be inversely proportional to the elasticity of the spring.

The observations above can be stated as a general rule for all oscillators, whether mechanical or electrical. Every oscillating system must have two elements, inertia and elasticity, which can store and release energy from one to the other at a natural frequency determined by the dimensions of the elements. In one form or another, inertia and elasticity are present in every oscillating system, and given an initial impulse they are sufficient to produce damped oscillations of the type described.

Many oscillators, such as the balance wheel of an ordinary watch or a radio-frequency generator, must be capable of producing continuous, undamped oscillations of the type shown in Fig. 80. In the spring pendulum this can be

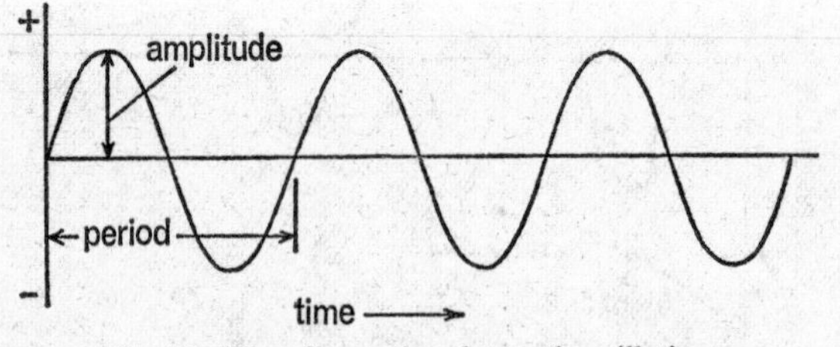

Fig. 80. Waveform of undamped oscillations

accomplished by pulling the weight each time it reaches bottom by just a sufficient amount to overcome losses due to friction. If the same result is to be achieved automatically, a mechanical source of energy must be supplied and some sort of synchronous trigger mechanism, which will release the energy to pull the weight at the moment it reaches the bottom. This is generally true for all oscillators producing continuous oscillations. An excellent example of a continuous mechanical oscillator is an ordinary clock or watch. Here, the balance wheel and hair spring represent the inertia and elasticity, respectively, of the oscillating system. The main spring is the source of energy, and the escapement is designed to release energy from the main spring at the proper instant during each oscillation of the balance wheel. An electrical oscillator, producing a continuous, undamped output, contains elements exactly analogous to those of a mechanical oscillator.

VALVE AND TRANSISTOR OSCILLATORS

Oscillators with a continuous output take energy from a unidirectional (d.c.) source and transform it into undamped oscillations or alternations. An electrical oscillator, therefore, acts as an energy converter which changes direct-current energy into alternating-current energy. Because of their ability to amplify, valves and transistors are very efficient energy converters, and are for this reason universally used as electrical oscillators. If the damped oscillations occurring in a resonant circuit containing inductance and capacitance are applied to the grid of a triode, the anode current varies in accordance with the grid signal, resulting in an amplified reproduction of the oscillations. Because of this amplification, more energy is available in the anode circuit than in the grid circuit. If part of this anode-circuit energy could be fed back by some means to the grid circuit in the proper phase to aid the oscillations of the resonant *L–C* circuit, its losses would be overcome and sustained, undamped oscillations would take place. This is in fact accomplished by a regenerative feedback circuit, which permits the combination of triode and resonant *L–C* circuit to function as a continuous self-sustaining oscillator.

ESSENTIAL PARTS OF TRIODE OSCILLATOR

To produce electrical oscillations with a valve or transistor, the following elements must be present:

(1) An oscillatory (resonant) circuit, containing inductance (*L*) and capacitance (*C*) to determine the frequency of oscillation. Such a circuit is called a tank circuit. (See Fig. 81.)

(2) A source of (d.c.) energy to replenish losses in the tank circuit.

(3) A feedback circuit for supplying energy from the source in the right phase (timing) to aid the oscillations; i.e. regenerative feedback. A valve or transistor can function as oscillator if it has sufficient amplification (μ) and if a sufficient amount of energy is fed back to the tank to overcome all circuit losses. If the losses in the tank and the anode circuit are completely overcome, the effective circuit resistance is zero, and oscillations take place. However, if either the amplification or the amount of energy fed back are insufficient to overcome the circuit losses and make the effective resistance zero, the valve or transistor will not oscillate. Both conditions must be fulfilled for sustained oscillations to occur.

OSCILLATIONS IN A TANK CIRCUIT

A tank circuit, consisting of a capacitor and inductance coil in parallel (Fig. 81), is the simplest type of electrical oscillating system. Its action is analogous to that of the simple spring pendulum, just described, and like the latter it can generate damped oscillations.

To understand how this comes about, assume that the capacitor (*C*) of

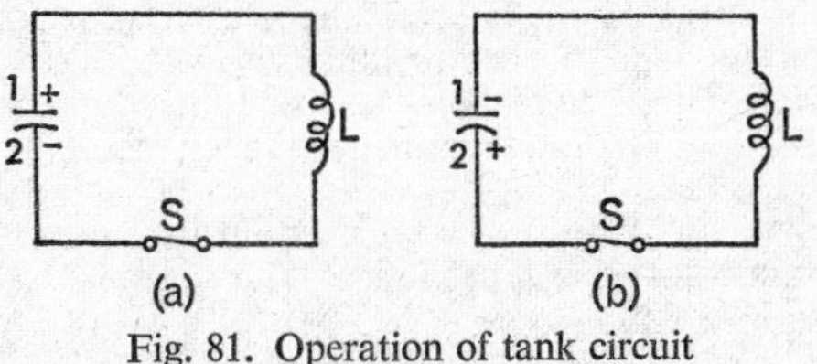

Fig. 81. Operation of tank circuit

Fig. 81a has been charged from a direct-voltage source, and is now discharged through the inductance coil *L*, by closing the switch *S*. Assume further that plate 2 of the capacitor is initially charged negative (that is, it has an excess of electrons), while plate 1 is charged positively (that is, it has a deficiency of electrons). Consequently, an electric field exists between plates 1 and 2.

When the switch is closed, electrons move rapidly from plate 2 through the inductance L to plate 1. As soon as the electron current flows through L, a magnetic field begins to be established around the coil. In building up the field, the magnetic flux lines cut across the turns of the coil, and induce a back e.m.f. which opposes the increasing current flow. This slows down the rate of discharge of electrons from the capacitor. As plate 2 loses its surplus of electrons by discharge, the current tends to die down, but is prevented from doing so by the inductance which now opposes the decrease in current flow, and keeps the electrons moving. Consequently, plate 2 not only loses its original excess of electrons, but gives up more electrons resulting in a deficiency, or positive charge. Simultaneously, an excess of electrons is pushed on to plate 1, so that it acquires a negative charge, as shown in Fig. 81b. The process stops momentarily when all the energy that was stored in the magnetic field is used up in pushing an excess of electrons to plate 1. At that moment the magnetic field collapses, but the energy has been stored again in the electric field of the capacitor, which is now charged in the opposite direction. The action then continues with electrons moving out of plate 1 through the coil to plate 2. Again, a magnetic field is built up, which stores the energy, and prevents the electron current from dying down. This time, however, the current is flowing in the opposite direction, from plate 1 to plate 2, and the 'flywheel' effect of the inductance now pushes an excess of electrons on to plate 2, recharging it negatively. A cycle is completed when the capacitor is again fully charged to the initial polarity.

The sequence of charge and discharge results in an alternating motion of electrons, or an oscillating current. The energy is alternately stored in the electric field of the capacitor and the magnetic field of the inductance coil. During each cycle, a small part of the originally imparted energy is used up as heat in the resistance of the coil and conductors. The oscillating current eventually dies down when all the energy has been transformed into heat. The waveform of these damped oscillations is exactly the same as that shown in Fig. 79 for the mechanical oscillator.

The capacitance of a tank circuit corresponds to the elasticity of a spring in a mechanical oscillator. An increase in capacitance, therefore, would be expected to lengthen the period of oscillations in a tank circuit, or equivalently, lower its frequency. This is indeed true. As the capacitance is increased, more charge must be transferred in and out of the capacitor during each cycle. Consequently, the period is lengthened, or the frequency is lowered.

We have pointed out that the greater the inertia element (weight) of a mechanical oscillator, the longer is the period of each oscillation. The self-inductance of a coil in a tank circuit corresponds to mechanical inertia. Accordingly, the greater the inductance of the coil, the greater is its opposition to any change in current flow and, hence, the longer is the time required for completion of each cycle. The greater the value of the inductance, therefore, the longer is the period, or the lower is the frequency of oscillations in the tank circuit.

The frequency of oscillations in a tank circuit is, thus, inversely related to both the inductance and the capacitance. The formula for the natural frequency of oscillations is

$$f = \frac{1}{2\pi\sqrt{LC}}$$

where $\pi = 3{\cdot}1416$
f = the frequency in cycles per second (hertz)
L = the inductance in henrys
C = the capacitance in farads

This expression shows that for any given frequency there is an infinite number of combinations of L and C. The same frequency of oscillations can be produced with a large inductor L and a small capacitor C, or with a small L and a large C. In practice, however, the choice of the L/C ratio for a particular frequency is restricted because the ultimate performance of the tank circuit depends to a large extent on this ratio. The impedance of the tank circuit varies directly in proportion to the L/C ratio. Since the desired impedance of the tank circuit in an electronic oscillator is determined to some extent by the valve or transistor operating conditions (i.e. anode or collector voltage and current), the choice of L/C ratios is limited.

TICKLER FEEDBACK CIRCUIT

Historically, one of the earliest electronic oscillator circuits is the so-called tickler feedback oscillator (Fig. 82).

The oscillatory tank circuit is made up of L_1 and C_1, the source of d.c. energy is the anode supply voltage, E, and the feedback circuit consists of the

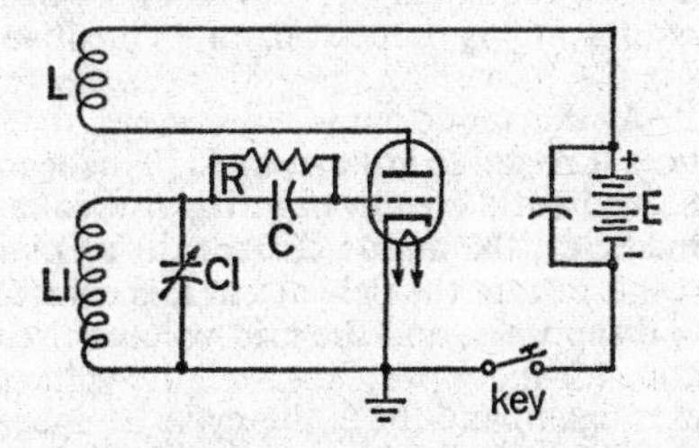

Fig. 82. Tickler feedback oscillator circuit

'tickler' coil L, which is coupled to L_1. Capacitor C_1 is made variable to adjust the frequency of the oscillations. The by-pass capacitor is placed across the battery to provide a low-reactance path for the alternating component of the anode current. The combination of R and C in the grid circuit is called the grid leak, and its function is to furnish a self-adjusting negative bias to the triode. We shall discuss its operation later on. The key serves to interrupt the oscillations into dots and dashes, for use in Morse-code telegraphy transmission.

Once oscillations have been started in the tank circuit L_1–C_1, they will appear in amplified form in the anode circuit. Part of the energy is fed back from tickler coil L to tank coil L_1 by mutual induction, thus overcoming losses and sustaining the oscillations. The phase of the feedback is correct, since the valve itself introduces a phase shift of 180° between grid and anode circuit. The combination of L and L_1 constitutes a transformer, and consequently, another phase shift of 180° takes place between the anode and the grid circuit. (All transformers introduce a phase shift of 180° between primary and secondary.) As a result, the voltage fed back is in phase with the voltage in the grid circuit, and regenerative or positive feedback takes place.

To understand the physical action you must remember that only a changing magnetic field is capable of inducing a voltage in a near-by coil. An expanding magnetic field induces a voltage in a certain direction in the coil, while a contracting magnetic field induces a voltage of the opposite polarity.

Assume the cathode of the triode is heated and electrons are being emitted. The moment the key is closed, anode current begins to flow through the valve and through the external circuit, consisting of the tickler coil L and the battery. This current sets up an expanding magnetic field around the tickler coil L and so induces a voltage in the tank coil L_1. Assume that the initial induced voltage is positive so that the upper end of the L_1–C_1 tank circuit is of positive polarity. This positive voltage charges the capacitor C_1 and places a positive charge on the grid of the valve which is connected to the upper end of L_1–C_1. Since the grid has initially no bias, this positive charge increases the anode current and thus further builds up the magnetic field around the tickler coil L. As a result, a still larger positive voltage is induced in L_1 and placed on the grid, and C_1 is further charged. Again the anode current keeps rising and the field of L expands, placing a still greater positive voltage on C_1 and the grid. Theoretically, this process continues until the anode current reaches its saturation point and tapers off. At that instant the field about coil L stops expanding and becomes static.

As the magnetic field about coil L stops expanding, the voltage induced in L_1 begins to drop and finally reaches zero. Now, however, capacitor C_1, which has been charged to the maximum positive voltage, begins to discharge in the opposite direction, making the upper portion of L_1–C_1 negative. As the voltage on the upper end of L_1–C_1 is reduced from its positive value to zero and then becomes negative, the voltage on the grid is equally reduced, thus lowering the anode current. As the anode current decreases, the field about L starts to contract and induces a negative voltage in L_1. This leads to a further discharge of capacitor C_1 and a still greater negative grid voltage. As the negative charge on the grid increases, the anode current drops more and more, until finally it is cut off. At this instant the field about L is completely collapsed, the voltage induced in L_1 disappears, and the grid voltage rises to zero. Since the grid is now less negative (or more positive) than the value it has at cutoff, the anode current increases again, and the entire cycle is repeated.

It must be understood that the entire process takes place very quickly, and is repeated thousands of times in an extremely brief period of time, at a rate determined by L_1 and C_1. You must further remember that oscillations will not take place if the amplification of the valve or the energy fed back from the tickler coil L to the tank circuit becomes insufficient to overcome circuit losses. The amount of energy fed back depends on the mutual inductance between L and L_1. As the coils are spaced farther apart, the mutual coupling between the coils decreases, and at a certain critical value of the mutual inductance, the coupling becomes too loose to sustain oscillations.

Grid-leak biasing. The grid-leak resistor R and grid capacitor C in Fig. 82 furnish the negative bias required for the triode operation. Grid-leak bias, rather than a resistor in the cathode circuit (cathode bias), is universally used in triode oscillators to ensure stable operation and make the oscillator self-starting. Capacitor C is large enough to provide a free (low-reactance) path to the grid for the excitation signal, and thus by-passes the high-resistance grid leak R. In operation, the grid is driven positive during positive half-cycles of the oscillations and, therefore, draws grid current. The electron current flows from the cathode to the grid and then through the external circuit consisting of L_1 and R, thus developing a voltage drop across R. The end of R connected to the grid is made more negative than the other end, and the grid is thus biased negatively by an amount equal to the voltage drop across R. The voltage present across R during grid-current flow charges capacitor C. If R and C are sufficiently large, the charge on C will leak off only very slowly during negative half-cycles, when no grid-current flows. Hence, for all practical purposes, the voltage across C remains constant throughout a complete cycle, and maintains a steady bias on the grid.

The use of grid-leak bias also makes the oscillator self-starting, since the grid bias is initially zero, making the anode current and the amplification large. Any initial impulse, such as the closing of the key, will thus be considerably amplified and start the building up of oscillations in the manner described before. Small random variations are always present in the circuit to start the oscillations.

HARTLEY OSCILLATOR

A modified version of the tickler feedback circuit using a tapped coil common to both anode and grid circuits, rather than two separate coils, is known as the Hartley oscillator (Fig. 83). Except for minor modifications in the manner in which coupling is obtained between the anode and grid circuits, the operation of the Hartley oscillator is identical with that of the tickler feedback oscillator just discussed.

The lower portion of the tank coil, L_2, is inductively coupled to the upper portion, L_1, the combination functioning as an autotransformer. Since the variable tank capacitor C is connected across both coil sections, capacitive coupling is present in addition to the mutual inductance. Adjustable taps are customarily provided to control the operating conditions.

Tap 1 together with the variable capacitor adjusts the frequency of oscillations. The inductance L of the tank can be decreased by moving the tap down. The frequency of the oscillator is $\frac{1}{2\pi\sqrt{LC}}$, where L is the total inductance connected across C.

Tap 2 adjusts the effective impedance of the anode-circuit coil L_2, and has only a slight effect on frequency. The more turns included between cathode and anode, the higher the anode-circuit impedance.

Tap 3 adjusts the grid excitation voltage to the proper value for maximum output. Sufficient excitation must be provided to overcome all losses in the

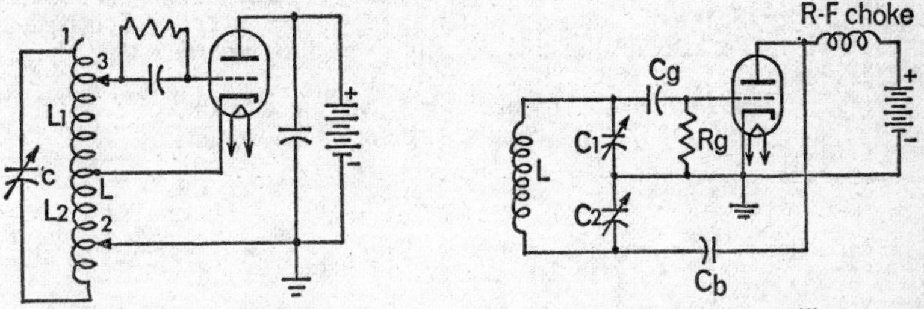

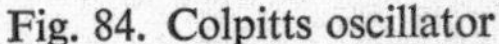

Fig. 83. Hartley oscillator

Fig. 84. Colpitts oscillator

tank and associated anode circuit and hence make the effective circuit resistance zero, as previously explained. At the same time, the excitation must not be so large as to overdrive the oscillator and thus distort the waveform of the output. The excitation is increased by moving tap 3 upward. All the taps are generally adjusted together, since their setting is to some degree interdependent.

Phasing of feedback. Since valves normally introduce a 180° phase shift between grid and anode, another phase shift of 180° must be provided in the feedback circuit to make the voltage being fed back in phase with the grid voltage. In this way positive feedback (regeneration) required for oscillation is obtained. This phase correction is obtained in the tank coil. You can readily confirm that the two opposite ends of the autotransformer L_1–L_2 are actually

180° out of phase, i.e. whenever one end of the transformer winding is positive the other end is negative. When the upper end of L_1 (connected to the grid) is positive, for instance, the lower end is at a minimum potential. However, the tap connected to the cathode is at an intermediate voltage, and so is negative with respect to the upper end and positive with respect to the lower end. Or, viewed from the tap, the upper end of the coil is positive and the lower end is negative. Since the grid and the anode are connected to opposite ends of the coil, they are of opposite polarity or phase. In this way, the feedback is properly phased to sustain oscillations.

COLPITTS OSCILLATOR

The Colpitts oscillator (Fig. 84) is similar to the Hartley circuit just discussed, except that two variable capacitors, C_1 and C_2, are used in the tank circuit instead of the tapped coil. The grid voltage is adjusted by capacitor C_1, instead of the tap used in the Hartley circuit. The tank is again common to both the anode and grid circuits, but the feedback is obtained by the relative voltage drops across the two capacitors. In Fig. 84, the anode is said to be shunt-fed, in contrast to the series-feed used in Fig. 83. Blocking capacitor C_b prevents the (d.c.) anode current from reaching the tank circuit while the radio-frequency choke keeps the r.f. currents out of the power supply. Grid-leak bias is used, but the grid leak R_g is connected in parallel with the grid circuit, rather than in series with it. This is necessary to provide a (d.c.) return path for the grid current, since otherwise the grid circuit would be completely isolated from the cathode for direct currents. Except for these minor modifications, the operation of the Colpitts oscillator is the same as that of the Hartley circuit.

Feedback. The two tank-circuit capacitors C_1 and C_2 act as a simple alternating-voltage divider. The tap between them fulfils the same purpose as the tapped coil in the Hartley circuit. It assures both the proper amount and the correct phase of the feedback voltage. As seen from the cathode-connected tap, whenever the top of C_1 is positive, the anode-connected end of C_2 is negative, or vice versa. Because of this polarity reversal, the correct 180° phase shift is obtained.

Frequency. The frequency of oscillations is determined by the coil L and the total capacitance C_T in the tank circuit. As before, the formula is

$$f=\frac{1}{2\pi\sqrt{LC_T}}$$

The capacitance C_T of two capacitors C_1 and C_2 in series is given by the formula $\frac{1}{C_T}=\frac{1}{C_1}+\frac{1}{C_2}$ or $C_T=\frac{C_1\times C_2}{C_1+C_2}$

Hence we can write

$$f=\frac{1}{2\pi}\sqrt{\frac{C_1+C_2}{L\times C_1\times C_2}}$$

TUNED-ANODE TUNED-GRID OSCILLATOR

The tuned-anode tuned-grid oscillator uses a tuned tank circuit in both the grid and the anode circuits, as shown in Fig. 85. The two coils L_1 and L_2 are not inductively coupled. Feedback takes place entirely through the grid-to-anode capacitance (C_{gp}) of the valve. In order to obtain oscillations of the correct frequency, both the grid and anode tank circuits must be tuned to a frequency somewhat higher than the desired operating frequency.

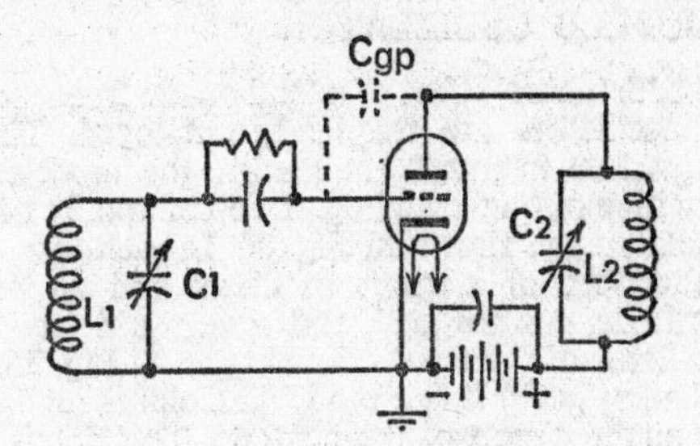

Fig. 85. Tuned-anode tuned-grid oscillator

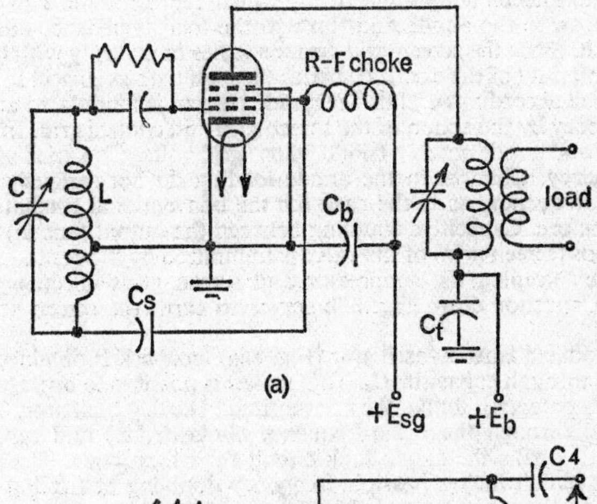

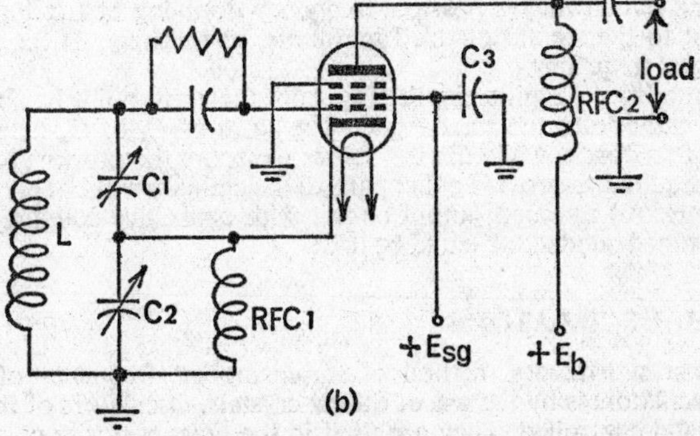

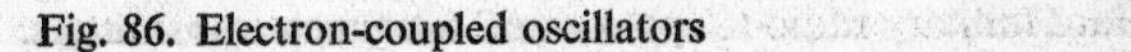

Fig. 86. Electron-coupled oscillators

ELECTRON-COUPLED OSCILLATOR

To minimize the 'detuning' effect of a load on the frequency of oscillations, electron-coupled oscillators are frequently employed. These substitute a common electron stream to couple the load to the anode circuit in place of inductive or capacitive output coupling. Two commonly used types of electron-coupled oscillators are shown in Fig. 86. In each case, a single pentode fulfils both the functions of a triode oscillator and an isolating or buffer amplifier. The cathode, control grid, and screen grid (acting as the anode) form a triode oscillator of the conventional type. In Fig. 86a, this is a modified Hartley oscillator; in (b) it is a modified Colpitts oscillator. The load is connected in the anode circuit of the pentode. The circuit derives its name from the fact that the oscillator and anode circuits are coupled solely by the stream of electrons between screen grid and anode.

Since the screen grid is at a positive voltage, electrons flow from cathode to screen grid, and oscillations are generated in the same manner as in the case of the conventional triode oscillator. The frequency is determined by the values of L and C in the grid tank circuit. Only a small number of electrons are intercepted by the screen, but they are sufficient to maintain oscillations in the tank circuit. The remaining electrons, which represent most of the space current, go on to the anode and through the load impedance connected in series with it. Since the screen-grid current varies in intensity with the oscillations, the intensity of the electron stream through the screen grid to the anode will be varied accordingly. Thus the anode current is modulated at the oscillator frequency by the action of the screen grid and control grid. In effect, the screen grid and anode act as a triode whose grid voltage is varied at the oscillator frequency. Changes in the anode loading do not seriously affect the oscillation frequency, as is the case for the conventional oscillator circuits described before. Capacitive coupling between the output (anode) and oscillator sections (screen grid) of the valve is minimized by the pentode construction, and by keeping the suppressor and screen grids effectively at earth potential. Capacitor C_3 in Fig. 86b serves to earth the screen grid for r.f. currents.

In the modified Hartley oscillator (Fig. 86a) feedback is obtained from the screen grid through capacitor C_s. This makes it possible to operate the cathode at earth potential, unlike the conventional Hartley oscillator. The screen is shunt-fed through the radio-frequency choke (r.f.c.) and capacitor C_b. Capacitor C_t earths the anode tank circuit for r.f. currents. The use of the anode tank circuit makes possible frequency doubling or tripling by tuning the output to the second or third harmonic, respectively, of the grid circuit fundamental frequency.

In the modified Colpitts circuit (Fig. 86b) the earth point has been shifted from the cathode to the screen grid. The ratio of C_1 to C_2 determines the amount of feedback, while the tap serves to secure the proper phase. Choke r.f.c.$_1$ is required to provide a d.c. path to the cathode without earthing it for r.f. currents. An untuned output circuit with capacitive coupling is shown, though a tuned anode tank could be used.

CRYSTAL OSCILLATORS

The most satisfactory method of stabilizing the frequency of radio-frequency oscillators is by the use of quartz crystals. Oscillators of this type are called crystal-controlled. They are used in the great majority of commercial and military radio telephone and telegraph transmitters.

Piezoelectric effect. The control of frequency by means of crystals is based upon the piezoelectric effect. When some crystals are compressed or stretched in certain directions, electric charges appear on the surfaces of the crystal that are perpendicular to the axis of mechanical strain. Conversely, when such crystals are placed between two metallic surfaces between which a difference of potential exists, the crystals expand or contract.

Thus, if a slice of a crystal is compressed along its width, or stretched along its length, so that it bulges inward, as in Fig. 87a, opposite electrical charges will appear across its faces, and a difference of potential is generated. If the crystal is squeezed or compressed lengthwise, so that it bulges outward, as in (b) of Fig. 87, the charges across its faces reverse. If alternately stretched and

Fig. 87. Expansion and contraction of a crystal

squeezed, a crystal slice becomes a source of alternating voltage. Conversely, if an alternating voltage is applied across the faces of a crystal wafer, it vibrates mechanically. The amplitude of these vibrations becomes very vigorous, if the frequency of the alternating voltage equals the natural mechanical frequency of vibration of the crystal, that is, if resonance takes place. If all mechanical losses are overcome, the vibrations at this natural frequency will sustain themselves and generate electrical oscillations of constant frequency. Accordingly, a crystal can be substituted for the tuned tank circuit in an electronic oscillator.

Types of crystal. Practically all crystals exhibit the piezoelectric effect, but only a few are suitable as the equivalent of tuned circuits for frequency-control purposes. Among these are quartz, Rochelle salts (sodium potassium tartrate), and tourmaline. Rochelle salt is the most active piezoelectric substance—that is, it generates the greatest voltage for a given mechanical strain. Rochelle salts are physically and electrically unstable, however, and therefore not suitable for frequency control. They have found applications in microphones and gramophone pickups. Quartz, although being much less active than Rochelle salt, is used universally for frequency control of oscillators because it is cheap, mechanically rugged, and expands very little with heat. Quartz is among the most permanent materials known, being chemically inert and quite hard physically.

Frequency of oscillation. Most crystals have at least two principal ways or modes in which they can vibrate. They can bulge in and out perpendicularly to their long parallel faces, as shown in Fig. 87, or they can stretch and contract along the width of these faces so that their short parallel faces bulge in and out. In the first case, called thickness vibration, pressure waves travel through the crystal from one long face to the opposite face, are reflected back from there to the first face and then reflected back again. At a particular thickness of the crystal, the reflected waves are in phase with the direct waves, and reinforce each other. As a result, standing waves are created between the two long faces of the crystal. The fundamental natural frequency of oscillation occurs at that particular thickness, where at least one complete wavelength (or cycle) can exist between the two long faces. However, for the same thickness, two or more shorter complete wavelengths can also exist between the two faces. Due to the inverse relationship between frequency and wavelength this means that the crystal also can vibrate at the second, third, or higher harmonic of this fundamental frequency. But for a given thickness, the fundamental frequency of the first principal mode is fixed.

The second principal mode of vibration (the short faces bulging in and out) is determined by the width of the plate, as measured along the long parallel faces. It is known as width vibration. Again, standing waves occur at the natural frequency of oscillation, and harmonics of this fundamental frequency are possible. But the fundamental frequency is determined by the width. Besides these principal modes and their harmonics, additional modes of vibration are possible, produced by various flexural and torsional tensions. Quartz crystals are produced commercially for frequencies from about 50 kilocycles per second to as high as 50 megacycles per second.

Crystal mounting. Crystals become practical circuit elements when they are associated with a crystal holder. In a holder (Fig. 88a), the crystal is placed between two metallic electrodes and forms a capacitor, the crystal itself being the dielectric. The crystal holder is arranged to add as little damping of the vibrations as possible, and yet it must hold the crystal rigidly in position. This

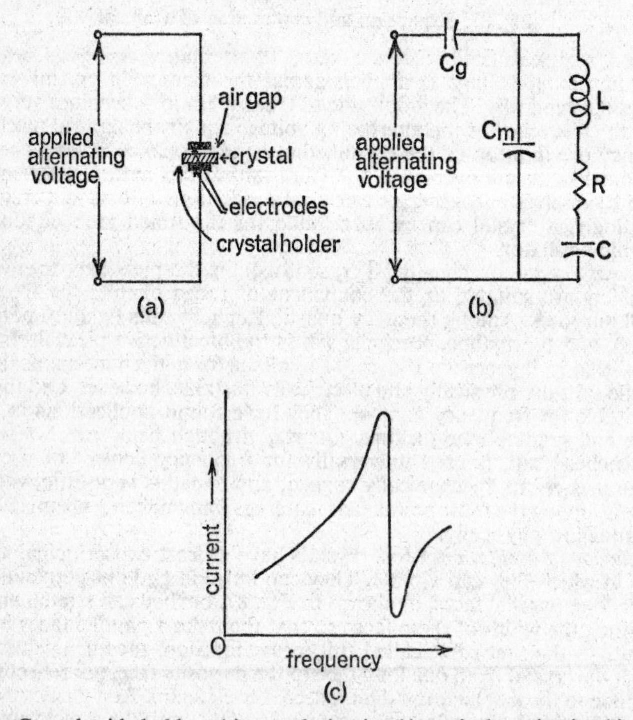

Fig. 88. Crystal with holder: (a) actual circuit; (b) equivalent circuit; (c) crystal resonance curve

is done in various ways. In some holders the crystal plate is firmly clamped between the metal electrodes, while others permit an air gap between the crystal plate and one or both electrodes. The size of the air gap, the pressure on the crystal, and the size of the contact plates affect the operating frequency to some degree. The use of a holder with an adjustable air gap permits slight adjustments of frequency to be made. For the control of appreciable amounts of power, however, a holder which firmly clamps the plate is generally preferred.

Equivalent circuit. At its resonant frequency a crystal behaves exactly like a tuned circuit, so far as the electrical circuits associated with it are concerned. The crystal and its holder can be replaced, therefore, by an equivalent electrical circuit, as shown in Fig. 88b. Here, C_m represents the capacitance of the mounting with the crystal in place between the electrodes, but not vibrating. C_g is the effective series capacitance introduced by the air gap when the contact plates do not touch the crystal. The series combination L, R, and C represents the electrical equivalent of the vibrational characteristics of the quartz plate. The inductance L is the electrical equivalent of the crystal mass effective in the vibration, C is the electrical equivalent of the mechanical compliance (elasticity), and R represents the electrical equivalent of the mechanical friction during vibration.

The frequency at which L and C are in series resonance is also the frequency of mechanical crystal resonance. Because of the presence of C_m, the circuit

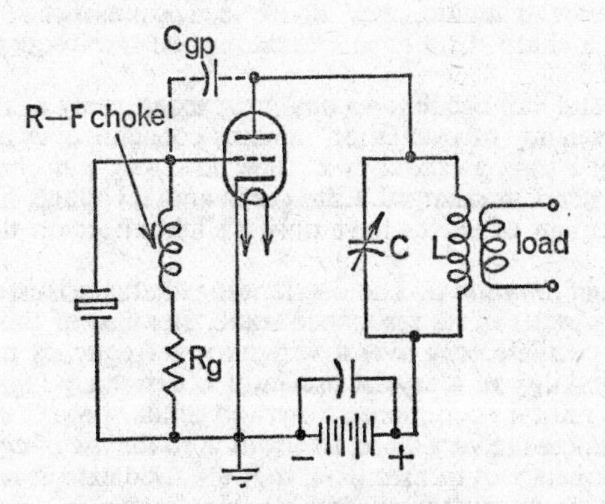

Fig. 89. Typical crystal-oscillator circuit

will also have a parallel-resonant frequency slightly above series resonance, when the series branch has an inductive reactance equal to the capacitive reactance of C_m. (A series L–C–R circuit is inductive above resonance.) Since C_m has a very low reactance (high C), only a small inductive reactance is required in the L–C–R branch to produce parallel resonance with C_m. The series and parallel resonant frequencies, therefore, are very close together. The presence of both resonant frequencies is clearly revealed by crystal resonance curve (Fig. 88c). This curve is extremely sharp, which means that the crystal will vibrate only within a very narrow band of frequencies: this provides high selectivity, or discrimination against unwanted frequencies.

CRYSTAL-OSCILLATOR CIRCUIT

A typical crystal oscillator is shown in Fig. 89. You can easily verify that this is the equivalent of the tuned-anode tuned-grid oscillator of Fig. 85, but with the crystal replacing the grid tank circuit. Feedback is obtained through the grid-to-anode capacitance C_{gp}. The choke (r.f.c.) keeps radio-frequency currents out of the grid-leak resistor, which provides the bias. The crystal functions in the same manner as the grid tank circuit of the tuned-anode tuned-grid oscillator. It stores energy in mechanical form during one half of the excitation-voltage cycle, and releases it in electrical form during the second half of the cycle. The rate of storage and release of energy depends on the natural frequency of the crystal and so determines the frequency of oscillations generated by the circuit. The losses in the crystal are overcome by the energy fed back through C_{gp}.

Frequency. The resonance curve of the crystal (Fig. 88c) is obtained by tuning the anode tank from a frequency below crystal resonance to one above crystal resonance. As the frequency is increased, series resonance of the crystal is reached first as indicated by the high current peak in the resonance curve. (The impedance is a minimum, and hence the current is a maximum for series resonance.) When the frequency of the anode tank circuit is increased slightly above this value, parallel resonance of the crystal is reached—that is, the inductive reactance of the crystal proper equals the capacitive reactance of the crystal holder. This is indicated by the sharp drop of the current in the resonance curve (Fig. 88c) and consequent high crystal impedance. Oscillations commence when the frequency of the tank circuit is again slightly increased to make the anode circuit inductive. The presence of oscillations can be detected by a sharp drop in anode current as the frequency of the tank circuit is raised.

Since the crystal can oscillate to any great extent only at its resonant frequency, the frequency of oscillation remains constant over a wide range of adjustment of the tuning capacitor C. The power output changes, however, substantially, when C is changed. Changes in anode voltage, filament voltage, and the replacement of valves have only a slight effect on the frequency of oscillations.

Advantages and limitations. The outstanding characteristic of the crystal is the extreme sharpness of its resonance curve. Because of this characteristic, the crystal can oscillate only over a very narrow frequency range and hence the frequency stability of a crystal oscillator is extremely high. This is taken advantage of in military communications and similar broadcasting, where the frequency tolerances are very close. In addition to the use of crystal oscillators in fixing the frequency of transmitters, they are used extensively as frequency standards for measurement purposes. If a low-frequency crystal is used in a circuit whose output is not tuned (a simple choke) a large number of harmonic frequencies (multiples of the basic frequency) are created. A great number of frequency calibration points can be obtained in this way with a single quartz crystal.

A crystal-controlled oscillator is a fixed-frequency oscillator. Its consequent disadvantage is that a different crystal must be used for each desired frequency. In many applications, especially military, it often is required to change the frequency of the transmitter rapidly and continuously. For these applications the ordinary variable-frequency oscillator is preferred, since it may be operated at any frequency within a band at the turn of a dial. Another limitation of the crystal oscillator is its low power output.

MICROWAVE OSCILLATORS

We have seen in Chapter 6 that conventional valves become progressively less effective as amplifiers and oscillators as the operating frequency is raised above 100 megacycles per second. The reason for this falling off in efficiency is essentially threefold:

1. The internal capacitances and inductances of conventional valves offer low reactances at very high frequencies, thus shunting out an appreciable part of the voltages applied to the electrodes. Moreover, at a certain critical fre-

quency, these internal reactances produce self-resonance, regardless of the tuned circuits connected externally to the valve. Valves cannot be operated above this resonant frequency.

2. The transit time of the electrons passing between the electrodes of the valve becomes an appreciable part of the cycle at high operating frequencies (between 300 and 3,000 Mc/s). As we have seen, this leads to a considerable radio-frequency power loss.

3. At very high frequencies the dimensions of the valve elements and associated circuit become comparable to the wavelength of the signal (wavelength = velocity/frequency) and, hence, considerable radio-frequency losses and power losses due to direct radiation from the valve and associated circuit take place.

We have further seen (in Chapter 6) that specially designed valves, such as the acorn and lighthouse types, are capable of functioning well in the u.h.f. region between 300 and 2,000 Mc/s. The generation of frequencies in the microwave region between 2,000 and 30,000 Mc/s, however, demands radically different techniques than have heretofore been described. These techniques make use of the very limitations in valve dimensions and electron transit time in achieving their normal operation. Among the most important valves using these novel principles to generate microwave frequencies are the so-called klystrons and magnetrons.

KLYSTRON OSCILLATOR

Klystrons make use of the transit time of the electrons to obtain a transfer of energy from the moving electrons to an electric field produced by an alternating voltage. Oscillations are produced by making most of the electrons pass through the electric field at such times that they deliver energy to the source of the alternating electric field. (Oscillations can occur only when more energy is delivered to the field than is taken from it.) This is accomplished by forming the electrons emitted from the cathode into compact bunches or groups, which deliver energy to the electric field on the grid at just the right time to synchronize with the alternations of the field. This principle is called velocity modulation. Klystrons are one type of velocity-modulated valve.

As in conventional oscillators, the frequency of the oscillations excited by a klystron is determined by a resonant circuit. However, the ordinary L–C tank circuit is replaced in microwave valves by a resonant cavity or chamber whose physical dimensions determine the frequency of its vibrations when it is excited by electron oscillations. A pulsating stream of electrons can excite oscillations in a resonant cavity in the same way as pulses of anode current excite a tank circuit to its resonant frequency. A valve which performs this action is shown schematically in Fig. 90a.

At one end of the klystron is an electron gun, similar to that used in the cathode-ray tube, consisting essentially of a cathode and accelerator grid. The electrons are emitted from the heated cathode and are attracted towards the positive accelerator grid. Most of the electrons pass through the grid wires to form a beam of electrons, all travelling at the same speed. The beam of electrons is then passed through a pair of closely spaced grids, called the buncher, each of which is connected to one side of a tuned (resonant) circuit. An alternating voltage exists across the resonant circuit, which speeds up or slows down the electrons, depending on the instant they enter the space between the buncher grids. An electron that passes the centre of the buncher at the instant when the voltage passes through zero, leaves the buncher with unchanged velocity. Electrons which pass through the buncher a little earlier, when the voltage is still negative, will be slowed down, while electrons which pass through the buncher a little later, when the voltage is positive, will be speeded up. This action causes the electrons to bunch together at some point beyond the buncher grids. Thus, the electron stream consists of bunches of electrons separated by regions in which there are few electrons.

These bunches of electrons pass through a second set of grids, called the catcher, which also are coupled to a resonant circuit. The polarity of the alternating field on the catcher grids is such that the electrons are slowed down, and thus give up some of their kinetic to the electric field and, consequently, to the tuned circuit. The spent electrons are removed from the circuit by a positive collector plate. The bunches pass through the catcher at the resonant frequency of the tuned circuit (i.e. once each cycle), and so maintain continuous oscillations in the tuned circuit. Some of the energy is fed back from the catcher to the resonant circuit of the buncher (see Fig. 90b) in the proper

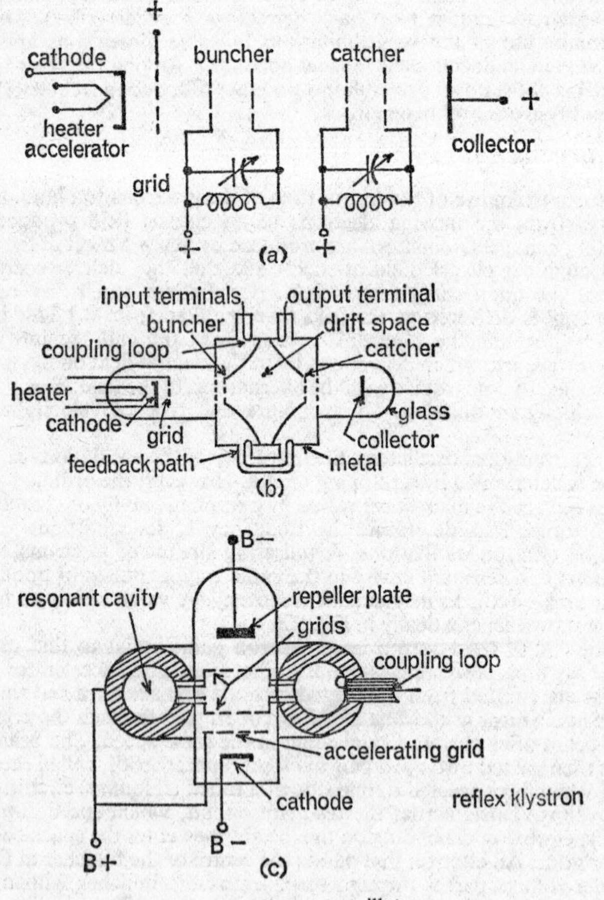

Fig. 90. Klystron oscillators

phase, thus producing self-sustained oscillations. This type of klystron can be used also as an amplifier or mixer.

As mentioned before, the tuned circuits of the buncher and catcher take the form of hollow metal chambers, called cavity resonators, with one of the grids attached to each side of the cavity (Fig. 90b). These cavities possess all the properties of conventional tank circuits, but are much more efficient at microwave frequencies. They are so tiny at these extremely high frequencies, that the entire cavity can be sealed inside the envelope of the valve, as shown. Energy is extracted from or coupled into the cavities by means of single-turn coupling loops, placed within the chambers.

For exclusive oscillator use, a simplified type of klystron that uses the same set of grids for both bunching and catching, has been developed (see Fig. 90c). In this so-called reflex klystron, a negative repeller plate is placed beyond the grids. This serves to repel the electrons that have been bunched on their first trip through the grids. The electrons then pass back through the grids, where energy is taken from them in the same manner as in the conventional catcher. The reflex klystron utilizes a single cavity resonator, which is more easily adjusted than the two-resonator klystron.

MAGNETRONS

The magnetron is a diode valve whose current is influenced by a magnetic field. As a microwave generator it can oscillate at frequencies from 300 to beyond 30,000 Mc/s, and produce peak powers of several thousand kilowatts. Early forms were of the split-anode type, as shown in Fig. 91a. The valve consists of a cathode in the form of a straight wire and a cylindrical plate or

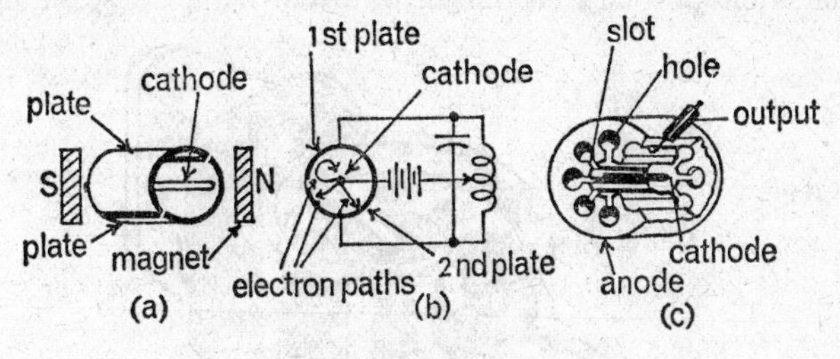

Fig. 91. Magnetrons

anode concentric with the cathode; the plate is split into an even number of segments (two are shown). A magnetic field is applied parallel to the valve's axis and is, therefore, perpendicular to the electric field existing between cathode and anode. A permanent magnet or an electromagnet can be used to provide the magnetic field.

In operation, the segments of the plate are kept at some positive potential with respect to the cathode. In the absence of a magnetic field, the electrons travel to the plate in a straight path. However, as a magnetic field is applied and increased, the electron path becomes more and more curved, and finally circular. (Magnetic and electric fields at right angles tend to bend the electron paths into circles.) At a certain critical value of the field strength, the electrons will miss the plate entirely and return to the cathode in a circular orbit, as shown in (b) of the figure. This is the point of anode-current cut-off. At higher values of the field strength, the radii of the circular orbits become smaller.

When the plate segments are made part of a resonant circuit (Fig. 91b) and the magnetic field is adjusted to anode-current cut-off, so that the electrons just fail to reach the plate, ultra-high-frequency oscillations can be produced. The action depends upon the transit time of the electrons. An alternating voltage is superimposed on the constant voltage present on the plate segments, causing the anode voltage to vary about its steady value. If the period of the alternating voltage is made equal to the transit time of an electron for a complete circular rotation, some electrons will be slowed down by the alternating field at the plate, and lose energy to it; others will be speeded up and, thus, gain energy from the electric field. The magnetron can be adjusted so that energy is extracted by the electric field from the majority of electrons grazing the plates. This energy sustains powerful oscillations in the associated resonant circuit. The frequency of oscillations is determined primarily by the transit time of the electrons. The magnetron also can be operated as a negative-resistance oscillator at frequencies that are low compared to the transit-time frequency. In this case, the frequency can be varied continuously by changing the constants of the external resonant circuit.

Modern super-high-frequency magnetrons are transit-time types made in the form of a multianode cavity resonator. The basic structure is shown in Fig. 91c. The assembly is a solid block of copper which assists in heat dissipation. The cathode is a cylinder of appreciable diameter located in a cavity in the centre of the structure. The plates or anodes are cut out of the block in the form of large circular holes and are divided into 4 to 16 or more segments. Radial slots lead out from the common cathode region to smaller circular holes, which act as resonant cavities. Each slot and terminating hole is electrically equivalent to a tuned resonant circuit, the slot having predominantly capacitive action, and the terminating hole being primarily inductive. Under proper conditions of voltage and magnetic field strength, energy is transferred

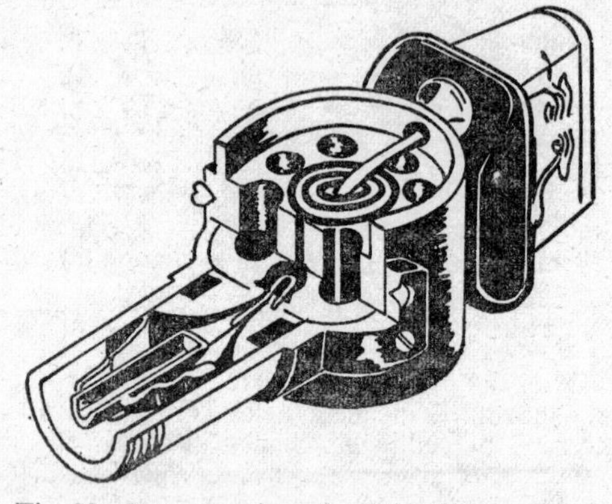

Fig. 92. Cut-away view of multicavity magnetron

from the swarm of gyrating electrons to the resonant cavities and powerful oscillations are sustained. The oscillations are coupled to the output by means of a wire loop, placed in one of the small circular holes. A cut-away view of a multicavity magnetron is shown in Fig. 92.

TRAVELLING-WAVE TUBES

The travelling-wave tube (t.w.t.) makes use of the interaction between an electron beam and a travelling electromagnetic wave to obtain highly efficient microwave amplification. Continuous-wave powers of hundreds of watts in the frequency range from 1,000 to 10,000 Mc/s are easily obtained, with voltage gains of 50 dB or better over a 1,000-Mc/s bandwidth being typical.

Fig. 93 illustrates the physical construction of a travelling-wave amplifier tube. One end of the tube is an electron gun from which a pencil-like beam of electrons is shot through a long, loosely wound helix and strikes the collector electrode, which is at the anode potential. The electromagnetic wave to be amplified is applied through a waveguide section at the input end of the helix and travels along the turns of the helix. As it does so, it produces an electric

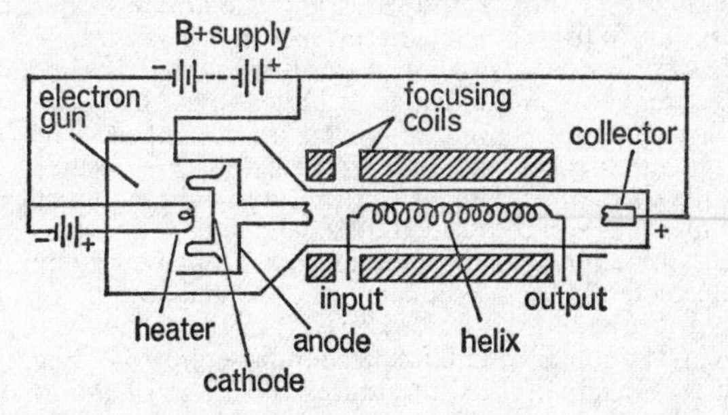

Fig. 93. Schematic drawing of travelling-wave tube

field at the centre of the helix which is directed along the axis of the helix. When the velocity of the electron beam approximates to the rate of advance of the axial field of the wave travelling along the helix, an interaction takes place in which the electrons impart energy to the travelling wave. As a result, a substantially amplified wave emerges from the output end of the helix. This wave is then coupled to another waveguide.

In practice, two electromagnetic focusing coils surround the glass envelope of the travelling-wave tube. One focuses the electron beam at the exit of the electron gun, while the other—arranged along the length of the tube—prevents the beam from spreading and guides it along the centre of the helix. Although the travelling-wave tube can be used only as amplifier, its essential operating principle has also been applied to an oscillator, known as backward-wave oscillator.

SUMMARY

One complete oscillation is a cycle; frequency (f) is the number of cycles completed in one second; period is the time required to complete one cycle and is equal to $1/f$; amplitude is the maximum displacement from the origin.

Damped oscillations die down because of internal losses. If losses are overcome, undamped, continuous oscillations take place.

Every oscillating system has two elements—inertia and elasticity—capable of storing and releasing energy from one to the other at a natural frequency determined by the dimensions of the elements.

Energy in a tank circuit is alternately stored in the electric field of the capacitor and the magnetic field around the coil.

The frequency of natural oscillations in a tank circuit is

$$f=\frac{1}{2\pi\sqrt{LC}}$$

An electrical oscillator acts as an energy converter which changes direct-current energy into alternating-current energy.

The essential parts of a triode oscillator are:

(1) An oscillatory tank circuit, containing L and C, to determine the frequency of oscillations.

(2) A source of (d.c.) energy to replenish losses in the tank circuit.

(3) A feedback circuit for supplying energy from the source in the right phase to aid the oscillations (regenerative or positive feedback).

The amplification factor of the valve and the amount of energy fed back must be sufficient to overcome all circuit losses.

In the tickler feedback oscillator, positive feedback occurs because of mutual induction between the tickler and tank coils.

The Hartley oscillator is a modified version of the tickler feedback circuit, using a tapped coil, rather than two separate coils.

The Colpitts oscillator is similar to the Hartley circuit, except that it uses a split capacitor in place of the tapped coil.

The tuned-anode tuned-grid oscillator uses a tuned tank circuit in both the grid and the plate circuits. Feedback in this oscillator takes place entirely through the grid-to-anode capacitance (C_{gp}) of the valve.

In an electron-coupled oscillator, a common electron stream couples the load to the anode circuit; this minimizes the effect of the load on the frequency of oscillation.

The control of frequency by means of crystals is based on the piezoelectric effect. Quartz is the most suitable piezoelectric crystal because it is cheap, mechanically rugged, chemically inert, and expands very little with heat.

Most crystals can vibrate in a thickness mode at a high frequency and in a width mode at a substantially lower frequency.

At its frequency of mechanical resonance, a quartz crystal behaves exactly like an electrical tuned circuit and can, therefore, be represented by an equivalent electrical circuit.

The basic crystal oscillator circuit is the equivalent of the tuned-anode tuned-grid oscillator, with the crystal replacing the grid tank circuit; its frequency of operation is the frequency of mechanical resonance of the crystal.

The advantages of crystal oscillators are extreme sharpness of resonance, high frequency stability, and possible use as frequency standards.

The limitations of crystal oscillators are their fixed frequency and relatively low power output.

The klystron is based on the principle of velocity modulation; klystrons form the electrons emitted by the cathode into compact bunches, capable of delivering energy to the electric field on the catcher grids. If feedback is provided from the resonant circuit of the catcher grids to the resonant circuit of the buncher grids, microwave oscillations can be generated. The klystron may be used also as microwave amplifier or mixer.

A reflex klystron uses the same set of grids for both bunching and catching, in conjunction with a negative repeller plate that reflects the electrons. Reflex klystrons can be used only as oscillators.

A magnetron is a diode whose anode current is influenced by a magnetic field at right angles to the electric field between anode and cathode.

Magnetrons can be operated either as negative-resistance oscillators, or as transit-time oscillators. In the first case, the frequency of oscillation is adjustable by means of the resonant circuit; in the second, the frequency is determined by the transit time.

A travelling-wave tube makes use of the interaction between a focused electron beam and a travelling electromagnetic wave to amplify microwave signals of thousands of megahertz over a large bandwidth and with high gain and low noise.

CHAPTER TWELVE

TUNED RADIO-FREQUENCY AMPLIFIERS

Most of the audio amplifiers we have discussed in earlier chapters will also work at radio frequencies—that is, above about 50 kc/s. However, the ordinary untuned, resistance-coupled amplifier becomes less efficient as the frequency goes up and the amplification falls off rapidly. Moreover, the audio amplifier amplifies a wide band of frequencies equally well and does not permit the selection of a particular desired frequency while discriminating against others. The ability to select a desired frequency (or band of frequencies) is quite important, however, inasmuch as radio and television transmissions are 'carried' on a specific radio frequency assigned to the station. The use of tuned circuits in conjunction with valve or transistor amplifiers makes possible the selection and efficient amplification of a specific radio frequency, or narrow band of frequencies.

TUNED (RESONANT) CIRCUITS

The tank circuit we discussed in the chapter on oscillators is a tuned or resonant circuit. We have seen that such an *L–C* tank circuit is capable of functioning as an oscillating system when d.c. pulses of energy are fed to it with the right timing to excite its natural frequency of oscillation. In this case a tank circuit functions as an energy converter from direct to alternating current. Oscillations in a tank circuit may also be excited by alternating current, provided that the frequency of the alternations is the same as the natural frequency of tank-circuit oscillations. In the latter case the tank circuit is said to be in resonance at the a.c. frequency.

The principle of resonance is best illustrated by ordinary physical objects. Every object has its own natural frequency of vibrations, depending on its size and mass. For example, when a certain note is struck on a piano, a near-by vase may begin to vibrate. This means that the natural frequency of oscillation of the vase has been excited by a piano tone of the same frequency, and energy is being transferred to the vase to sustain the vibrations. Similarly, soldiers marching across a bridge in step may cause it to vibrate at its natural frequency. If the constant small impulses from the marching soldiers take place at the same frequency as the natural frequency of oscillation of the bridge, resonance occurs and the bridge starts to vibrate. This effect is cumulative and the amplitude of vibrations can become so large that the bridge may be destroyed.

Resonance in a tank circuit is analogous to mechanical resonance. When an alternating voltage is impressed across a tank circuit, electrical resonance occurs at that a.c. frequency at which the tank circuit breaks into natural oscillations. The circuit then draws just enough energy from the a.c. supply to overcome its internal resistance losses. This, indeed, is the fundamental meaning of resonance. At the resonant frequency, the external supply releases just enough energy with the proper timing to sustain the natural self-oscillations of the tank circuit. With these ideas in mind, let us now look more closely at the familiar parallel-resonant or tank circuit, which forms the basis of all tuned circuits.

PARALLEL-RESONANT (TANK) CIRCUIT

As shown in Fig. 94, a parallel-resonant circuit consists of two branches in parallel, one branch containing capacitance (*C*), the other inductance (*L*) and resistance (*R*). A source of alternating voltage (*E*) of a certain frequency (*f*) is applied to the two branches. The resistance of the inductive branch is always present, although sometimes not shown, and it represents that of the coil and associated conductors.

We know from basic electricity that the current in a capacitor leads the applied voltage by a quarter cycle, or 90° in phase, while the current through an inductor (coil) lags the applied voltage by a quarter cycle, or 90°. When both capacitance and inductance are present in a circuit, therefore, the current

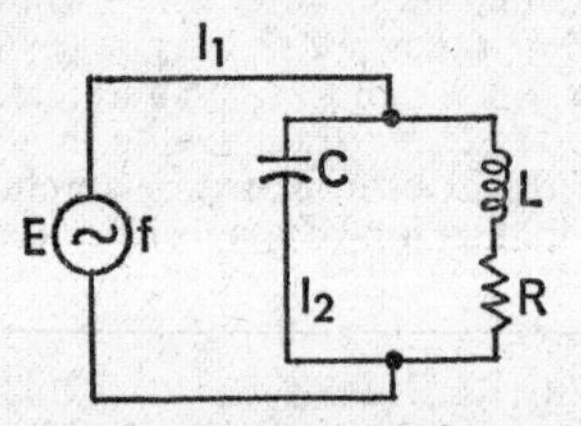

Fig. 94. Parallel-resonant circuit

through the capacitor leads that through the inductor by one half-cycle, or 180°. In effect, the two currents are in phase-opposition and tend to cancel each other. Similarly, the voltage developed across the inductor by the inductive current is in phase opposition to that developed across the capacitor. The voltage (E_L) developed across an inductance is the product of the current (*I*) and the inductive reactance (X_L), where the latter is defined as

$$\text{inductive reactance } X_L = 2\pi \times f \times L$$

Here *f* is the applied frequency and *L* is the inductance in henrys. Further, the voltage E_c developed across a capacitor is the product of the current (*I*) and the capacitive reactance (X_c), where the capacitive reactance

$$X_c = \frac{1}{2\pi \times f \times C}$$

(Here *C* stands for the value of capacitance in farads.) Of course, the voltage developed across a resistor (E_R) is by Ohm's law simply the product of the current (*I*) and the resistance (*R*), and this voltage is always in phase with the applied voltage.

By comparing the above relations you will note that the inductive reactance, X_L, increases directly with the applied frequency, while the capacitive reactance is inversely proportional to the frequency. It stands to reason that there must be some frequency, where the inductive and capacitive reactances are equal in value and, hence, present equal opposition to current flow. At that frequency the currents through the two branches in Fig. 94 must be about equal in magnitude (except for the small current through the resistance), but they are, of course, in phase opposition in respect to each other. As far as the external line supplying the two branches is concerned, the branch currents are equal and opposite at this frequency and, hence, cancel each other out. Under these conditions only a tiny current (necessary to supply the losses in the resistance) flows. The line current is then a minimum, as shown in Fig. 95.

Now, parallel resonance is precisely defined as the frequency where the inductive and capacitive reactances of the two branches are equal and the line current is a minimum. It is interesting to note that these two conditions take place at the same frequency only if the resistance is small, which is generally the case.

At resonance, by definition, the inductive reactance equals the capacitive reactance.

$$X_L = X_c$$

hence,

$$2\pi f L = \frac{1}{2\pi f C}$$

solving for the resonant frequency,

$$f = \frac{1}{2\pi\sqrt{LC}}$$

This turns out to be the same as the natural frequency of oscillations of a tank circuit, discussed in the last chapter. You should not be surprised by this, since at resonance a tank circuit breaks out into natural or self oscillation.

Although the line current (I_1 in Fig. 94) is very small at resonance, the current circulating in the two branches (I_2 in Fig. 94) can be very large and may, in fact, be several hundred times the value of the line current. As evident

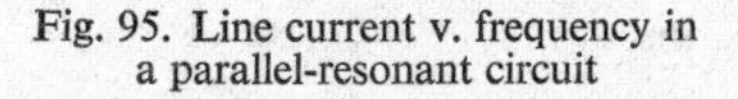

Fig. 95. Line current v. frequency in a parallel-resonant circuit

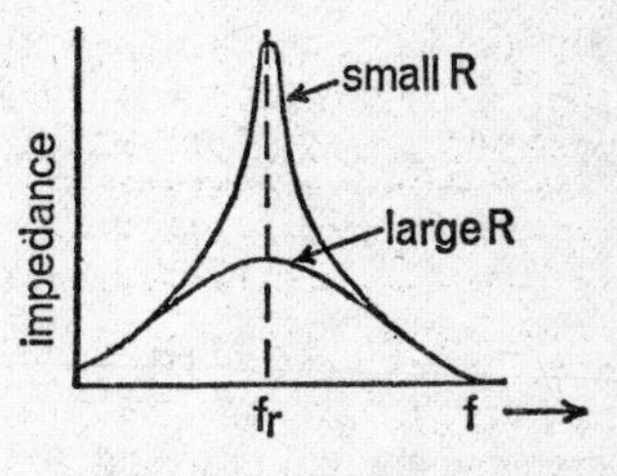

Fig. 96. Impedance curve of parallel-resonant circuit

from Fig. 95, the line current is a minimum at resonance and increases above and below resonance. The impedance presented by the parallel branches is simply the ratio of the applied voltage to the line current (E/I_1), and since the line current is small at resonance, the impedance is correspondingly large. This is graphically shown in Fig. 96 by the impedance resonance curve. The impedance rises to a steep peak at resonance.

'Q' factor. As is apparent from Fig. 96, the impedance diminishes rapidly when the frequency is varied in either direction from resonance. Note also that the impedance peak is much less pronounced when the resistance in the parallel-resonant circuit is large than when it is small. This is due to the fact that a large resistance consumes a considerable amount of power and draws a relatively large line current. In general it is desirable to have the resonance curve

as sharp as possible, in order to provide the necessary selectivity to discriminate between different radio frequencies. The sharpness of the resonance curve (Figs. 95 and 96) is determined by a quality factor called '*Q*'. The *Q* is defined as the ratio of the reactance of either the coil or the capacitor at the resonant frequency to the total resistance of the circuit. Mathematically,

$$Q=\frac{X_L}{R}=\frac{2\pi fL}{R}, \text{ or } Q=\frac{X_c}{R}=\frac{1}{2\pi fCR}$$

Since $X_L = X_c$ at resonance, both definitions result in the same value for *Q*.

Q is also a measure of the ratio of the reactive power stored in the tank circuit to the actual power dissipated in the resistance. The higher the *Q*, the

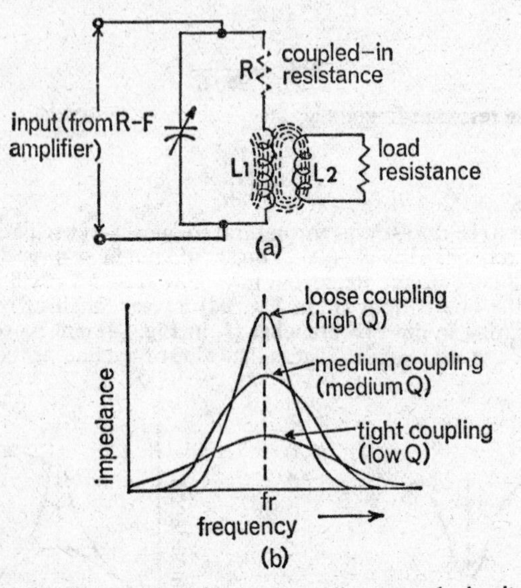

Fig. 97. Effect of variable load coupling on tank circuit

greater the amount of energy stored in the circuit compared with the energy lost in the resistance during each cycle. Consequently, the higher the *Q*, the greater is the efficiency.

The quantitative performance of a parallel-resonant circuit is easily evaluated by means of the *Q* factor. The higher the *Q* of a parallel-resonant circuit, the greater is its resonant impedance and circulating current (I_2 in Fig. 94), and the smaller is the line current. (The circulating current is approximately *Q* times the line current.)

COUPLED CIRCUITS

Fig. 97a shows a circuit frequently used for coupling a radio-frequency amplifier to a resistive load. Here the radio-frequency energy from the tank circuit (in a transmitter, for instance) is coupled to the load by means of an air-core transformer, consisting of coils L_1 and L_2. By changing the mutual inductance between the coils, the impedance of the tank circuit can be matched to the load resistance. The impedances must be made comparable in value to obtain the greatest possible energy transfer from the tank circuit to the load. The easiest way to change the mutual inductance and, thus, obtain the required impedance match is to vary the coupling between the coils by changing the distance between them.

When coil L_2 is coupled to L_1, a portion of the load resistance is coupled into the primary (tank) circuit and affects the primary circuit in exactly the same manner as though a resistor had been added in series with the coil. This is shown dotted in Fig. 97a. The closer the coupling between the two coils, the greater is the amount of series resistance coupled into the primary (tank) circuit. We can consider this apparent series resistance coupled into the tank circuit as being reflected from the load (secondary) circuit into the tank (primary) circuit. Increasing the coupling by bringing the coils closer together increases the reflected series resistance in the tank circuit and, hence, lowers the Q in the same manner as shown in Fig. 96 for a primary resistance.

Fig. 97b shows the effect of various degrees of coupling on the shape of the impedance resonance curve. When the coils are loosely coupled (large distance between them), the reflected resistance is small and, hence, the Q is high and the resonance curve is sharp. When the coupling is increased to an intermediate value, the coupled-in resistance is larger, the Q is lowered and the resonance curve is broader. Finally, when the coupling between the coils is very tight (coils close together), the reflected resistance is large, the Q is low, and the resonance curve is very broad, as shown by the bottom curve in Fig. 97b.

Coupled resonant circuits. Two resonant tank circuits are frequently used to couple the output of one stage of a radio-frequency amplifier to the input of the following stage. Fig. 98 shows such a circuit used primarily in the intermediate-frequency (i.f.) amplifier stages of superheterodyne receivers.

The primary tank circuit in Fig. 98 is connected between the positive terminal of the anode supply voltage and the anode (output) of the valve, while

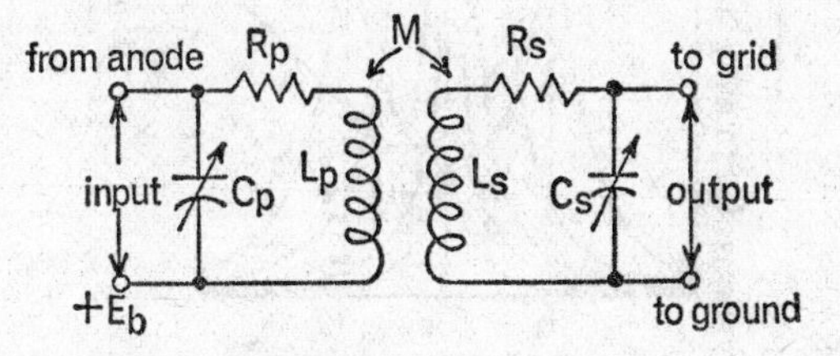

Fig. 98. Coupled tuned circuits in intermediate-frequency amplifier

the secondary tank circuit is connected between earth and the grid (input) of the following stage. The primary and secondary circuit resistances, R_p and R_s, respectively, are chiefly the resistances associated with the coils L_p and L_s. Both the primary and secondary tank circuits are tuned to resonance at the same frequency (possibly the frequency of a radio signal). The combined resonant response of such a coupled circuit depends primarily on the degree of coupling—that is, the amount of mutual inductance (M) between the primary and secondary circuits.

Fig. 99 illustrates some typical resonance curves for two coupled resonant circuits for various degrees of coupling. These curves were obtained by plotting the current in the secondary circuit against frequency for a constant input voltage applied to the primary circuit. The resonance frequency of both circuits is designated by f_r. When the coupling between the primary and secondary is quite loose, the secondary current is small, but the resonance curve is sharply peaked (curve 1). As the coupling is increased slightly, the secondary peak becomes larger and the resonance curve becomes somewhat broader (curve 2). This tendency continues with increased coupling until the secondary current becomes a maximum for a critical degree of coupling (curve 3).

Critical coupling is said to occur when the resistance reflected back into the primary from the secondary is equal to the primary resistance (R_p). The coefficient of critical coupling, k, for this condition is found to be

$$\text{critical } k = \frac{1}{Q}$$

provided both the primary and the secondary Q-factors are the same. Since the Q is usually high (above 100), the coefficient of critical coupling is generally very small.

When the coupling is increased beyond the critical value defined above, the secondary resonance curve begins to show two resonance humps, a condition known as double-peaking (curve 4). As the coupling is increased still further, the two peaks begin to spread apart in frequency and the valley or dip between the peaks becomes more pronounced (curve 5). For extremely tight coupling between the two circuits, the response between peaks may go almost to zero.

The double-peaked resonance characteristic shown in Fig. 99 is taken advantage of in the transformers of intermediate-frequency amplifiers to

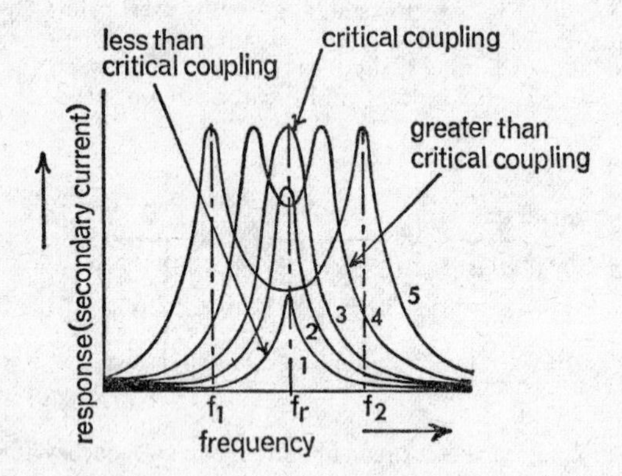

Fig. 99. Resonance curves of coupled tuned circuits for various degrees of coupling

obtain the required bandpass characteristic (i.f. amplifiers in radio sets must pass a band of frequencies of about 10 kc/s on both sides of the desired radio frequency). By slightly overcoupling the tuned circuits of an i.f. transformer beyond the critical k, the secondary current will be approximately constant near resonance over a range of frequencies between the two peaks. Beyond the peaks, however, the response falls off rapidly, so that the discrimination or selectivity against adjacent unwanted channels is excellent. Various resistance-loading methods are used to smooth out the dips between peaks at the extreme frequencies of the pass band. The greater the required bandwidth, the tighter must be the coupling to spread the two peaks sufficiently far apart.

TUNED AMPLIFIERS

Radio-frequency amplifiers can be divided into two main types: those used in radio and TV receivers, where small voltages are to be amplified, and those used in transmitters, where large voltages are to be amplified. The first type is always a Class A voltage amplifier, which means that anode current flows during the entire cycle of the input voltage. The second type is generally a Class C voltage or power amplifier; in Class C relatively large power output and efficiency are obtained by biasing the valve beyond cutoff, so that anode current flows for less than one half-cycle of the input voltage. The distortion attendant with Class C amplification is largely eliminated by use of a resonant anode tank circuit, as we shall see later. Either of these two basic types may also be operated untuned, but most r.f. amplifiers are tuned to amplify one frequency or a narrow band of frequencies and reject all other frequencies. This is accomplished by means of the parallel-resonant circuits, which—as we have seen—have frequency-selective properties.

A circuit for the initial stages of a tuned Class A radio-frequency amplifier in a radio receiver is shown in Fig. 100. The input signal from the aerial is

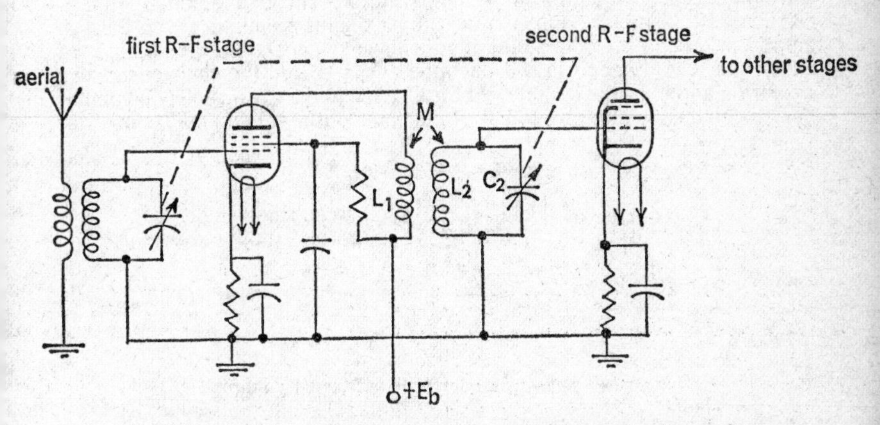

Fig. 100. Class A tuned radio-frequency input stages of radio receiver

applied to the primary coil of an inductively coupled circuit. The secondary coil of this circuit is tuned to resonance at the frequency of the desired radio station. The voltage across the tuned secondary is then applied to the grid of the first r.f. amplifier. The anode circuit of this valve is coupled to the grid of the second stage by a similar coupled circuit. Both of these tuned coupled circuits serve to select a voltage of the desired frequency and reject all others. Maximum input voltage to each stage is developed across the tuned secondary of the coupling transformer when its impedance is a maximum. As you can see from the resonance curve of a parallel-resonant circuit (Figs. 96 and 97), the impedance is a maximum at the resonant frequency and falls off sharply on either side of resonance. The tuned coupled circuits are, therefore, highly effective in selecting the desired frequency and rejecting all others.

The maximum voltage gain of a tuned radio-frequency amplifier is obtained when the secondary of the coupling transformer is tuned to the frequency of the incoming stage and the coupling between primary and secondary is adjusted to the critical value. (See curve 3 in Fig. 99.) The amplification under these conditions is then given by

$$\text{amplification at resonance} = 2\pi f_r g_m M Q$$

where f_r is the resonant (incoming) frequency
g_m is the transconductance of the amplifier valve
M is the mutual inductance between coils
and Q is the effective 'Q' ($=2\pi f_r L_2/R$) of the secondary.

Tuned r.f. voltage amplifiers always employ pentode valves of the types used with resistance-coupled and video amplifiers (Chapters 8 and 9). Triodes are not suitable for use as r.f. voltage amplifiers for two reasons. First, triodes have a low anode resistance compared with pentodes. This results in a low value of the mutual inductance (M) for critical coupling and, hence, a low gain per stage. Secondly, triodes have a large grid-to-anode interelectrode capacitance, which results in a feedback of energy from the anode circuit to the grid circuit. Since this feedback is positive, instability and oscillations may result.

Class C amplifiers. Class C radio-frequency amplifiers are used in radio transmitters for voltage and power amplification. The grid excitation voltage required in transmitters is many times larger than is permissible with Class A amplification and the power output and efficiency in Class C are higher than with either Class A or Class B operation. Apart from the changes in valve operation and bias, the circuit used for Class C may be a single-ended amplifier, like that shown in Fig. 100, or it may be a push-pull amplifier, similar to

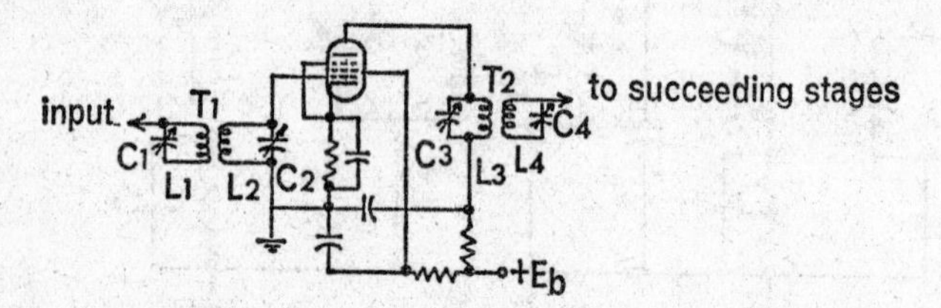

Fig. 101. Single-stage intermediate-frequency amplifier

that shown in Fig. 73 (Chapter 10). However, the important point of any Class C amplifier is that a tuned tank circuit must be used in the anode circuit of the valve to avoid distortion. As was explained in Chapter 10, the anode current in a Class C amplifier is zero during most of the input cycle and flows only in short pulses during the intervals when the grid voltage is near the positive peak of its cycle. Because of this, the anode current is far too distorted to be used directly. However, by feeding an anode current pulse once during each cycle to the anode tank circuit, continuous sinewave oscillations can be maintained in the tank circuit and large amounts of undistorted power may be withdrawn.

INTERMEDIATE-FREQUENCY AMPLIFIERS

Let us now consider the use of two coupled resonant circuits in an intermediate-frequency (i.f.) amplifier. These amplifiers are high-gain circuits used in superheterodyne receivers. As will be explained later (Chapter 15), i.f. amplifiers are permanently tuned to a fixed frequency, which is the difference between the incoming radio signal and a locally generated frequency. By using a fixed intermediate frequency, the tuned circuits of an i.f. amplifier may be adjusted for maximum amplification and selectivity. The i.f. transformers used in these amplifiers consist of pairs of coupled resonant circuits, as shown in Fig. 101. They are tuned either by varying the capacitance in the circuit, or by moving a powdered iron core in or out of the coils to change the inductance of the tuned circuit.

Fig. 101 illustrates the circuit of a single-stage i.f. amplifier with two i.f. transformers (T_1 and T_2). Transformer T_1, the input transformer, has its primary winding connected to the anode circuit of the previous stage and is tuned to the selected intermediate frequency (usually 455 kc/s). The secondary circuit, L_2–C_2, is inductively coupled to the primary and is in series with the grid of the amplifier pentode. Transformer T_2 in the output circuit is identical with transformer T_1. For use in wideband and high-fidelity receivers, the transformers are usually overcoupled to produce a resonance curve similar to curve 4 of Fig. 99. This provides a small band of (audio) frequencies in addition to the radio 'carrier' frequency.

SUMMARY

Electrical resonance occurs in a tank circuit at the frequency of natural oscillations. The circuit then draws just enough energy to overcome its internal resistance losses.

Parallel resonance in a tank circuit takes place at the frequency where the inductive and capacitive reactances of the two branches are equal and the line current is a minimum. The impedance is then a maximum. The resonant frequency

$$f_r = 1/2\pi\sqrt{LC}$$

The quality factor Q is the ratio of the reactance of the coil or the capacitor at the resonant frequency to the total resistance. Q is also a measure of the ratio of the reactive power stored in the tank circuit to the actual power dissipated in the resistance.

When a tuned resonant circuit is coupled to a load resistance, an apparent series resistance is reflected from the load (secondary) circuit into the tank (primary) circuit. This reflected load resistance lowers the Q and thus decreases the sharpness of resonance.

When two resonant tank circuits are coupled to each other, the resonant response depends on the degree of coupling between primary and secondary. As the coupling is increased, the secondary current peak becomes larger and the resonance curve becomes broader. Maximum response takes place at critical coupling ($k = 1/Q$). When the coupling is increased beyond the critical value, the secondary resonance curve shows two humps, which move apart with tighter coupling. This characteristic is used in wideband r.f. and i.f. amplifiers.

Radio-frequency amplifiers make use of resonant circuits to amplify a desired frequency and reject all others. Class A voltage amplifiers are used for receivers, while Class C power amplifiers are used in transmitters.

I.F. amplifiers with two coupled resonant circuits are used in superheterodyne receivers.

CHAPTER THIRTEEN

MODULATION AND DETECTION

Radio frequencies generated by oscillators and sent out as radio waves from transmitting aerials are by themselves mute messengers. You can neither see nor hear them, although you may detect their presence by various electrical devices. If a radio wave is to convey a message, some feature of the wave must be varied in accordance with the information to be transmitted. This superimposition of some sort of intelligent information on a radio wave is known as modulation. The reverse process, the extraction of this information from the radio wave at a receiver, is called demodulation or detection.

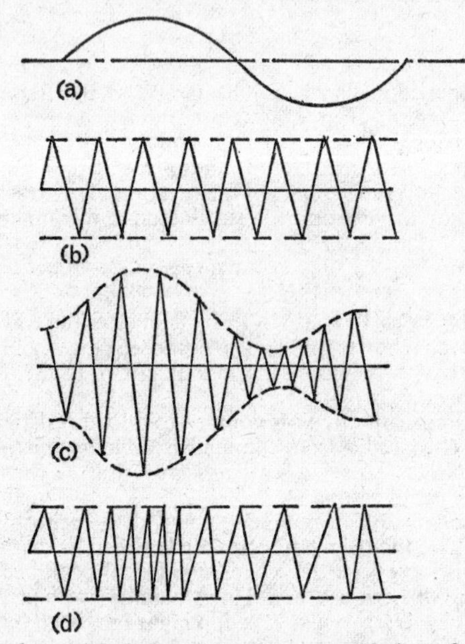

Fig. 102. Amplitude modulation and frequency modulation of a radio-frequency carrier by an audio-frequency wave

There are various ways for conveying information by means of radio waves. You can turn a radio transmitter on or off in accordance with some prearranged code, which may be the dots and dashes of the telegraph (Morse) code. This system of radio-telegraphy by means of continuous waves (c.w.) permits the transmission of written messages, but it can neither reveal the sound of a voice nor the appearance of a face. The latter, more complicated task can be performed in several ways. The amplitudes of the radio-frequency waves can be varied in accordance with the pressure changes of sound waves (after conversion to electrical audio frequencies) or they may be varied in accordance with the light intensity of a portion of a picture to be transmitted. This process is called amplitude modulation (a.m.) of a radio wave. It is used in radio-telephony for the transmission of sounds, in facsimile for the transmission of still pictures, and in television for the transmission of moving pictures or actual scenes.

A second way of modulating a radio wave is to vary its frequency (number of alternations per second) in accordance with the pressure of a sound wave, the light intensity of a picture, or some other form of intelligence. This is called frequency modulation (f.m.). Fig. 102 illustrates the difference between amplitude and frequency modulation—the two most important methods of modulation. The illustration shows the modulation of a radio-frequency 'carrier' wave (so called because it carries the modulation) by a single audio tone.

Fig. 102a shows the electrical equivalent of a single musical tone of a particular pitch. This is seen to be a simple sinewave of the corresponding frequency. A radio-frequency 'carrier' wave of constant frequency and amplitude is shown in (b). Since the carrier completes ten complete cycles during the time of one audio cycle, its frequency must be ten times that of the audio tone. In (c) of Fig. 102 the radio-frequency carrier is being amplitude-modulated by the audio-frequency wave. Note that the amplitudes of both the positive and negative half-cycles of the r.f. wave vary in accordance with the audio signal. The detector of a radio receiver simply eliminates the radio-frequency carrier, while retaining its audio-frequency amplitude variations. Finally, in (d) the frequency of the carrier is varied in accordance with the amplitude of the audio signal. The greater the positive amplitude of the audio wave, the higher is the frequency of the radio-frequency carrier.

There are other methods of modulating a radio wave in addition to amplitude and frequency modulation. For instance, the radio signals can be sent out as a series of sharp discontinuous pulses and the timing or spacing of these pulses may be varied in accordance with some information to be conveyed. This is known as pulse-time modulation and it has the advantage of permitting many different messages to be sent out over the same radio-frequency channel. Still other ways of conveying information via radio waves exist, but we shall be concerned primarily with the modulation and detection of radio waves, which have been modified either in their amplitude or frequency in accordance with the desired information.

Need for a radio-frequency carrier. You may well wonder why we cannot simply talk into a microphone and then transmit the electrical equivalents (audio frequencies) of the sound waves directly from a radio transmitter. As a matter of fact, this is not impossible, but quite impractical. As you will recall, the audio frequencies corresponding to sound waves range from under 20 c/s to over 15 kc/s. Since wavelength = velocity/frequency, and the velocity of radio waves is 186,284 miles per second, we see that the wavelength at 20 c/s is about 9,000 miles and at 15 kc/s is about 12½ miles. To transmit a radio wave, the aerial of the transmitter must be approximately of the same size as the waves to be radiated. It is obviously impossible to construct aerials thousands of miles long. The wavelength of a 1,000 kc/s radio wave, on the other hand, is only about 328 yards, and aerials of this order of size are easily constructed. The only practical solution, therefore, is to modulate a radio-frequency carrier with audio (sound) or video (picture) signals.

AMPLITUDE MODULATION

We have defined amplitude modulation as the process of modifying the amplitudes of a radio-frequency carrier wave in accordance with the strength of an audio (sound) or video (picture) signal. For a pure tone, or single audio-frequency, amplitude modulation looks as illustrated again in Fig. 103. Assume, for example, that (a) represents a 1,000 kc/s carrier and (b) a pure 1 kc/s note. The effect of impressing the carrier with the audio tone simultaneously across a resistor is shown in (c). Note that the amplitude of the carrier does not vary at all, but the instantaneous polarity of the radio-frequency

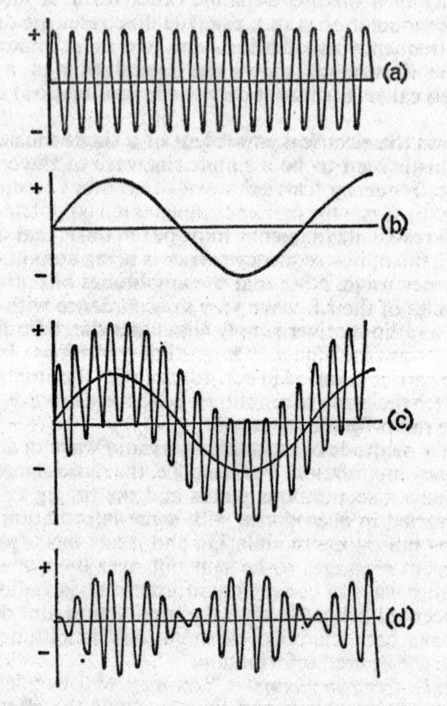

Fig. 103. Impressing an r.f. carrier (a) and audio tone (b) simultaneously across a resistor results in waveform (c), which is not amplitude-modulated; to obtain amplitude modulation (d) a non-linear device must be used

cycles varies continuously. This, obviously, is not amplitude modulation. A radio receiver has no means for examining the momentary polarity of this signal and, hence, for extracting the audio signal.

It has been found that the desired amplitude modulation, shown in (d), can be obtained only when the audio signal and carrier are impressed on a circuit where the current is not directly proportional to the applied voltage; that is, a circuit that does not obey Ohm's law. Because the graph of current *v.* voltage in such a circuit is a curve, it is called non-linear. The necessity for this can be proved mathematically, but you may simply accept it as a fact that an amplitude modulator must be a non-linear circuit.

It will be somewhat easier to appreciate the significance of this fact if you recall our discussion on amplitude distortion in the chapter on power amplifiers (Chapter 10). We demonstrated there that the non-linearity of valve characteristics results in amplitude distortion. In a sense, you may consider amplitude modulation as a form of intentional amplitude distortion. To produce this distortion, or modulation, a non-linear characteristic is required. Furthermore, valves make ideal modulators when they are operated in the non-linear portions of their characteristic.

Sidebands. We further pointed out in Chapter 11 that amplitude distortion of a waveform results in the production of new frequencies not present in the original waveform. The same thing happens with amplitude modulation. Although we intended to change only the amplitude of the waveform in Fig. 103, the process of amplitude modulation has also resulted in the production of some additional frequencies. These new frequencies are equal to the sum and the difference of the carrier frequency and the modulating frequency. In the example of Fig. 103 only two new frequencies result. One is 1,001 kc/s frequency equal to the sum of the 1,000 kc/s carrier wave and the 1 kc/s audio signal; this is called the upper side frequency. The other frequency is 999 kc/s or the difference between the carrier and audio tone; this is called the lower side frequency.

If it is desired—as in radio broadcasting—to modulate a carrier with the greater part of all audible frequencies, say up to 10 kc/s, each of these audio frequencies will produce upper and lower side frequencies during modulation, resulting in upper and lower sidebands. Thus, for the 1,000 kc/s carrier, the upper sideband will extend to 1,010 kc/s (for a 10 kc/s modulating frequency), while the lower sideband will reach to 990 kc/s. To broadcast audio frequencies up to 10 kc/s on a 1,000 kc/s carrier, therefore, a transmitting channel must be provided that has a bandwidth of 20 kc/s (from 990 to 1,010 kc/s), or twice the

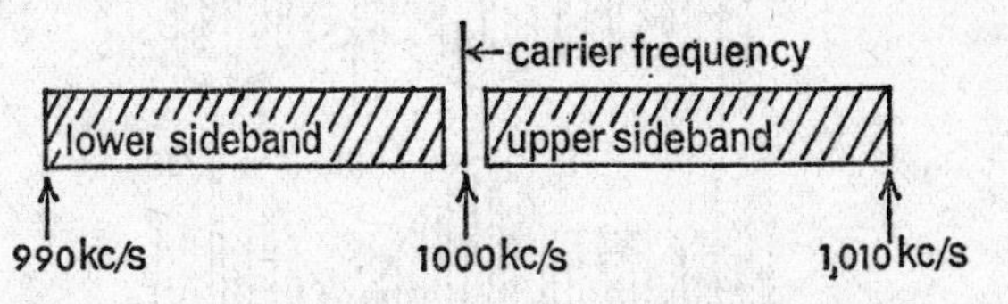

Fig. 104. Sidebands of 1,000 kc/s carrier produced by 10 kc/s amplitude modulation

highest audio modulating frequency (see Fig. 104). This is not only true for audio but also for video (picture) transmission, and in general the overall bandwidth of a channel must be twice the band of frequencies included in the information (sound or sight) to be transmitted. There is an exception to this rule, called single-sideband transmission, which utilizes only one of the two sidebands, but this system is chiefly used for trans-oceanic radio-telephony and amateur radio.

We now understand why the tuned amplifiers in transmitters and radio receivers must be able to pass a whole band of frequencies, rather than only the frequency of the radio carrier wave. In order to extract the information contained in the amplitude modulation of a carrier, it is clearly necessary that all tuned transmitter and receiver circuits pass the bandwidth of the entire channel with its upper and lower sidebands. While passing the desired channel, the tuned circuits should nevertheless be sufficiently selective to discriminate against unwanted adjacent channels on either side.

Depth of modulation. So far we have described the principle of modulating the amplitude of a carrier, but we have not said how much. The degree of modulation is a very important consideration, since it determines the strength and quality of the transmitted signal. Fig. 105 illustrates a radio-frequency carrier modulated to various degrees by an audio-frequency signal. The audio signal is shown in (a) of the figure and the unmodulated carrier in (b). The waveform of the modulated carrier for a low degree of depth of modulation is shown in (c). While the carrier varies faithfully in accordance with the audio signal (a), the amount of variation is rather small. As we shall explain later, the detector in a receiver responds only to the variations in the amplitude of the carrier and not to its absolute magnitude. When the carrier is modulated

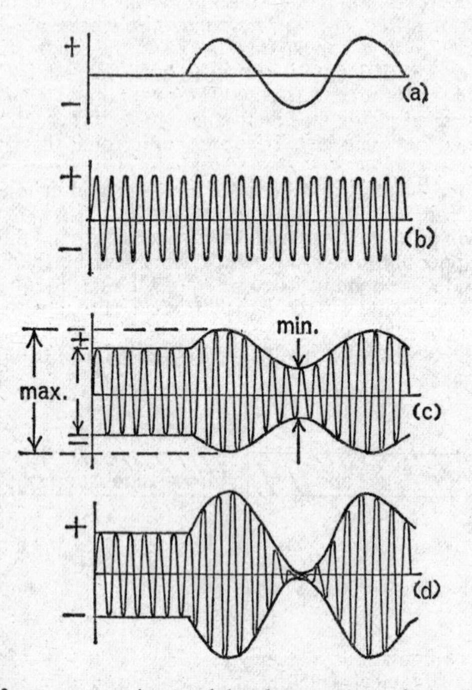

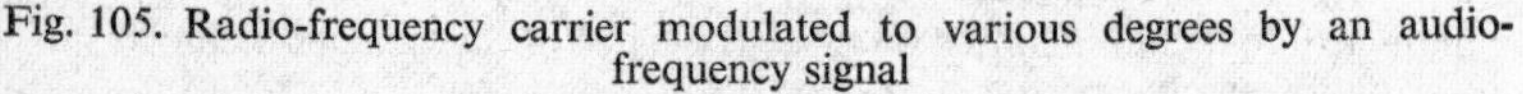
Fig. 105. Radio-frequency carrier modulated to various degrees by an audio-frequency signal

only to the small degree shown in (c), the audio signal will not be very strong and it may possibly be drowned out by extraneous noise. The greater the depth of the modulation, the stronger and clearer will be the audio signal.

In (d) of Fig. 105 the r.f. carrier has been modulated to the maximum possible extent, namely, to twice its normal amplitude during the positive peak of the modulating signal and to zero amplitude on the negative peak of the modulating signal. This is known as 100% modulation. Any further increase in the amplitude of the modulating signal will result in distortion during reception.

The depth of modulation of a carrier is conveniently expressed as a percentage of the normal (unmodulated) carrier amplitude. Knowing the relative amplitudes of the modulating signal and the carrier, as indicated in Fig. 105, the percentage of modulation is easily computed from the following relation: (See Fig. 105.)

$$\text{percentage modulation} = \frac{max. - Y}{Y} \times 100, \text{ or } \frac{Y - min.}{Y} \times 100$$

where Y represents twice the carrier amplitude.

If the modulating signal is a pure sinewave, as in Fig. 105a, both relations will result in the same percentage; if the modulating signal is distorted, however, this is not the case and the larger of the two values is generally used.

For the example of 100% modulation, shown in Fig. 105d, the *max.* value is twice the value of Y, or $2Y$, while the *min.* value is zero. Thus, substituting in the above relation, we verify:

$$\text{percentage of modulation} = \frac{2Y - Y}{Y} \times 100 = 100\%$$

or

$$\frac{Y - 0}{Y} \times 100 = 100\%$$

TYPES OF MODULATOR

To obtain amplitude modulation of a radio carrier, the modulating signal must be injected in some way into the radio-frequency power-amplifier stages of a transmitter. Depending upon where the audio signal is inserted, different

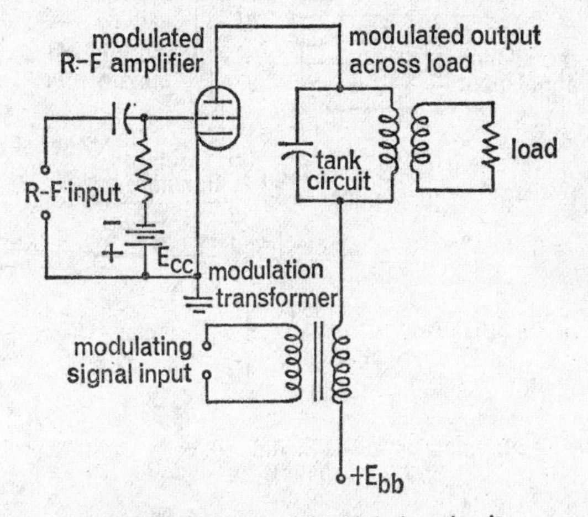

Fig. 106. Basic anode modulation circuit

types of modulating circuit result. If the modulating signal is inserted in series with the anode voltage supply of a transmitter, anode modulation results; if it is injected into the control grid of a transmitter stage, control-grid modulation occurs; if injected into the screen or suppressor grids, screen grid or suppressor grid modulation, respectively, result. Because of its efficiency and ease of adjustment, anode modulation is the most widely used modulating method.

Anode modulation. A basic form of anode modulation circuit is illustrated in Fig. 106. Here the r.f. carrier signal is applied to the grid of a tuned r.f. amplifier stage and the audio-frequency modulating signal from the output of an audio amplifier is inserted in series with the anode circuit.

In the absence of the modulating signal, the r.f. stage amplifies the carrier signal and the output voltage appears across the anode tank circuit, from which it is transformer-coupled either to the next stage or to the aerial. When the modulating signal is applied through the modulation transformer, the amplitude of the anode-circuit signal is made to vary in accordance with the audio signal, and the modulated carrier appears across the secondary of the r.f. output transformer. You can easily see how this happens. During one half-cycle of the audio signal, a positive voltage is induced in the secondary of the modulation transformer that adds to the anode supply voltage and, thus, causes an increase in the r.f. voltage across the tank. During the next, negative half-cycle of the audio signal, the voltage induced in the secondary of the modulation transformer subtracts from the anode supply voltage, thus causing a decrease in the r.f. voltage across the tank circuit. Since the valve and associated circuit contains non-linear elements, true amplitude modulation takes place.

To attain maximum (100%) modulation, the peak amplitude of the audio modulating signal must be made equal to the anode supply voltage. If it is smaller than this value, the percentage of modulation will be less than 100%;

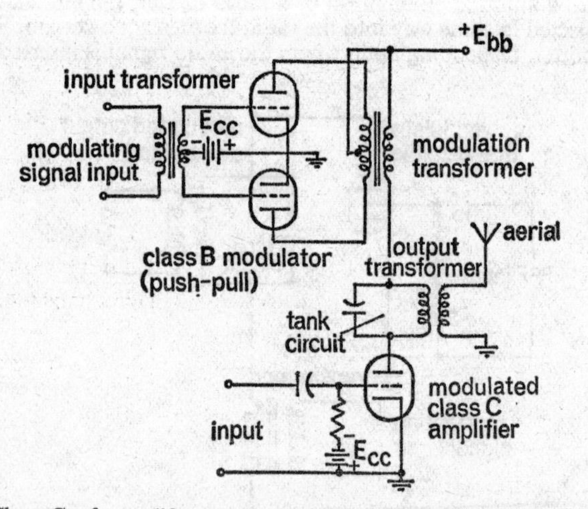

Fig. 107. Class C r.f. amplifier anode-modulated by a Class B push-pull audio amplifier

if it is larger, the anode current will be cut off during part of the audio cycle and the tank output will be zero during this portion of the cycle. This so-called overmodulation results in excessive distortion of the transmitted modulated carrier signal.

A very popular circuit for anode modulation is shown in Fig. 107. Here the final r.f. stage of a transmitter is anode-modulated by a Class B push-pull audio amplifier. This is known as high-level modulation. When the modulating signal is injected into one of the earlier stages of the transmitter, where the r.f. power is small, the system is called low-level modulation. In Fig. 107 the r.f. carrier signal is applied to the grid of a Class C amplifier and the output voltage is developed across the anode tank circuit. The modulating signal is obtained through a push-pull Class B power amplifier stage (the modulator). The output of the modulator is applied in series with the r.f. anode tank circuit through the secondary of the modulation transformer. The d.c. plate power for both modulator and r.f. amplifier is obtained, in this case, from the same anode supply ($+E_{bb}$), but this is not necessarily so.

You may have wondered how the power of a modulated wave compares with that of an unmodulated r.f. carrier. Well, with 100% modulation each sideband is one-half the amplitude of the carrier wave, as we have seen (Fig. 105). Since power is proportional to the square of the amplitude, the power associated with each of the two sidebands is one-quarter of the carrier power. The total audio-modulating power for two sidebands, therefore, is one-half

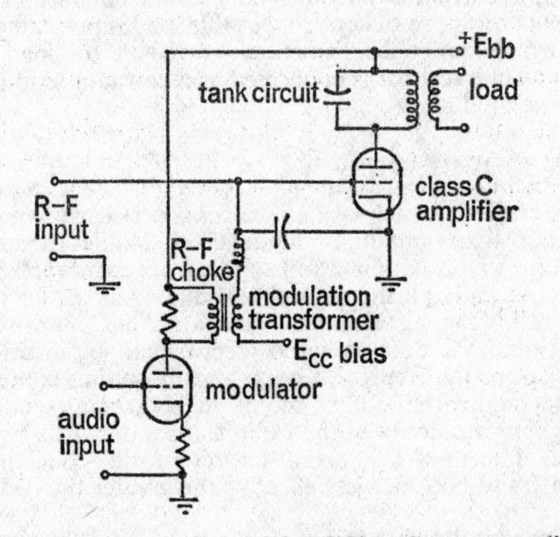

Fig. 108. Control-grid modulated Class C power amplifier

the carrier power and, hence, the average output of the transmitter increases by 50% with 100% modulation compared with the power of an unmodulated carrier. This additional power in the modulated wave must be supplied by the modulator in an anode-modulated system. The modulator must, therefore, be designed to furnish 50% of the desired transmitter output power without overloading and resulting distortion. Stated in another way, the maximum transmitter power that can be utilized with 100% modulation is twice the power available from the modulator. Since this places quite a burden of expense on the modulator, other methods of amplitude modulation are occasionally employed.

Control-grid modulation. If it is desired to eliminate the expense of a high-power modulator, the modulating signal can be injected into the grid of the Class C r.f. power output stage of a transmitter and control-grid modulation results. However, regardless of how the modulation is applied, the transmitter output power is still required to increase by 50% with 100% modulation, and this power must come from somewhere. In control-grid modulation, the modulator power requirements are very light and the required additional output power is generated within the r.f. amplifier stage itself. Grid modulation differs from anode modulation primarily in that the anode supply voltage remains constant, and the extra output power is obtained by causing the anode current and anode efficiency of the r.f. amplifier to vary in accordance with the modulating signal. The anode current is made to double at the peak of the modulation swing, which produces the required extra amount of output power.

Fig. 108 illustrates a typical control-grid modulated r.f. power amplifier circuit. The r.f. carrier signal is applied to the grid of the Class C amplifier, which is biased well beyond cutoff by a fixed direct-voltage source. The audio modulating signal is applied in series with the bias lead through the secondary of the modulation transformer, thus superimposing the a.f. signal on the bias voltage. Consequently, the anode current of the r.f. stage also varies in accordance with the audio modulating signal.

As you will remember, the grid of the r.f. stage is driven positive in Class C operation and grid current flows. Since the grid current varies with the amplitude of the input voltage to the grid, a variable load is presented to the modulator output, which can cause distortion. To reduce the load variation and resulting distortion, a resistor is connected as a constant load across the output circuit of the modulator.

FREQUENCY MODULATION

Historically, frequency modulation (f.m. for short) arose from the search for something better than amplitude modulation. While theoretically highly effective, amplitude modulation (a.m.) suffers from two practical defects. The first is noise. Practically all the natural and man-made radio noises, such as atmospheric static, razors, electrical machines, etc., consist of electrical amplitude disturbances. Since a radio receiver cannot distinguish between amplitude variations that represent noise and those that contain the desired sound (i.e. the modulation), a.m. reception is generally noisy anywhere, except within close proximity of the radio transmitter. Raising the power of the transmitter improves the 'signal-to-noise ratio', but this brute-force method is costly and becomes less effective the greater the distance from the transmitter.

The second practical defect of a.m. is the lack of fidelity or audio quality. To attain high-fidelity reception, all audio frequencies within the span of human hearing up to about 15,000 c/s must be reproduced. This necessitates a channel bandwidth of 30,000 c/s, or 30 kc/s, since both sidebands must be reproduced. But a.m. broadcasting stations are assigned channels only 20 kc/s wide and most of them use only 15 kc/s to avoid interference with adjacent channels. This means, in practice, that the highest audio modulating frequency can be only 7,500 c/s, which is hardly enough to reproduce music realistically. The reason for the narrow width of the a.m. channels is that at the time of construction of the early broadcast stations, which were exclusively a.m., only the medium frequencies from about 500 kc/s to 1,600 kc/s were suitable for the existing electronic equipment. To accommodate as many broadcast stations as possible within this narrow broadcast band of 1,100 kc/s, the width of each channel was deliberately restricted. Hi-fi was a remote consideration at the time these channel assignments were made. Outside of this arbitrary restriction, there is nothing in the nature of a.m. that would not permit full high-fidelity reproduction of speech and music.

Frequency modulation, which came into practical use just before the Second World War, dramatically removes these limitations of amplitude modulation. By impressing the modulation on the carrier through a variation in its frequency, while keeping the carrier amplitude constant, all the amplitude-sensitive noises are immediately eliminated, since variations in amplitude (i.e. noise) are simply not reproduced. This means that the signal-to-noise ratio for noise-free f.m. reception can be much lower than for a.m. and, hence, the f.m. transmitter power may be lower for the same quality of reception. Furthermore, f.m. is high fidelity, since the full audio band from 20 to 15,000 c/s is transmitted. As we shall see later, this requires a channel bandwidth even wider than that required for a.m. By the time f.m. was developed, it became feasible to transmit in the very wide v.h.f. frequency band from 30,000 kc/s to 300,000 kc/s (30 to 300 Mc/s). F.M. broadcasting was assigned a band from 88 Mc/s to 108 Mc/s (i.e. 20,000 kc/s wide) and if each station is allotted a channel width of 200 kc/s, this permits simultaneous operation of 100 stations in the same area.

BASIC PRINCIPLES

The appearance of a frequency-modulated carrier wave is illustrated in Fig. 109. A pure sinewave audio-modulating tone is shown in (a) and an un-modulated carrier wave in (b). When this tone is frequency-modulated on to the carrier wave, the result is as shown in (c) of the figure.

As you will note, the amplitude of the carrier in Fig. 109c remains constant, but its frequency is continuously varied in accordance with the instantaneous

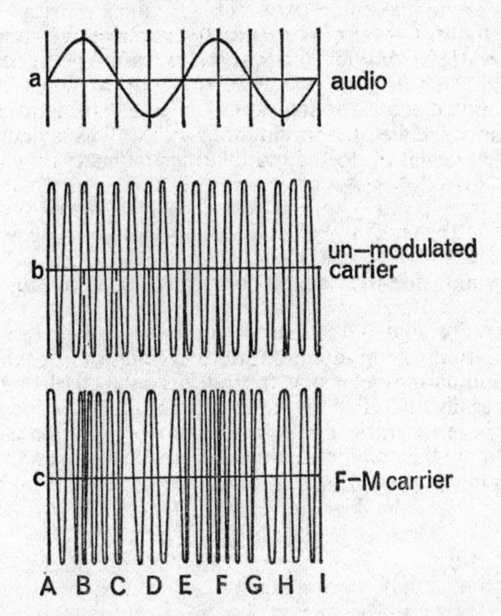

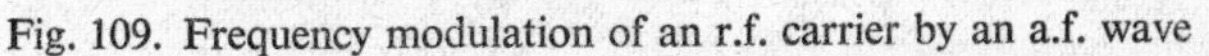

Fig. 109. Frequency modulation of an r.f. carrier by an a.f. wave

amplitude (strength) of the audio signal. During the times when the audio voltage is zero, as at *A*, *C*, *E*, *G*, and *I*, the carrier frequency is not modulated and, hence, remains the same as in (b). When the audio signal approaches its positive peaks, however, as in *B* and *F*, the frequency of the carrier is increased towards its maximum positive deviation, as shown by the closely spaced cycles in (c). When the audio signal approaches its negative peaks, on the other hand, as in D and H, the carrier frequency is deviated towards its minimum frequency, as shown by the wider than normal spacing of the cycles in (c).

Assume, for example, that the frequency of the carrier without modulation, called the centre frequency, is 100,000 kc/s (100 Mc/s) and the audio-modulating frequency is 10,000 c/s, or 10 kc/s. Say that the frequency deviation of the carrier is 50 kc/s during the peaks of the audio signal. The carrier is then deviated to 100·05 Mc/s during the positive peaks of the audio signal, and to 99·95 Mc/s during the negative peaks of the audio signal. The total range of the f.m. carrier frequency, called frequency swing, is then 100 kc/s (from 99·95 Mc/s to 100·05 Mc/s), or twice the deviation in either direction.

Modulation index. The depth of modulation in a.m. depends on the change in carrier amplitude caused by the modulating signal, 100% modulation taking place when the carrier amplitude goes to zero during the negative peak of the modulating signal. Since the amplitude of an f.m. carrier is constant, the depth of modulation depends on the amount of frequency deviation during the peaks of the audio signal. Theoretically, maximum or 100% modulation could be attained if the carrier frequency goes to zero during the negative peak of the audio-modulating signal. This is obviously not practical, since a zero-frequency signal could not be radiated and in any case would require a prohibitive channel bandwidth. Maximum or 100% modulation is, therefore, arbitrarily defined as the maximum permissible frequency swing. This is 150 kc/s ($\pm$75 kc/s deviation) for f.m. broadcast stations throughout Europe and America and varies from 15 to 25 kc/s for various f.m. communications services.

In practice, the degree of modulation of an f.m. wave is specified by a more significant quantity, called the modulation index. This is defined as the ratio of the frequency deviation to the modulating frequency:

$$\text{Mod. index } (m) = \frac{\text{carrier frequency deviation}}{\text{modulating frequency}}$$

The modulation index varies, of course, with the audio-modulating frequency. Thus for a frequency deviation of $\pm$75 kc/s and a modulating frequency of 1,000 c/s (1 kc/s), the modulation index is 75/1, or 75; for 5,000 c/s (5 kc/s) modulation, the modulation index is 75/5, or 15; while for a 10,000 c/s (10 kc/s) modulating frequency, the index is only 75/10, or 7·5. A particular f.m. system is easily identified by the limiting modulation index, called deviation ratio. This is the ratio of the maximum permissible carrier frequency deviation to the highest audio-modulating frequency. For f.m. broadcasting the maximum deviation is 75 kc/s and the highest audio frequency is 15,000 c/s, or 15 kc/s; hence the deviation ratio is 75/15 = 5.

F.M. sidebands. The significance of the modulation index arises out of the curious fact that it determines the f.m. bandwidth. You might think that the f.m. bandwidth is simply the frequency deviation on both sides, that is, the total frequency swing of the carrier. But this is not so. An inspection of Fig. 109c reveals that a frequency-modulated carrier consists of distorted sine-waves. Distortion, however, results in new upper and lower side frequencies, and frequency modulation by a single audio tone sets up a whole group of upper and lower side frequencies, rather than just one pair, as for a.m. These side frequencies are spaced at intervals equal to the modulating frequency, and their number and intensity depends directly on the modulation index. A useful rule of thumb states that the total bandwidth of an f.m. channel is approximately equal to the total frequency swing plus twice the highest audio modulating frequency.

REACTANCE-VALVE MODULATOR

To produce f.m., the frequency of an r.f. oscillator must be varied in accordance with the amplitude of an audio signal. The best way to do this is by means of a reactance-valve modulator. The basic circuit of such a modulator together with the input circuit of a conventional Hartley oscillator is shown in Fig. 110.

An alternating-voltage divider *R–C* is connected across the oscillator tank circuit between anode and cathode of the pentode. The tap of the voltage divider is connected to the control grid of the valve. The resistance *R* is made large compared to the reactance of *C* by choosing a large resistor and a large capacitor. (Capacitive reactance varies inversely with capacitance.) With the reactance almost negligible compared to the resistance *R*, the current through the voltage divider is practically in phase with the voltage across the tank

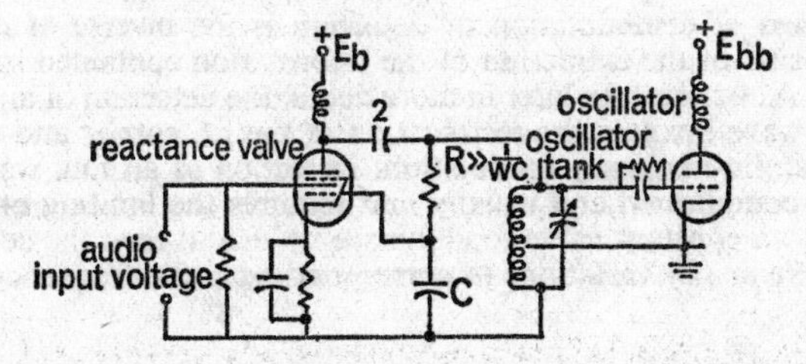

Fig. 110. Frequency modulation of oscillator by a reactance valve

circuit. However, the current through a capacitor always leads the voltage across it by 90°, or equivalently, the voltage lags the capacitive current by 90°. The voltage across *C*, which is the input voltage to the control grid of the valve, thus lags the current through *C* by 90°. Since the anode current is in phase with the control-grid voltage, the r.f. anode current also must lag behind the current through *C* by 90°, or equivalently, it lags the oscillator tank voltage by 90°. The effect is the same as if an additional inductance were connected across the tank. This shunt inductance decreases the total tank circuit inductance and, consequently, raises the oscillator frequency.

To accomplish frequency modulation, we must raise the tank circuit frequency in proportion to the amplitude of the modulating frequency. Now, the larger the lagging anode current of the reactance valve, the greater will be the apparent shunt inductance of the tank circuit and, hence, the higher will be the oscillator frequency. The amplitude of the reactance valve anode current is controlled by the audio input voltage on the screen grid. Evidently, then, the amplitude of the lagging anode current varies directly with the voltage on the screen grid and, hence, the tank circuit frequency is proportional to the audio modulating voltage applied to the screen grid. True frequency modulation is thus accomplished.

The frequency-modulated oscillator is usually operated at a relatively low frequency to attain stability of the carrier frequency. The oscillator frequency is then raised to the desired carrier frequency by a number of frequency multipliers. The output frequency of an r.f. amplifier may be doubled or tripled by tuning the anode tank circuit to twice or three times the input frequency, respectively. When the output frequency of a frequency-modulated oscillator is thus multiplied, the deviation increases by the same factor as the frequency is raised. For example, if a 12·5 Mc/s oscillator frequency is raised to a 100 Mc/s carrier frequency, an original frequency deviation of, say, 2 kc/s will be increased to 16 kc/s deviation at the output. This fact is sometimes taken advantage of by connecting a reactance-valve modulator to a relatively low-frequency crystal oscillator, thus obtaining exceptional frequency stability. The reactance valve cannot vary the crystal frequency by more than a small amount, which may be less than one cycle (360°). When the variation is less than a cycle, it is in effect only a phase shift, and the modulation is then called phase modulation, rather than frequency modulation. The two types of modulation are essentially the same. The phase modulation of a crystal oscillator can easily be converted to frequency modulation by multiplying the crystal frequency, and hence the frequency deviation, to the desired value through a series of frequency doublers or triplers.

DETECTION

The process of demodulation or detection is the inverse of modulation, since it consists of the extraction of the information contained in the modulated wave. As we shall see later in more detail, the detection of an amplitude-modulated wave involves the rectification of the r.f. carrier and the filtering out of the audio-frequency modulation. Detection of an f.m. wave is somewhat more complicated and usually first requires the limiting of the carrier amplitude to a constant value to eliminate noise and then the conversion of the carrier frequency variations to corresponding audio-frequency amplitude variations.

A.M. DETECTORS

The modulation of an a.m. wave consists of amplitude variations of the carrier by the audio signal. An a.m. detector, consequently, must extract these a.f. amplitude variations of the carrier, while eliminating the r.f. carrier itself. The first step in this process is rectification of the carrier wave to eliminate the

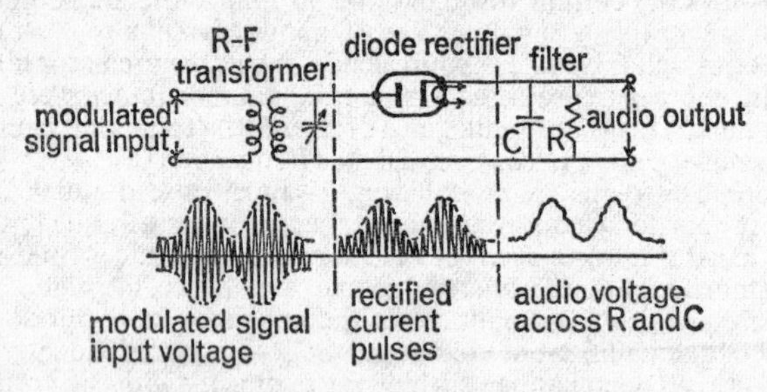

Fig. 111. Basic action of diode detector

negative half-cycles. This is necessary, since the negative and positive peaks of the carrier tend to cancel each other. The second step in the detection process is to remove the carrier-frequency variation so as to leave only the audio modulation. This is accomplished by a filter circuit. We shall consider three popular types of a.m. detector out of the multiplicity of existing circuits. These are the diode detector, the grid-leak detector, and the infinite-impedance detector.

Diode detector. The basic circuit and waveforms of the diode detector are illustrated in Fig. 111. Although a valve is shown in the figure, semiconductor diodes made of silicon or germanium may equally well be used.

The action of the diode detector is essentially that of a half-wave rectifier. (See Chapter 14.) The modulated r.f. signal is coupled to the anode of the diode through the tuned r.f. input transformer. As we know, the valve conducts only when its anode is positive in respect to the cathode. Consequently, a current pulse flows through the valve and the load resistance, R, whenever the signal voltage at the anode is positive. No current can flow during the negative half-cycles of the carrier signal. The output of the diode, thus, consists of a series of positive half-cycles of the carrier wave, as shown.

The rectified current pulses are still much too rapid to be made audible in a loudspeaker. The remaining r.f. carrier variations are, therefore, smoothed out by a filter, consisting of filter capacitor C and load resistor R. The value of the capacitor must be sufficiently large to present a low reactance (opposition) to the r.f. current pulses, while presenting a relatively high reactance to audio frequencies. Under these conditions, the remaining r.f. carrier frequencies are by-passed around R, while the audio-frequency currents (the modulation) develop a voltage across R.

Let us consider the filtering action in more detail. When the valve conducts during the positive half-cycles, the capacitor quickly charges up to the peak value of each voltage pulse. The values of the capacitor and resistor are chosen so that the capacitor is able to discharge only very slowly through the resistor. The capacitor thus tends to hold its charge between successive positive peaks, when the applied signal voltage drops to zero and cuts off the anode current. As a result the voltage across the load resistance R does not drop to zero between pulses, but decreases slowly as the charge on the capacitor leaks off. During each successive r.f. pulse the capacitor charges up again to the peak value of the new pulse and the voltage across the load rises to the new peak value. The upshot is that the load voltage is unable to follow the rapid r.f. carrier variations, but follows only the peak values of the applied r.f. signal voltage, as shown in Fig. 111. But the changes in the peak value of the r.f. carrier contain the original audio modulation, that is, the desired information. By following the peaks, the filter thus recovers the original audio-frequency signal.

Grid-leak detector. Although having excellent fidelity and little distortion, diode detectors are not very sensitive and need a fairly strong signal voltage for efficient operation. Their use is therefore chiefly confined to the highly sensitive superheterodyne receivers, where large amplification takes place before demodulation. When extreme sensitivity for weak-signal reception is required, the grid-leak detector is generally used. A typical triode grid-leak detector circuit with its associated waveforms is shown in Fig. 112. A pentode circuit could be used to obtain added amplification.

The circuit is highly sensitive because it provides amplification, as well as detection of the modulated wave. Detection of the modulated signal takes place in the control-grid-to-cathode portion of the valve, while the grid-to-anode portion amplifies the signal. The grid and cathode are operated like a diode rectifier. Resistor R_g (the grid leak) represents the load for the rectifier, while grid capacitor C_g acts as by-pass for radio frequencies. The modulated r.f. signal is coupled to the grid of the valve through transformer T_1.

The valve has initially zero grid bias. Whenever the signal voltage at the grid becomes positive, grid current flows through R_g and the secondary of T_1 back to the cathode. This grid current produces a negative voltage drop across R_g and C_g, which serves to bias the valve negatively. A large enough capacitor is chosen so that it holds its charge, and hence the bias, during the negative half-cycles of the r.f. input signal. Since the bias builds up during the positive half-cycles, the valve is eventually driven to cut-off, and anode current flows only during positive half-cycles. This results in rectified r.f. grid current pulses, as shown in Fig. 112b.

As in the diode detector, the rectified r.f. current pulses develop a load voltage across the combination of R_g and C_g that varies at the audio-modulation frequency rather than at the frequency of the r.f. carrier. Again capacitor C_g is chosen large enough so that its charge can leak off only slowly and, hence, the voltage across it can follow only the peaks of the r.f. carrier, just as in the diode detector. The peaks of the r.f. carrier contain the desired audio modulation, as shown in the bottom waveform in Fig. 112b. But in contrast to the diode detector, the r.f. component of the grid voltage also is amplified by the

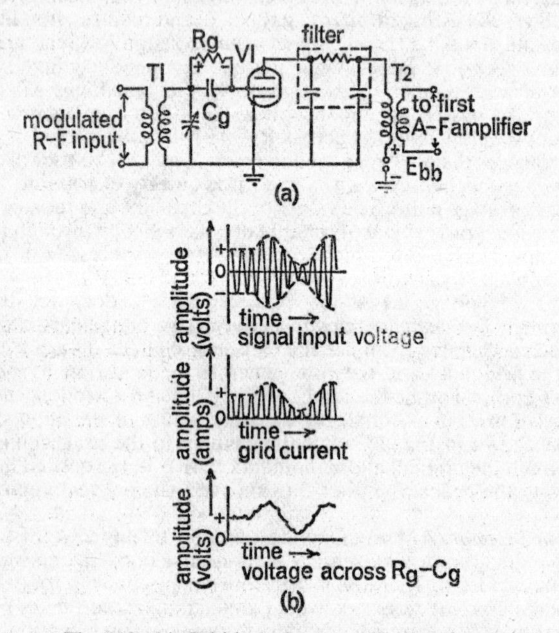

Fig. 112. (a) Grid-leak detector, (b) Waveforms

valve and appears at the anode. The audio-modulation component of the anode current, therefore, must be separated from this r.f. component by a suitable filter. The filter may consist of a resistance–capacitance combination, as shown, or an inductance–capacitance filter may be used with even greater effectiveness. In either case the r.f. component is by-passed to earth and the audio component of the anode current is coupled through transformer T_2 to the input of an audio-frequency amplifier.

For the detection of weak signals, the grid resistor R_g is usually between 1 and 5 megohms and C_g has a value between 100 and 300 micromicrofarads. Since a weak input signal operates the valve near the bottom, curved portion of the characteristic, the detection is non-linear and considerable distortion takes place. Furthermore, with a strong input signal the positive grid current overloads the valve and additional distortion results. Because the curved portion of the characteristic follows a square law (that is, the anode current increases as the square of the grid voltage), this type of detector is sometimes referred to as a square-law detector.

Infinite-impedance detector. The infinite-impedance detector shown in Fig. 113 is used frequently in high-fidelity a.m. receivers because of its ability to handle relatively large signal voltages at low distortion and with excellent fidelity and selectivity. No amplification takes place, however, since the output is taken from the cathode of the valve; the sensitivity of the circuit is therefore low. This is of no importance, however, in high-gain receivers with plenty of r.f. amplification.

The circuit of Fig. 113 derives its name from the fact that its grid input has theoretically infinite impedance. The grid cannot be driven positive by the input signal and, hence, no grid current can flow. The cathode-load resistor, R_1, serves several purposes. First, it provides automatic grid bias near the anode-current cut-off of the valve. Because of the cut-off bias, anode current

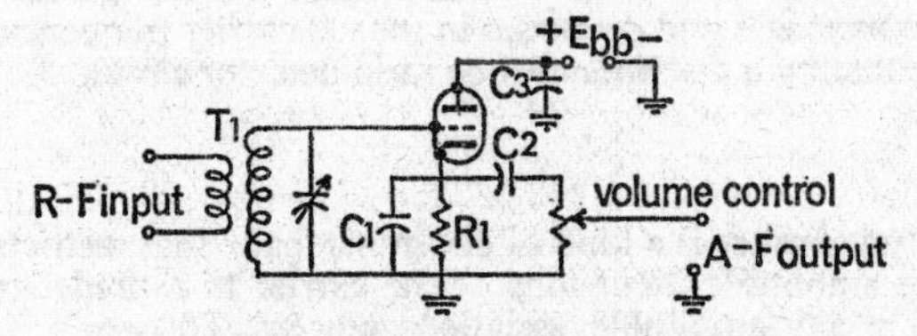

Fig. 113. Infinite-impedance detector

flows only during the positive half-cycles of the r.f. signal, the negative half-cycles being cut off. The rectified current pulses appear across R_1, which thus acts also as a load for the rectified audio signal. Capacitor C_1 filters out the remaining r.f. carrier component, just as in the diode detector. The value of the capacitor is chosen so that it by-passes the r.f. fluctuations to earth, but does not shunt out the audio-frequency signal appearing across load resistor R_1. This audio output voltage is then coupled through capacitor C_2 and a volume control (a potentiometer) to the grid of the first audio amplifier stage. Since the load resistor R_1 is common to both the anode and grid circuits of the valve, it has still another effect. It feeds a portion of the audio output signal back to the grid input circuit. As you can easily verify, this feedback is negative and, hence, results in further reduction of distortion and improvement of linearity. It also prevents any gain from being realized.

This detector operates somewhat differently from those discussed before, since it does not respond to the peak values of the r.f. carrier wave. What happens is this: the series of rectified anode-current pulses result in a certain average value of the anode current, which would be the current measured by a d.c. ammeter. When the amplitude of the signal at the grid increases, the rectified current pulses become larger and the average value of these pulses also increases. The average value of the anode current, rather than the peak value, will thus faithfully follow changes in the carrier signal amplitude and, in this way, reproduce the original modulation. Furthermore, when the average anode current increases for a large input signal, the bias developed across R_1 also increases, and prevents the grid from being driven positive and drawing current.

F.M. DETECTORS

The function of an f.m. detector is to recover the audio modulation of the r.f. carrier wave, just like an a.m. detector. But because of the nature of the f.m. signal, the process of f.m. detection differs quite substantially from a.m. detection. We can divide f.m. detection conveniently into two main tasks. The first is to smooth out or limit any variations in the amplitude of the carrier wave, to obtain a signal that varies in frequency only, as shown in Fig. 109c. This is necessary because the actual f.m. signal arriving at the aerial is likely to vary quite a bit in amplitude owing to noise voltages superimposed on it and the effects of signal fading, reflections, and absorption by various objects. If these extraneous amplitude variations were allowed to pass through the detector, noisy and distorted reception would result. The task of limiting the f.m. signal to constant amplitude is carried out by an f.m. limiter circuit. As we shall see, however, some f.m. detectors do not require a limiter.

The second task to be performed in f.m. detection is to obtain a detector output voltage that varies instantaneously as the frequency of the carrier wave. In other words, the detector output voltage should be zero or a fixed steady value for the unmodulated carrier, and it should rise and fall proportionally with increases and decreases in the r.f. carrier frequency. This task is performed either by a discriminator or ratio detector circuit.

LIMITERS

An amplitude limiter is a kind of electronic 'gate' that restricts the positive and negative amplitude excursions of the carrier to a predetermined value, thus removing any amplitude variations present. To obtain limiting in an ordinary triode or pentode, advantage is taken of the fact that the characteristic

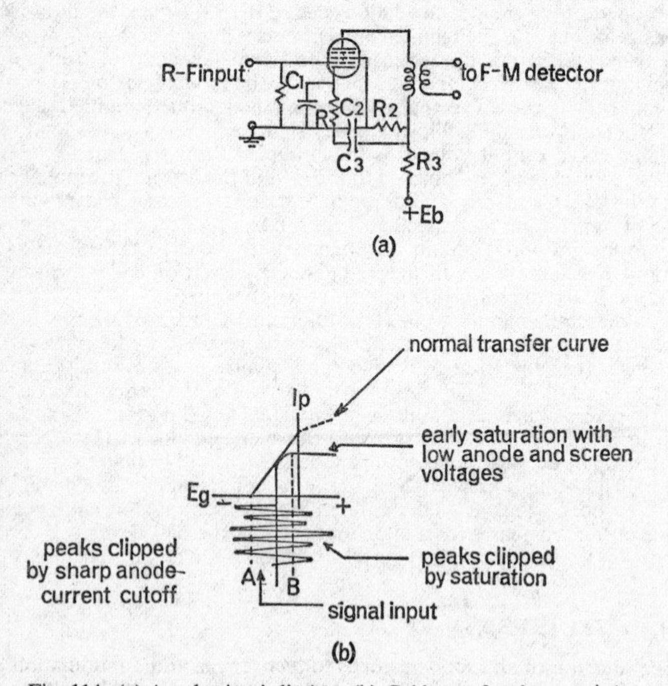

Fig. 114. (a) Anode circuit limiter, (b) Grid-transfer characteristic

curves of valves show a levelling off or saturation when the input voltage at the grid exceeds a certain value. You may further remember that anode-current saturation sets in for relatively small values of the grid input voltage if the anode and screen voltages are kept low. Such early saturation effectively limits the positive swings of an r.f. input signal. To limit the negative signal swings, the valve is simply biased in the lower portion of the characteristic, so that negative amplitudes beyond a set level will produce anode-current cut-off.

Anode-circuit limiter. A typical limiter circuit using a sharp-cut-off pentode is illustrated in Fig. 114. Resistors R_2 and R_3 are chosen large enough to drop the anode and screen voltages to a low value, of the order of 30 V. The grid-transfer characteristic is thereby flattened out in the manner shown in (b) of Fig. 114. Because of the early saturation, the amplitudes of relatively small input voltages are cut off above a certain positive value. Thus, all positive amplitude peaks exceeding line B in the figure are clipped off. At the same time the sharp anode-current cut-off of the pentode eliminates all negative signal peaks that exceed line A. The anode-circuit limiter achieves effective limiting action, but very little amplification is obtained because of the low anode and screen grid voltages.

Combined anode- and grid-circuit limiter. Fig. 115 shows a circuit that combines both anode and grid circuit limiting. By adding the grid-limiting action,

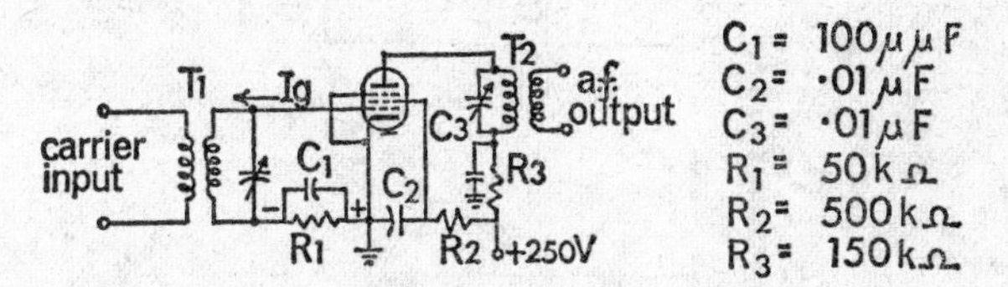

Fig. 115. Combined anode- and grid-circuit limiter

the anode and screen voltages may be raised somewhat, thus producing higher gain without a loss in limiting action. The values are typical of a practical receiver circuit.

The input transformer T_1 couples the f.m. carrier wave to the grid of the pentode. Anode-circuit limiting is achieved in the same way as in the circuit of Fig. 114, except that the anode and screen voltages need not be quite as low. Resistors R_2 and R_3 drop the screen and anode voltages to the desired value and together with capacitors C_2 and C_3 provide by-passing action for the radio frequencies.

Grid-circuit limiting takes place through the combined action of R_1 and C_1, which act essentially as a grid-leak biasing circuit. The input signal must have sufficient amplitude to drive the grid of the valve positive and, hence, draw grid current. The grid current charges capacitor C_1 to nearly the peak voltage of the positive signal half-cycle. When the signal falls away from its peak value, the capacitor discharges through resistor R_1 and produces a negative bias. Since most of the input voltage appears across R_1 rather than at the grid, the flow of grid current tends to flatten the positive peaks of the signal, i.e. limiting results.

PHASE-SHIFT DISCRIMINATOR

One of the most popular and effective f.m. detectors is the phase-shift discriminator sometimes referred to as Foster–Seeley discriminator. (Fig. 116.)

The primary and secondary tank circuits are tuned to the frequency of the incoming carrier signal and are inductively as well as capacitively coupled. Capacitor C_1 connects the frequency-modulated voltage from the primary coil to the centre of the secondary and also blocks the steady anode voltage from the diodes V1 and V2. A choke coil L, connected from the centre of the secondary to the junction of diode load resistors R_1 and R_2, serves as a return path for the rectified current of the two diodes.

The two equal load resistors R_1 and R_2 are by-passed for the r.f. carrier frequency through capacitors C_3 and C_2, respectively. When the carrier is at the centre frequency in the absence of modulation, the rectified voltages across R_1 and R_2 are equal and opposite in polarity; hence, no audio output voltage appears between point 'X' and earth. As the carrier frequency increases or decreases with modulation, however, the voltage across one of the

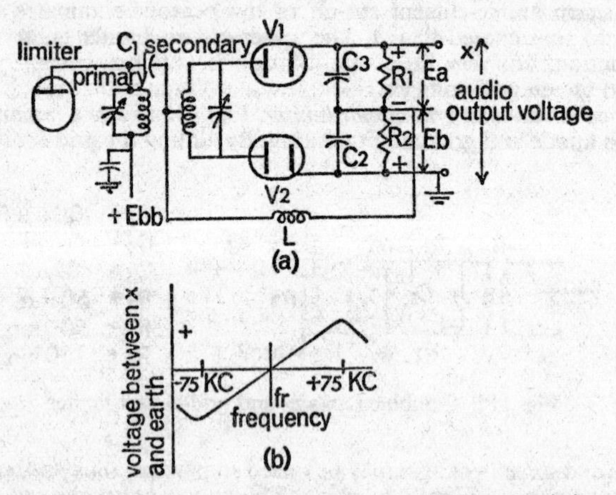

Fig. 116. (a) Phase-shift discriminator, (b) Characteristic

load resistors increases or decreases with respect to the other and a net audio output voltage appears at 'X'. The variation in carrier frequency is translated into an equivalent voltage change by means of the changing phase relationship between the voltages across the primary and secondary tuned circuits. The process is rather complicated and is explained in detail in more advanced texts. Essentially, the voltages across the top and bottom half of the secondary coil add vectorially to the primary voltage. When the secondary tank circuit is resonant to the carrier (centre) frequency, the top and bottom secondary coil voltages are exactly 90° out of phase with the primary voltage. The vector addition thus produces equal and opposite voltages across the diodes, and the rectified diode output voltages, E_a and E_b, cancel each other out. When the applied frequency of the input wave is either higher or lower than the resonant frequency of the secondary tank, the secondary voltages are shifted in phase with respect to the voltage across the primary coil. Vector addition will then no longer result in equal and opposite voltages across the top and bottom diodes, but one will be larger and the other smaller, depending on the direction of the frequency deviation. The difference between the unequal, rectified voltages developed across load resistors R_1 and R_2, appears as audio output voltage between point 'X' and earth (the chassis). The output voltage thus varies in accordance with the instantaneous frequency of the input signal.

A typical discriminator characteristic is shown in Fig. 116b. The characteristic is a plot of the discriminator output voltage versus frequency. At the resonant frequency (f_r) the output voltage is zero, as explained. For frequencies above resonance, the output voltage increases positively and becomes maximum for a frequency deviation of 75 kc/s. Below resonance the output voltage becomes negative and reaches a minimum at the maximum negative deviation of −75 kc/s. The voltage/frequency characteristic is seen to be linear between the two resonant peaks at the extremes of the frequency swing.

RATIO DETECTOR

A variation of the phase-shift discriminator, called the ratio detector, is insensitive to carrier amplitude variations and, hence, permits eliminating the limiter circuit. The basic circuit, shown in Fig. 117, is seen to be very similar to the discriminator. As a matter of fact, the circuit behaves in roughly the same way as the discriminator as far as the detection of frequency-modulated

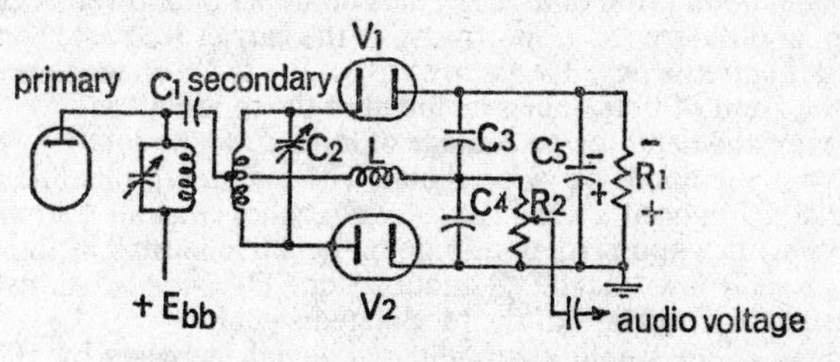

Fig. 117. Basic ratio-detector circuit

waves is concerned. The only physical difference is that one of the diodes, V1, has been reversed in polarity and that the audio output voltage is taken off across one of the capacitors, rather than across both of them. Slight though these changes may be, they add up to a significant difference, as far as possible amplitude variations in the carrier wave are concerned.

Because of the inversion of V1, the two diodes conduct on the same r.f. half-cycle, whenever the bottom half of the transformer secondary is positive. A current then flows in series through the two diodes, the secondary coil and resistor R_1, and produces a voltage across R_1 with the polarity shown. This voltage charges capacitor C_5 to the peak value. The capacitor is chosen large enough to hold the voltage across R_1 constant during the non-conducting half-cycles. When frequency modulation is present, unequal rectified diode output voltages are developed across capacitors C_3 and C_4. However, since C_3 and C_4 are in series across C_5, the sum of their voltages must remain the same as the voltage across C_5 regardless of the ratio between them. Since the sum of the two output voltages across R_1 remains a fixed value, amplitude variations cannot occur. The ratio between the individual diode output voltages, however, changes continuously in accordance with the instantaneous frequency of the carrier wave. The audio output voltage cannot be taken off across the output resistor (R_1), since that voltage remains constant, but must be picked off from one of the capacitors, C_3 or C_4, as shown. The output voltage-frequency characteristic is practically the same as that shown in Fig. 116b for the discriminator.

SUMMARY

Modulation is the process of superimposing information on a radio carrier wave. Detection or demodulation is the extraction of the original information from the carrier wave at the receiver.

Amplitude modulation modifies the amplitudes of a radio-frequency carrier wave in accordance with the strength of an audio (sound) or video (picture) signal.

In frequency modulation, the amplitude of the carrier remains constant, but its frequency is continuously varied in accordance with the instantaneous amplitude of the audio or video signal. The number of periodic frequency changes of the carrier (per second) equals the modulating frequency. By keeping the amplitude constant, frequency modulation eliminates static and electrical interference.

In pulse-time modulation the radio signal is transmitted in the form of sharp, discontinuous pulses and the timing or spacing of these pulses is varied in accordance with the information to be conveyed.

To produce amplitude modulation a non-linear modulator circuit must be used. Valves have the required non-linear characteristic.

Amplitude modulation of a carrier sets up upper and lower sidebands equal to the sum and difference, respectively, of the carrier frequency and the highest modulating frequency. The bandwidth of the radio channel must therefore be twice the band of frequencies included in the modulation.

The greater the depth or percentage of amplitude modulation, the stronger and clearer is the audio or video signal. The maximum possible modulation without distortion occurs when the carrier reaches twice its normal amplitude on positive swings and zero amplitude on negative swings of the modulating signal. This is known as 100% modulation. Over-modulation (more than 100%) results in distortion of the modulated signal.

The power of an amplitude-modulated signal increases by 50% with full (100%) modulation, compared to the unmodulated wave.

In anode modulation the modulating signal is injected in series with the anode voltage supply of the transmitter; in control-grid modulation the modulating signal is injected into the control grid, and in screen grid or suppressor grid modulation the signal is injected into either the screen or suppressor grid.

The maximum frequency change of a frequency-modulated carrier from its centre frequency in either direction is its frequency deviation; it occurs during the positive and negative peaks of the modulating signal.

The frequency swing is the total frequency range of the f.m. signal and it is equal to twice the frequency deviation.

The modulation index determines the degree of modulation of an f.m. wave; it is defined as the ratio of the frequency deviation to the modulating frequency. The limiting modulation index, called the deviation ratio, is the ratio of the maximum deviation to the highest audio modulating frequency.

Frequency modulation sets up groups of upper and lower sidebands, spaced at intervals equal to the modulating frequency; their number and intensity depends on the modulation index.

The bandwidth of an f.m. channel is approximately equal to the total frequency swing plus twice the highest audio modulating frequency.

A reactance-valve modulator produces f.m. by varying the frequency of an oscillator in accordance with the amplitude of an audio signal.

Detection of an amplitude-modulated wave requires the rectification of the r.f. carrier and the separation (filtering) of the modulation.

In a diode detector the r.f. carrier is rectified and an *R–C* filter develops a load voltage that follows the peak values of the rectified pulses; these peak values contain the audio modulation. The detector has good fidelity and low distortion, but is not sensitive.

A grid-leak detector is highly sensitive, but tends to overload for large signals; it also follows the peak values of the r.f. signal.

An infinite-impedance detector handles large input signals with low distortion; it follows the average values of the rectified anode current.

A limiter confines the amplitudes of an f.m. carrier to a predetermined, constant value. Limiting is obtained by anode-current saturation and cut-off, by grid-current flow, or by a combination of both.

In a phase-shift discriminator the variations in carrier frequency are translated into equivalent voltage changes by means of the changing phase relationships between the primary and secondary tank circuit voltages.

A ratio detector eliminates the carrier amplitude variations, while demodulating the f.m. carrier in nearly the same way as a discriminator circuit.

CHAPTER FOURTEEN

POWER SUPPLIES

Most electricity supplies provide alternating current rather than direct current; a.c. is more easily generated and transmitted over long distances. Valves, on the other hand, require direct current for all their electrodes, except the filaments, which may be heated either by a.c. or d.c. The most convenient way to change alternating to direct current is by means of a rectifier. A rectifier is capable of changing alternating current into a pulsating form of direct current; to obtain smooth d.c. power, additional filter circuits are required. A complete power supply also contains a voltage divider for providing d.c. at various desired voltages, and sometimes a voltage regulator to keep the output voltage at a relatively constant value. We shall be primarily concerned with rectifiers and filters, which are the main elements in all power supplies.

RECTIFIERS

Rectifiers change a.c. into pulsating d.c. by eliminating the negative half-cycles of the alternating voltage. Thus, only a series of sinewave pulsations of positive polarity remain. An ideal rectifier may be thought of as a switch that closes a load circuit whenever the alternating current is positive, and opens the circuit whenever the alternating current is of negative polarity. Such a switch would have in effect zero resistance when the circuit is closed during positive a.c. half-cycles, and infinite resistance for the time when the circuit is open during negative half-cycles. Practical rectifiers do not attain this goal, but come close to it. The resistance during the non-conducting interval (called back resistance) of valve rectifiers is extremely high—for practical purposes infinite—but the resistance during the conducting interval (called forward resistance) is never zero or even constant. In any case, all rectifiers must provide a substantially one-way path for electric current; that is, conduction must take place primarily in one direction only. This is called unilateral conduction, or a unidirectional characteristic. You will remember that diodes have such a unidirectional characteristic.

Diode as rectifier. A diode is any electronic device consisting of two elements, one being an electron emitter or cathode, the other an electron collector or anode. Since electrons in a diode can flow in one direction only, from emitter to collector (or cathode to anode), the diode provides the unilateral conduction necessary for rectification. This is true for all diodes regardless of type. As we have already seen, diodes come in many forms. They may be thermionic valves of the vacuum or gas-filled type; crystals or semiconductors made of germanium or silicon; or metallic types, such as copper-oxide and selenium rectifiers. Although the rectifier circuits that follow will illustrate primarily diode valves, any of these other types may equally well be substituted, with the additional advantage that no filament supply is required, since no filaments are present. The latter is one of the main reasons that selenium rectifiers and crystal diodes have become increasingly popular in television and radio receivers.

HALF-WAVE RECTIFIER

Since a diode will permit current to flow only during the positive half-cycles of the applied alternating voltage, a single diode is known as a half-wave rectifier. A half-wave rectifier circuit with its input and output waveforms is shown in Fig. 118. As you can see this is a very simple circuit. The a.c. supply voltage is applied through a power transformer in series with the diode valve and load resistance, R_L. Anode current I_b flows through the diode every other half-cycle, during positive alternations of the input voltage; it is

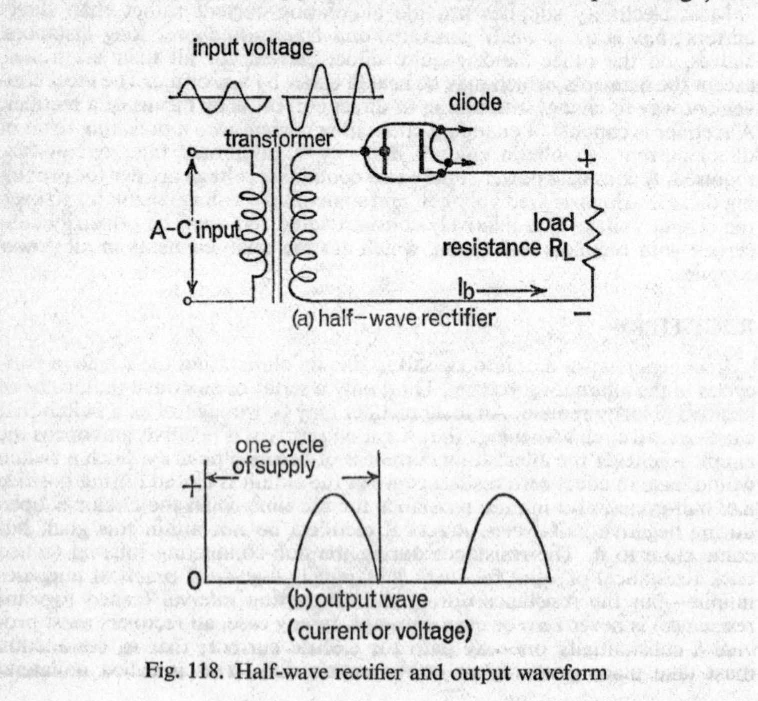

Fig. 118. Half-wave rectifier and output waveform

blocked during the negative half-cycles. The current consequently flows through the load always in the same direction so as to make the cathode-connected end of R_L positive. Although unidirectional, the current is not direct (d.c.), because of its continuous changes in amplitude, or pulsations. It can be shown mathematically that these pulsations contain both a d.c. component and an alternating (a.c.) component, known as ripple. The current can be converted into a steady d.c. by filtering out the a.c. ripple with a suitable smoothing filter, as we shall see later on.

It is evident from Fig. 118 that during the time the anode current flows, its instantaneous amplitude follows exactly the changes in the applied voltage. The shape of the anode-current waveform (Fig. 118b), therefore, is an exact

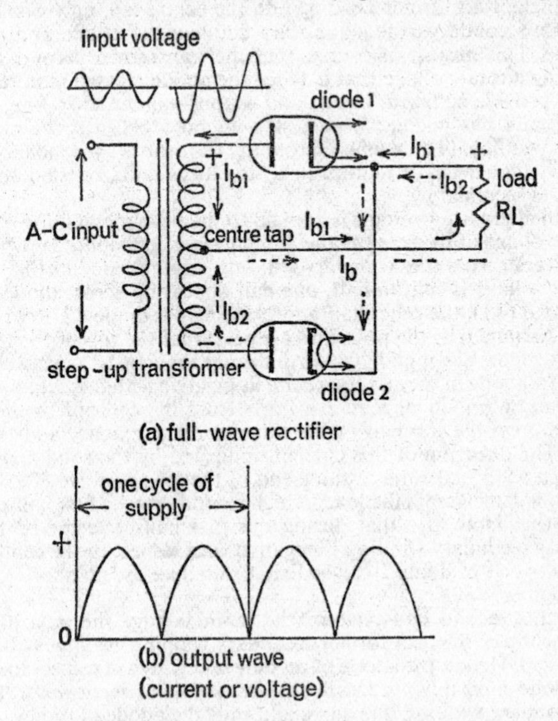

Fig. 119. Full-wave rectifier and output waveform

replica of the input voltage waveform during positive half-cycles. The anode current flowing through the load resistance develops a d.c. output voltage, whose waveform is exactly the same as that of the anode current and is also represented by Fig. 118b. But since only the positive half-cycles of the input voltage are reproduced, one half of the input voltage is in effect lost. The efficiency of the half-wave rectifier is therefore low and is used only for applications requiring a small current drain. Another disadvantage of the half-wave rectifier is that the pulsations of the output current and voltage are at the same frequency as the a.c. power supply. Elaborate filter circuits are required to eliminate this low-frequency ripple (usually 50 c/s) and produce smooth direct current.

FULL-WAVE RECTIFIER

By employing two diode half-wave rectifiers in a so-called full-wave rectifier circuit, anode current can be made to flow during the full cycle of the alternating supply voltage. As shown in Fig. 119, the two diodes alternately supply rectified current to the load during both halves of the input voltage cycle, and always in the same direction.

Note in Fig. 119a that the cathodes of the two rectifier valves are 'tied' together and the common junction is connected to one side of the load resistor, R_L. The other end of R_L is connected to the centre tap of the secondary winding of the power transformer. Since each diode is connected between one end of the transformer winding and the centre tap, only one half of the transformer secondary voltage appears between the anode and cathode of each diode. This means, of course, that the transformer secondary winding must supply a total voltage that is twice the anode voltage required for each valve. To provide sufficient anode and output voltage, therefore, the transformer usually has a considerable step-up ratio between the primary and secondary winding. (The voltage across the transformer secondary winding is the primary input voltage multiplied by the turns ratio, or 'step up', between primary and secondary.)

When an alternating voltage is applied to the primary winding of the transformer, a voltage of the same shape, but enlarged in amplitude by the step-up ratio, appears across the secondary winding, as illustrated in Fig. 119a. This secondary voltage is split in half, one half appearing across diode 1 in series with the load (R_L), the other half appearing across diode 2 also in series with the load. Assume that the polarities are such that the top of the transformer secondary winding is initially positive during the first half-cycle of the input voltage. The anode of diode 1, therefore, is positive with respect to the cathode junction, and a anode current, I_{b1}, flows from the cathode to the anode of diode 1 through the top half of the transformer secondary and through the load R_L. The direction of this current, indicated by the solid arrows in Fig. 119, is such as to make the cathode-end of the load positive. The current I_{b1} develops a voltage across the load, which reproduces the first half-cycle of the input voltage. Note also that during this first half-cycle the bottom of the transformer secondary winding is negative with respect to the centre tap and, hence, the anode of diode 2 is negative. Consequently, no anode current can flow through diode 2.

During the second half-cycle of the input voltage the polarities reverse, making the top of the transformer secondary winding negative with respect to the centre tap. Hence, the anode of diode 1 is negative in respect to its cathode and no anode current flows. During this same half-cycle, however, the bottom of the secondary winding is positive and thus the anode of diode 2 is positive with respect to its cathode. Consequently, an anode current, I_{b2}, flows from the cathode to the anode of diode 2, through the bottom half of the transformer secondary and through the load, R_L. As indicated by the dotted arrows, the current flows through the load in the same direction as the previous half-cycle so that rectification is obtained. The current I_{b2} flowing across the load develops an output voltage, which reproduces the second half-cycle of the input voltage. Thus, two positive half-cycles appear across the load during one complete cycle of the supply voltage, as is illustrated by the output waveform in Fig. 119b.

During successive half-cycles of the input voltage, diodes 1 and 2 will continue to conduct alternately, each permitting current to flow during one half-cycle whenever its anode is positive with respect to the cathode junction. The resulting output current is a series of unidirectional pulses, as shown in Fig.

119b. Since there are two output pulses for each complete cycle of the input voltage, the output frequency is twice that of the a.c. supply frequency; that is, for a 50 c/s supply, the a.c. ripple in the output is 100 cycles per second. With the ripple frequency being twice the a.c. supply frequency and the current much less discontinuous than that of the half-wave rectifier, the pulsations are easily smoothed out by a suitable filter circuit. Furthermore, since both halves of the a.c. input cycle are rectified, the efficiency of the full-wave rectifier is far better than that of the half-wave type.

The full-wave rectifier has another important advantage. You may have noted in Fig. 119 that the individual valve currents flow in opposite directions through the secondary winding of the power transformer. The rectified current pulses, therefore, cancel each other out and no direct current flows through the secondary winding. This avoids d.c. magnetization of the transformer core and the resultant saturation, which is one of the serious defects of the half-wave rectifier. Since saturation of the core reduces the inductance of the transformer winding, a transformer that is subject to d.c. saturation must be much larger for the same power rating than a unit that has no d.c. flowing through its winding. This advantage—combined with the high efficiency, high permissible current drain, low ripple at twice the a.c. frequency, and relatively high output voltage—makes the full-wave rectifier suitable for a wide variety of applications in electronics. It is the standard circuit for low power applications.

FULL-WAVE BRIDGE RECTIFIER

The need for a centre-tapped power transformer is eliminated by the bridge rectifier circuit, in which four diodes are used. (See Fig. 120.) The a.c. input to the bridge circuit is applied to diagonally opposite corners of the network, while the output to the load is taken from the remaining two corners. The circuit is a full-wave type because both halves of the a.c. input cycle are utilized. As we shall see, two valves in series carry the load current on alternate a.c. half-cycles.

Assume as before that the top of the transformer secondary winding is initially positive during the first half-cycle of the a.c. supply. We can consider the transformer secondary voltage during this half-cycle to be applied across a series circuit, consisting of diode 1, the load resistor, and diode 2. Since the cathode of diode 1 is at maximum negative potential and the anode of diode 2 at maximum positive potential, an electron current flows through diode 1, the load resistor, diode 2, and the transformer secondary, in the direction indicated by the solid arrows. During this first half-cycle the anodes of diodes 3 and 4 are more negative than their cathodes and, hence, these valves do not conduct. This is indicated by the dotted positive half-cycles for diodes 3 and 4, in Fig. 120. Thus, the first half-cycle is reproduced by the conducting valves (diodes 1 and 2) and across the load resistor, as illustrated by the solid waveforms in Fig. 120.

One half-cycle later the top of the transformer secondary is negative and the bottom is positive, so that diodes 1 and 2 cannot conduct. This is indicated by the dotted negative half-cycles for diodes 1 and 2. The anodes of diodes 3 and 4 are now positive, however, with respect to their cathodes and an electron current flows through diode 3, the load resistor, diode 4, and the transformer secondary, in the direction indicated by the dotted arrows. Thus, the second half-cycle of the a.c. supply voltage is reproduced by conducting diodes 3 and 4 and also across the load resistor, as shown by the solid waveforms in Fig. 120. Note that the current pulses flow through the load resistor in the same

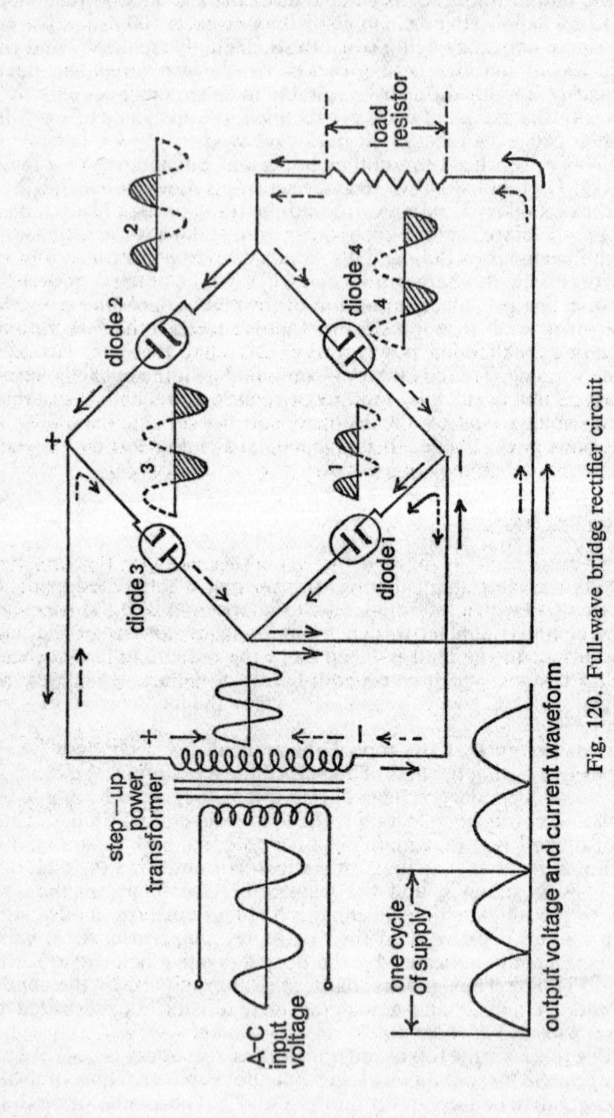

Fig. 120. Full-wave bridge rectifier circuit

direction during both a.c. half-cycles, each pulse flowing from the anode junction of diodes 1 and 3 through the load resistor to the cathode junction of diodes 2 and 4. This makes the anode-end of the load resistor negative and the cathode-end positive, since electrons flow from negative to positive. The current pulses are, therefore, unidirectional and their ripple frequency is twice that of the a.c. supply frequency, just as for the conventional full-wave rectifier.

One advantage of the bridge rectifier over the conventional full-wave rectifier (Fig. 119) is that the bridge circuit produces a voltage output nearly twice that of the conventional circuit for a given power transformer. This is so because in the bridge circuit the full voltage of the transformer secondary winding is applied across two conducting valves during each half-cycle, in contrast to the circuit of Fig. 119, where the secondary voltage is split in two. The bridge circuit, however, has the disadvantage of requiring four valves, which makes for uneconomical operation. It is not possible to combine two of the valves in one envelope (a double diode) because the cathodes are not at the same potential and hence cannot be connected in parallel. Hence, separate filament transformers are required to heat the cathodes of the individual diodes, which also adds to the expense of the circuit. The last disadvantage does not apply to metallic-oxide rectifiers and semiconductor diodes, which do not require any cathode heating power. The bridge circuit is therefore widely used with selenium, copper-oxide, and crystal rectifiers.

VOLTAGE DOUBLER

Diode rectifiers can be made to deliver direct voltages that are twice or several times the peak amplitude of the alternating voltage supplied. Moreover, by multiplying the supply voltage to the desired value, it becomes possible to omit the expensive step-up power transformer required in conventional rectifiers. Voltage-multiplying circuits are therefore especially useful in high-voltage circuits, where a transformer with a sufficiently high secondary

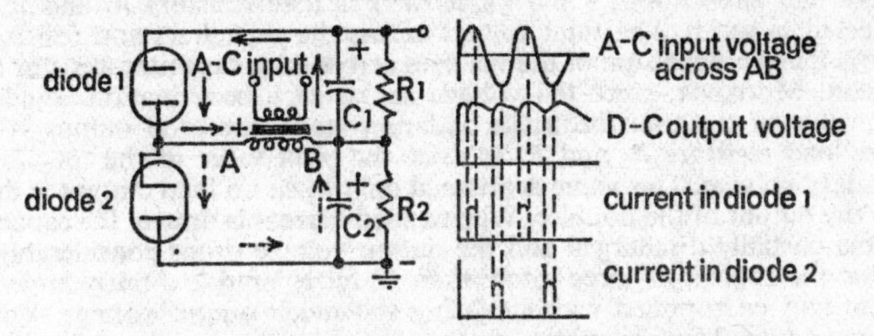

Fig. 121. Full-wave voltage-doubler circuit and waveforms

voltage would be expensive and inconvenient, and in transformerless radio and TV receivers, operated directly from the a.c. supply line. The voltage doubler described in the following paragraphs illustrates the principles involved in all voltage multiplier circuits.

The full-wave voltage doubler illustrated in Fig. 121 is capable of delivering a d.c. output voltage of twice the peak value of the applied a.c. input voltage. Basically, the circuit combines the output voltages of two half-wave rectifiers (diodes 1 and 2) in series. It is a full-wave circuit because each diode passes current to load resistors R_1 and R_2 on alternate half-cycles of the input voltage.

The full-wave doubler operates as follows. When point *A* of the transformer secondary winding is instantaneously positive during positive a.c. half-cycles, diode 1 passes a current in the direction of the solid arrows, which charges up capacitor C_1 so that its upper plate becomes positive. (The transformer could be omitted, as mentioned.) You can see that the upper plate becomes positively charged, since electrons flow out from it to the cathode of diode 1, leaving an excess positive charge. The lower plate of C_1 becomes negatively charged, since electrons from the anode of diode 1 flow through the transformer secondary into the lower plate of C_1, thus giving it an excess negative charge. As we have previously explained, a capacitor always charges up to the peak value of the voltage applied to it and hence capacitor C_1 charges up to the peak value of the transformer secondary voltage. This is shown in the output-voltage waveform of Fig. 121. Note also that only a brief charging current pulse flows through diode 1 during the first half-cycle.

During the next half-cycle, point *B* of the transformer secondary becomes positive and point *A* becomes negative, so that diode 1 cannot conduct. Since the valve is an open circuit, the charge (and hence voltage) on capacitor C_1 remains essentially constant, except for a small amount that leaks off through R_1. This is shown in Fig. 121 by the slow dropping off of the output voltage between conducting half-cycles.

Since point *A* of the transformer secondary is negative during the second half-cycle, the cathode of diode 2 also becomes negative with respect to its anode. A current flows, therefore, through diode 2 in the direction of the dotted arrows, which charges up the lower plate of capacitor C_2 negatively to the peak value of the transformer secondary voltage. The waveform of the brief charging pulse flowing through diode 2 is illustrated in Fig. 121. Again, the voltage across capacitor C_2 remains essentially constant during the next non-conducting half-cycle since the capacitor tends to hold its charge. A small amount of charge, however, leaks off through resistor R_2, as shown by the dropping off in the output voltage waveform between the second and third half-cycles.

Note that capacitors C_1 and C_2, as well as load resistors R_1 and R_2, are connected in series. The total voltage across the capacitors and resistors in series is therefore the sum of the voltages across each capacitor–resistor combination. Moreover, since the voltage across each combination equals the peak value of the transformer secondary voltage, the total output voltage across load resistors R_1 and R_2 is twice the peak value of the transformer secondary voltage. This value is attained only when no load current is drawn from the output of the doubler. When a load current is drawn, the capacitors become partially discharged and the output voltage drops considerably. By making the capacitors large (more than 10 microfarads), a fairly large load current can be supplied without losing too much output voltage. Voltage triplers and quadruplers operate on these same basic principles.

FILTER CIRCUITS

Although the rectifier circuits we have discussed deliver an output voltage that always has the same polarity, this voltage is not suitable as a d.c. supply for valves because of the pulsations in amplitude, or ripple, of the output voltage. These pulsations must be smoothed out before the output voltage can be applied to the anode, screen, and grid circuits. The required smoothing action is obtained by filter networks, consisting of choke coils and capacitors. (See Figs. 122 and 123.)

The filtering action of coils and capacitors depends on basic electrical principles. A capacitor opposes any change in the voltage applied across its terminals by storing up energy in an electrostatic field whenever the voltage tends to rise, and converting this stored energy back into voltage or current flow whenever the voltage across its terminals tends to fall. Thus, if some of the energy of the rectifier output pulsations could be stored in the electric field of a capacitor, and the capacitor would then be allowed to discharge between current pulses, the fluctuations in the output voltage could be considerably reduced. This is exactly what happens when a capacitor is connected in parallel with the rectifier output and load.

An inductor coil (choke, for short) opposes any change in the magnitude of the current flowing through it by storing up energy in a magnetic field when the current through it tends to increase; and by taking energy away from this field to maintain the current flow when the current through the inductor tends to decrease. Hence, by placing a choke coil in series with the rectifier output and load, abrupt changes in the magnitude of the load current and voltage are minimized. Another way of looking at the action of a series choke coil is to consider that the coil offers only the low resistance of the winding to the passage of d.c. while offering at the same time a high impedance to the passage of fluctuating or alternating currents. Thus, the d.c. passes through, while the a.c. ripple is largely reduced.

CHOKE-INPUT FILTER

Practical filter circuits are derived by combining the voltage-stabilizing action of shunt capacitors with the current-smoothing action of series choke coils. If the first component of the filter is a shunt capacitor connected across the rectifier, a capacitor-input filter results; if the first component of the filter consists of a choke coil connected in series with rectifier output, a choke-input

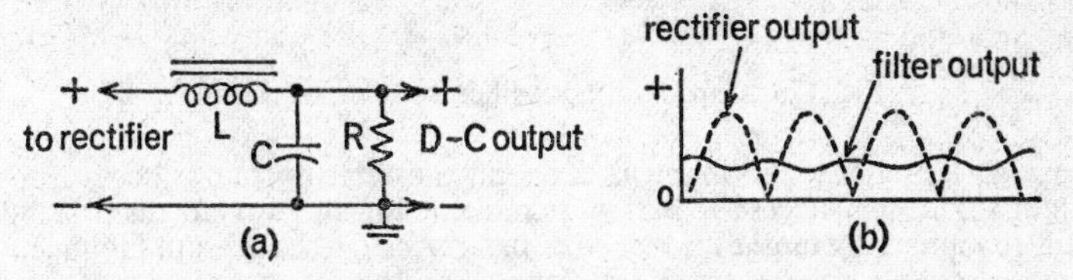

Fig. 122. Choke-input filter and output waveform

filter results. A typical choke-input filter is illustrated in Fig. 122. Only one filter section is shown, but several identical sections are often used to improve the smoothing action.

The choke coil *L* at the input of the filter readily passes the direct current from the rectifier, but opposes the a.c. pulsations, or ripple. Any fluctuations in the current that remain after it passes through the choke are largely bypassed around the load by the shunt capacitor, *C*, in the output of the filter. However, a small ripple is still present in the filter output, as shown in Fig. 122b. This is considered negligible if it is less than 1% of the steady voltage. A filter with a 1% ripple may be obtained, for example, with a 10-henry choke and an 8-microfarad capacitor.

Note in Fig. 122b that the output voltage of a choke-input filter is not equal to the peak values of the pulsations. This is so because the series choke prevents the capacitor from charging to the peak voltage when a load current is drawn. In the absence of a load current, the output voltage of the filter is nearly equal to the peak value of the applied alternating voltage. As soon as even a small load current is drawn, however, the voltage drops off to some lower value, as shown. Beyond this initial drop, the output voltage of the choke-input filter changes little with changes in load current and the filter is said to have good voltage regulation.

You may wonder why a resistor, *R*, is connected across the output of the filter. This resistor is known as a bleeder resistor and its main purpose is to place a minimum load across the rectifier during the time when the receiver valves (or other load) are heating up and do not draw any current. Without the resistor there would be an initial high-voltage surge when the rectifier is turned on, which might damage the rectifier and other valves. The bleeder also helps to maintain a constant output voltage for changing loads by drawing a minimum current at all times. Moreover, the bleeder resistor discharges the capacitors after the rectifier has been turned off and so helps to prevent dangerous shocks. For these reasons a bleeder resistor is usually present in the output of any filter.

CAPACITOR-INPUT FILTER

The action of the capacitor-input filter, illustrated in Fig. 123, is slightly different from the choke-input filter. Here the rectifier output voltage first charges the capacitor to the peak value of the pulsations. The capacitor tends to hold this charge between successive pulses, though discharging slowly

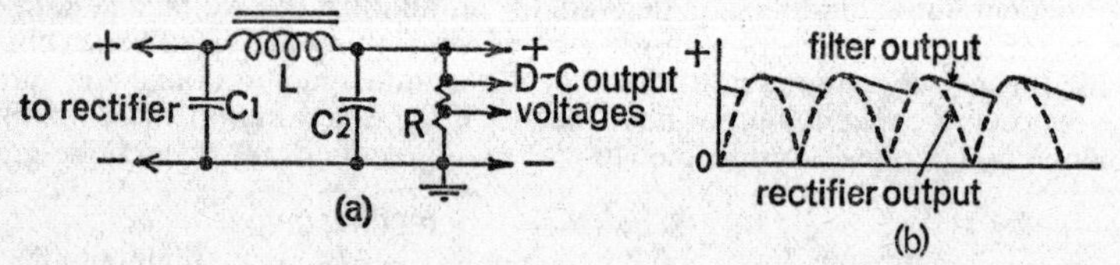

Fig. 123. Capacitor-input filter and output waveform

through the choke and load. As a result, the filter output voltage drops off slightly between successive pulses, as indicated in the waveforms (Fig. 123b), but remains substantially near the peak value. The output voltage of a capacitor-input filter is therefore higher than that of a choke-input filter for the same input voltage.

The remaining fluctuations of the rectifier output current are opposed by the series choke and by-passed to earth by the output capacitor, C_2. A small a.c. ripple remains which may be further reduced by adding additional, identical filter sections in series. The output voltage of a capacitor-input filter falls off rapidly with increasing load current and, hence, the voltage regulation of this filter is considerably poorer than that of the choke-input type.

Note that the bleeder resistor in the filter of Fig. 123 has been provided with taps for supplying the proper voltages to various valve electrodes. The resistor, therefore, acts as a voltage divider, as well as providing the required bleeder current.

COMPLETE POWER SUPPLY

Now that we have learned something about rectifiers and filters, let us look at a complete d.c. power supply, which combines these basic elements. A typical high-quality power supply, such as might be found in a high-fidelity radio receiver or amplifier, is illustrated in Fig. 124.

Note that the power supply consists of a full-wave rectifier circuit and a two-section capacitor-input filter with a 15,000-ohm bleeder (voltage divider) in the output. The two diodes are housed in a single envelope of a full-wave rectifier valve, which supplies a (d.c.) load current of approximately 200 mA maximum. The step-up power transformer has four secondary windings: an 800 V winding to supply the anodes of the rectifier, a 5 V winding for the heater of the rectifier, and two 6·3 V filament windings for the heaters of the valves in the set. Since an 800 V anode-to-anode voltage is used at the input of the rectifier, the first filter capacitor theoretically charges up to a peak voltage of one-half this value, or about 565 V, when no load current is drawn. With the maximum load current of 200 mA being drawn, however, the rectifier output voltage at the input of the filter drops to 400 V because of the voltage lost in the valve and transformer windings. This value is further reduced to about 350 V at the output of the filter owing to additional voltage drops occurring in the two filter chokes, which have about 125 ohms resistance each.

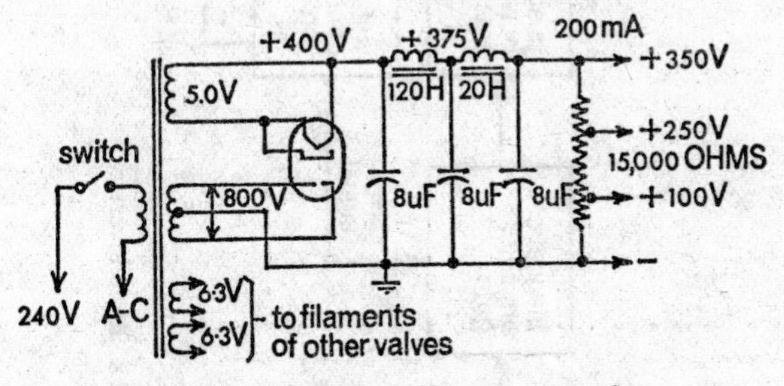

Fig. 124. Typical rectifier power supply

The upshot is that the filter output provides a maximum output voltage of 350 V at a load current of 200 mA. This output voltage would probably be used for the anodes of the final power amplifier valves in the audio amplifier. Reduced voltages of 250 V and 100 V are provided at the taps of the voltage divider for the anodes and screens of the earlier amplifier valves. Sometimes an additional tap is provided to supply a fixed negative bias voltage. The voltage divider provides a minimum load current (bleeder current) of about 23 mA.

The two-section capacitor-input filter employs large (20-henry) chokes and (8-microfarad) capacitors and, hence, is extremely effective. The ripple voltage at the output of the filter is only about 0·01% of the output voltage, or about 0·035 V a.c.

D.C. TO A.C. INVERTERS

We have spent a considerable amount of time describing power supplies that furnish smooth d.c. power from the a.c. mains. In mobile and field applications frequently the opposite need exists—namely, to furnish a ready source of a.c. power for lighting and equipment. We are all familiar with mobile, petrol-driven motor generators which furnish a.c. power directly for emergency and military uses. We also take the valve-type car radio for granted without ever questioning how the 6–12 V car battery is capable of providing the 100 V d.c. or so needed to operate the valves. In the age of the transistor radio, this is no longer a problem, since transistors will operate directly and efficiently from the car battery. In valve-type radios, however, the low-voltage

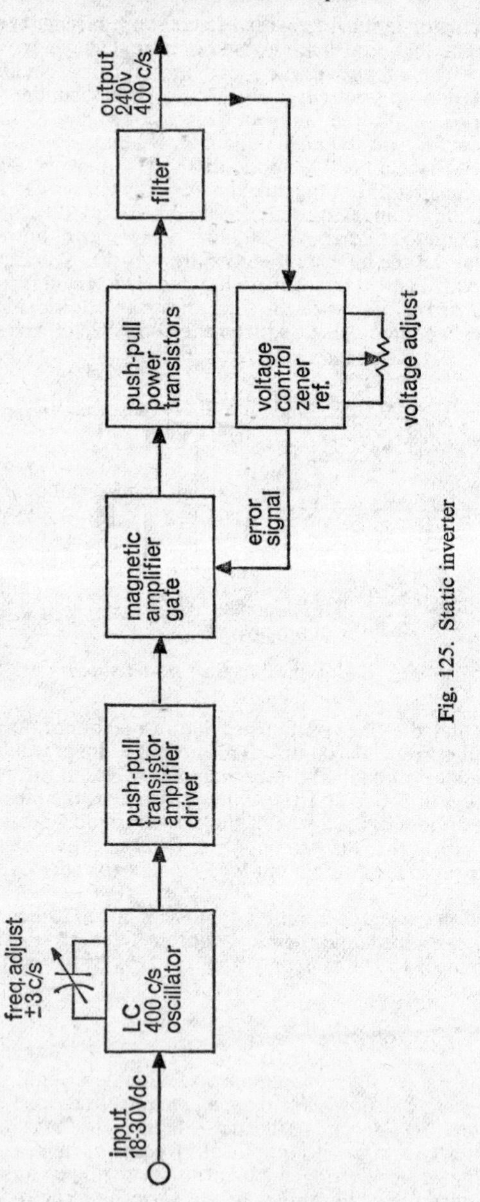

Fig. 125. Static inverter

d.c. from the battery first had to be converted to a.c., then transformed to a suitable high voltage, and finally rectified again to direct current. This formidable task used to be accomplished by means of mechanical 'vibrator'—to chop up the direct current to alternating current—and the resulting a.c. was then transformed, rectified, and filtered by a conventional rectifier power supply.

In general, any electrical or electronic device for converting direct current into alternating current is known as an inverter. How d.c. motors coupled to a.c. generators (motor generators) and vibrator power supplies accomplish this feat should be fairly obvious. The majority of high-power inverters are still of the motor-generator type. However, the advent in recent years of transistors and other solid-state components capable of operating from a low-voltage d.c. source has made it possible to design all-electronic inverters—without any moving parts. These are called static inverters.

Static Inverter. The functional operation of a typical static inverter used in military aircraft is portrayed in the block diagram (Fig. 125). This particular inverter will provide 80 volt-amperes of 400 c/s, 115 V alternating current from the standard 24 V direct-current primary aircraft power supply. The entire package comes in a compact box that weighs approximately nine pounds.

The conversion from d.c. to 400 c/s a.c. is accomplished in the first stage by means of a transistorized 400 c/s *L–C* oscillator. A push-pull transistor amplifier strengthens the output of the oscillator and drives the succeeding magnetic amplifier. A magnetic amplifier is essentially an a.c. inductor (reactor) whose inductance—and hence output—can be varied by means of a control winding that changes the saturation of the iron core. The purpose of the magnetic amplifier is twofold. First, it provides amplification of the 400 c/s square-wave output of the transistor amplifier. In addition, and more important, it regulates the voltage output of the inverter in accordance with a feedback signal derived from a voltage control circuit. The latter consists of a Zener diode, which 'samples' the output of the inverter and compares it with a Zener reference voltage. The comparison provides an 'error signal', which is used to 'gate' the magnetic amplifier to provide more or less amplification to compensate for variations in the output voltage. A 'voltage adjust' potentiometer permits manual control of the output.

The magnetic amplifier drives the power output stage, consisting of two push-pull power transistors. The transistors are 'gated' (i.e. switched on and off) by the magnetic amplifier to provide the proper output power and voltage. A bandpass filter smoothes the output.

SUMMARY

Rectifiers change alternating current into pulsating direct current by eliminating the negative half-cycles of the alternating voltage.

An ideal rectifier acts like a switch that closes a load circuit during the positive a.c. half-cycles and opens the circuit during the negative a.c. half-cycles. All rectifiers must provide a one-way path for electric current; this is called unilateral conduction.

Diodes provide unilateral conduction since the current can flow in one direction only, from emitter (cathode) to collector (anode). Diodes may come in the form of evacuated or gas-filled valves, germanium or silicon crystals (semiconductors), and copper-oxide or selenium metallic rectifiers. Crystal diodes and metallic rectifiers do not have filaments and hence do not require heating power.

A single diode is called a half-wave rectifier because it permits anode current flow only every other half-cycle during positive alternations of the input voltage. Since only positive half-cycles are reproduced, one half of the input voltage is lost and the efficiency is low. Furthermore, a half-wave rectifier has a large ripple voltage at the same frequency as the a.c. power supply.

A full-wave rectifier rectifies both halves of the a.c. input cycle by employing two diodes back to back. The conventional full-wave rectifier requires an anode-to-anode voltage twice the value of the anode voltage for each valve. The circuit has a small a.c. ripple at twice the a.c. supply frequency, high efficiency and high permissible current drain.

A full-wave bridge rectifier uses four diodes, but produces a voltage output nearly twice that of the conventional full-wave circuit. Separate transformers are required to heat the filaments of the valves.

Voltage multipliers (doublers, triplers, quadruplers) deliver rectified voltages that are several times the peak amplitude of the applied alternating voltage; this is done by charging series-connected capacitors through diode rectifiers on alternate a.c. half-cycles so that the direct voltages across the capacitors add up in series.

Filter circuits smooth out the pulsations in amplitude (a.c. ripple) of the rectifier output voltage. Filter capacitors oppose any change in the voltage applied across them by storing up energy in an electric field, while choke coils oppose any change in the magnitude of a current flowing through them by storing up energy in a magnetic field.

Choke-input and capacitor-input filters combine choke coils with shunt capacitors to obtain increased filtering action.

An inverter is any electrical or electronic device that converts direct current into alternating current. Motor generators and vibrator power supplies were once commonly used inverters. A static inverter utilizes only electronic components and has no moving parts. Transistor oscillators and amplifiers and magnetic amplifiers frequently constitute static inverters.

CHAPTER FIFTEEN

RADIO COMMUNICATION

Radio communication was one of the first important uses electronics was put to. Simple radio transmitters and receivers had existed long before the application of electronics made them what they are today; the existence of electromagnetic (radio) waves was predicted as early as 1865 in a mathematical treatise by James Clerk Maxwell, a professor at King's College, London. The German scientist Heinrich Hertz had verified the existence of these mysterious waves in 1887, using a spark between two electrodes as his transmitter, and a similar pair of electrodes as his detector. A somewhat more sophisticated system, using an aerial, an earth, and an improved detector, sent out the first dot-dash radio-telegraphy messages in 1896. The range of this communication system was two miles, just beyond shouting range. A milestone in radio communication occurred in December 1901, when the young Italian Guglielmo Marconi and his associates succeeded in spanning the Atlantic with three faint buzzes, the letter 'S' of the Morse code. Further attempts, in 1902, to institute a commercial transatlantic radio telegraph service failed, however, because of the seasonal variations in signal strength, which nobody had suspected.

Radio telegraphy and telephony continued to make slow and erratic progress in the following years, but it was not until the invention of the crystal detector and, in 1907, the triode valve, that it finally came to fruition. A regular broadcasting service, as we know it, did not appear until 1920.

BASIC ELEMENTS

Modern radio communication and broadcasting may consist of the periodic interruptions of continuous radio waves (c.w.) in accordance with a telegraphic code, called radio telegraphy, or it may consist of the modulation of

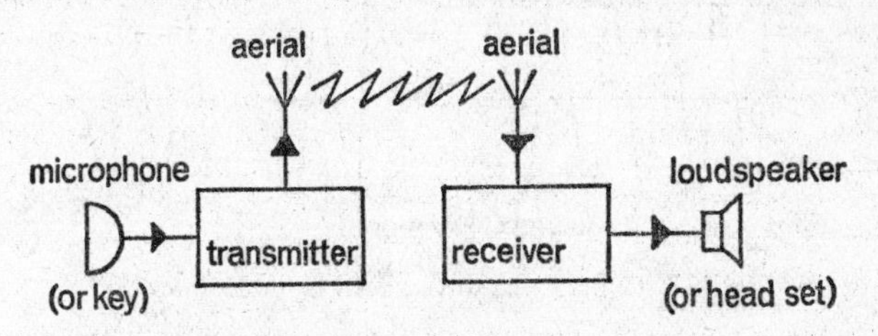

Fig. 126. Elements of radio communication

radio waves by speech or music—radio telephony. Fig. 126 illustrates the basic elements required to transmit messages, speech or music from one location to another by radio.

These elements are:

1. A radio transmitter to generate the radio-frequency waves.
2. A telegraph key or a microphone to control the radio waves in accordance with the information to be transmitted.
3. A transmitting aerial to radiate the waves into space.
4. A receiving aerial to intercept a portion of the radiated waves.
5. A radio receiver to select and amplify the desired transmitter signal and to demodulate the information contained in the radio waves.
6. A loudspeaker (or headphones) to convert the demodulated electrical waves into sound and, thus, reproduce the original information.

This picture is rather crude and we shall have to fill in many important details to gain a significant understanding of the process of radio communication.

RADIO TRANSMITTERS

Radio transmitters generate energy at a definite frequency and convey this energy to the transmitting aerial for radiation. To obtain a useful radiated signal, information must be superimposed on the radio waves. In the continuous-wave (c.w.) transmitters used for radio telegraphy, the desired information is added by interrupting the radio-frequency oscillations in accordance with a telegraphic code. In radio-telephone (modulated) transmitters the information is added by modulating either the amplitude or the frequency of the radio-frequency carrier wave with the speech or music to be transmitted.

CONTINUOUS-WAVE (C.W.) TRANSMITTERS

Any of the oscillators described in Chapter 11 are capable of generating radio frequencies and, when connected to a suitable aerial, can radiate a useful signal. The radio-frequency output of such a simple one-stage transmitter may be interrupted in accordance with a coded telegraphic message by turning the anode-supply power on and off with a switch, or key. The output of the transmitter then consists of interrupted continuous r.f. waves, which may be intercepted by a distant receiver and made audible. A typical Morse code

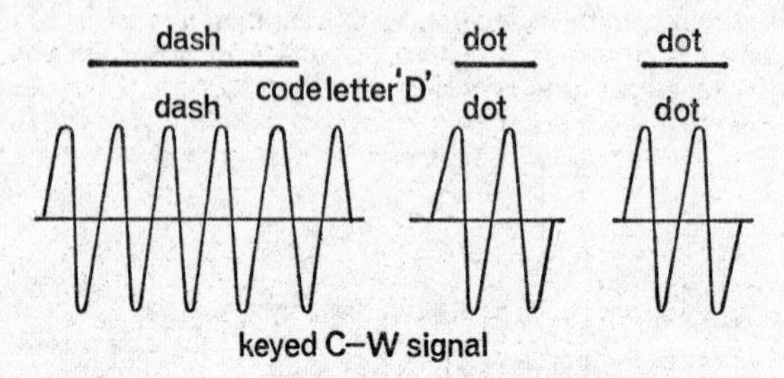

Fig. 127. Continuous wave keyed by code letter 'D'

symbol, the letter 'D' consisting of a dash and two dots, is shown in Fig. 127.

Master-oscillator power-amplifier. While a simple one-stage transmitter illustrates the principles involved, most practical transmitters use several stages to obtain additional r.f. amplification or to multiply the oscillator frequency. A popular combination that provides more power than is possible with a simple oscillator is known as the master-oscillator power-amplifier (m.o.p.a.) transmitter. As its name implies, this consists of an oscillator stage, which may be crystal-controlled, and one or more r.f. power amplifier stages coupled to an aerial. The block diagram of a m.o.p.a. transmitter and a typical schematic circuit diagram are shown in Fig. 128.

Note that in the block diagram (Fig. 128a) the key controls the power to the oscillator stage, while in the circuit diagram (Fig. 128b) the key has been inserted in the anode-supply circuit of the power amplifier stage. (Batteries have been shown for convenience.) Either method of keying is practicable, but anode-circuit keying of the power amplifier is preferred. In addition to

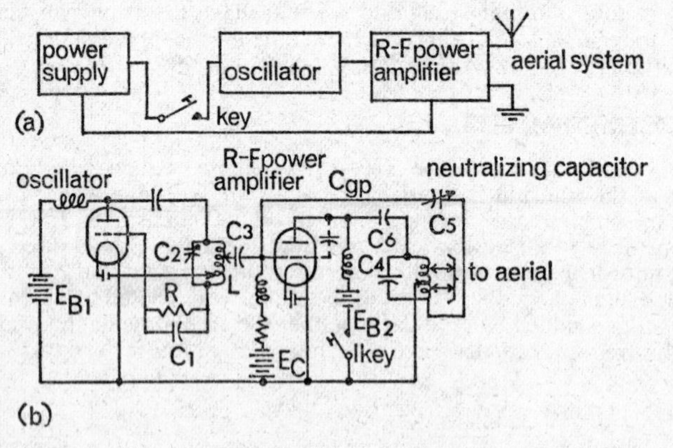

Fig. 128. Master-oscillator power-amplifier transmitter: (a) block diagram; (b) schematic circuit diagram

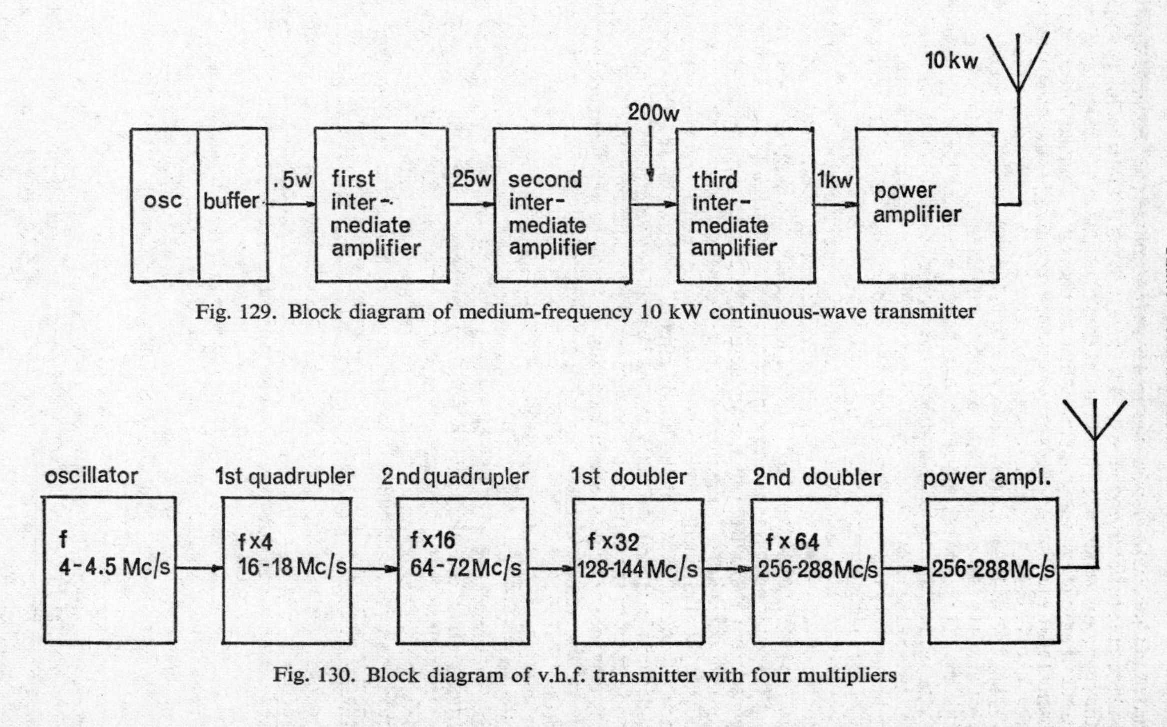

Fig. 129. Block diagram of medium-frequency 10 kW continuous-wave transmitter

Fig. 130. Block diagram of v.h.f. transmitter with four multipliers

providing increased power, the m.o.p.a. transmitter has the advantage of isolating the aerial from the oscillator stage through the intervening r.f. power amplifier. An aerial may change its effective resistance and capacitance as a result of swaying in the wind and other influences. If coupled directly to the oscillator, these changes in impedance cause detuning of the oscillator tank circuit and, hence, frequency instability.

The typical m.o.p.a. transmitter illustrated in Fig. 128b consists of a conventional Hartley oscillator and a Class C power amplifier, both being shunt-fed through r.f. chokes and capacitors. The output of the oscillator is coupled to the grid of the power amplifier through capacitor C_3, which also prevents the bias voltage, E_c, of the amplifier from being short-circuited. The anode-tank circuit of the amplifier is tapped, as shown, and coupled to an aerial.

Neutralizing capacitor C_5 in the anode-to-grid circuit of the power amplifier is used to balance out or neutralize the effects of the triode's grid-to-anode interelectrode capacitance, C_{gp}, shown dotted in (b). You will recall that the grid-to-anode capacitance of a triode feeds back energy from anode to grid with a resulting loss in power output. By feeding back a voltage in opposite phase from the bottom of the anode-tank circuit to the grid of the valve, through capacitor C_5, the interelectrode feedback voltage may be cancelled out and its detrimental effects are overcome. When this is done through a neutralizing capacitor in the anode circuit, as shown in (b), the process is called anode neutralization. When the neutralizing voltage is developed in the grid circuit of the valve, it is known as grid neutralization. In either case, the capacitance of the neutralizing capacitor (C_5) will be approximately equal to the grid-to-anode capacitance of the valve, when the stage is properly neutralized.

Buffer Amplifier. A two-stage m.o.p.a. transmitter can supply only a limited amount of output power. When high output power is required, several intermediate amplifiers are inserted between the oscillator and the final power amplifier that feeds the aerial. A sufficient number of intermediate stages must be inserted so that the output of the last intermediate amplifier is capable of driving the final power amplifier to the required transmitter power. Furthermore, an additional buffer amplifier is generally inserted between the oscillator and the first intermediate amplifier to isolate the oscillator from the following stages and minimize changes in the oscillator frequency due to variations in coupling and aerial loading. The buffer stage is a voltage amplifier operated Class A; it therefore draws no grid current, requires no input power and does not load down the oscillator stage. Fig. 129 illustrates the block diagram of a typical medium-frequency (300 kc/s to 3 Mc/s) 10-kilowatt c.w. transmitter, which employs a buffer stage and three intermediate amplifiers. The combined oscillator-buffer stage has the correct transmitter frequency, but supplies only 0·5 W output. This is multiplied to 1 kilowatt by three intermediate r.f. power amplifiers, operating Class C, and to 10 kW by the final power amplifier stage.

High-frequency and v.h.f. transmitters. When the transmitter must operate in the high-frequency band between 3 Mc/s and 30 Mc/s, or in the v.h.f. band between 300 Mc/s and 3,000 Mc/s, frequency multipliers must be used to obtain the required output frequency. Generally, crystals cannot be obtained that will oscillate well in the v.h.f. band, and other oscillators are far too unstable for direct frequency control in these bands. The obvious way around this difficulty is, of course, to employ a suitable low-frequency oscillator and multiply its output frequency to the required value by a series of multiplier stages whose output tank circuits are tuned to multiples of the oscillator frequency. When the anode-tank circuit of the multiplier is tuned to twice the input (oscillator) frequency, the stage is called a doubler; when tuned to three

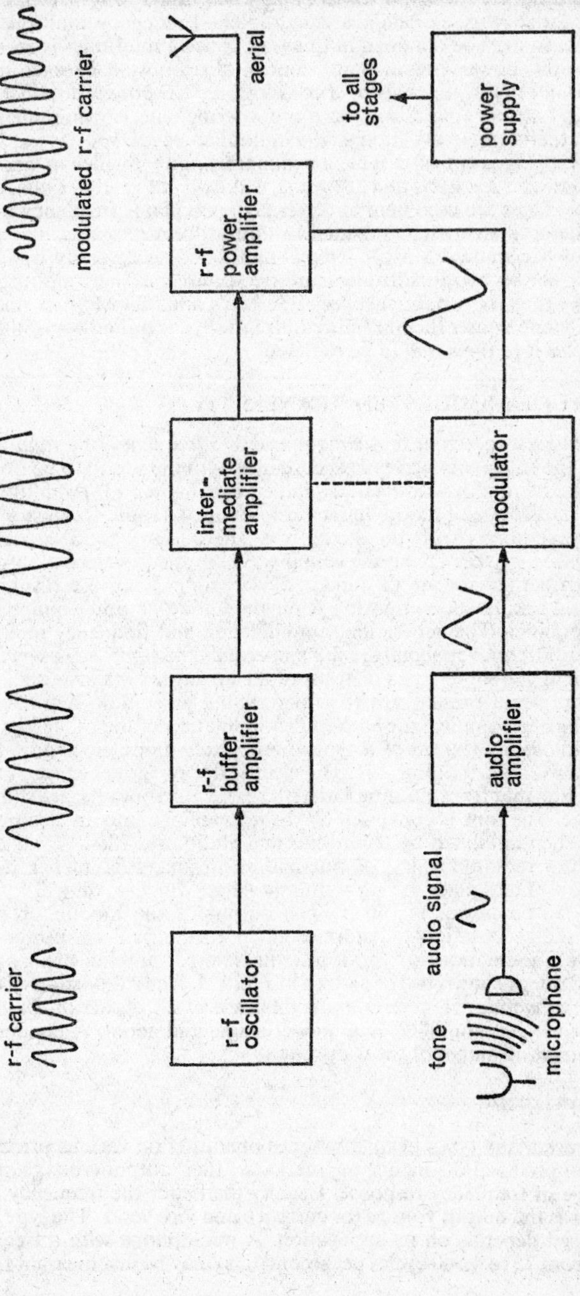

Fig. 131. Block diagram of amplitude-modulated radio-telephone transmitter and waveforms

times the input frequency, a tripler; and when tuned to four times the input frequency, the stage is called a quadrupler. Frequency multiplications of several dozen may be obtained in this way. Since a multiplier is an r.f. power amplifier, the power output is, of course, also boosted by each additional stage, though not as much as by a conventional r.f. power amplifier. In high-power v.h.f. transmitters, one or more intermediate r.f. amplifiers may be required, therefore, in addition to the multiplier stages.

The block diagram of a typical v.h.f. transmitter, which is continuously tunable between 256 Mc/s and 288 Mc/s, is shown in Fig. 130. Four frequency-multiplier stages are used here to increase the oscillator frequency by a factor of 64. The oscillator itself provides an output frequency that may be varied between 4 Mc/s and 4·5 Mc/s. This is multiplied successively by four, four, two, and two by two quadruplers and two doublers. The transmitter output frequency, thus, is variable between 256 Mc/s and 288 Mc/s. A final power-amplifier stage boosts the multiplier output to the required transmitter power and couples it to the aerial to be radiated.

AMPLITUDE-MODULATED TRANSMITTERS

To broadcast speech or music requires either frequency or amplitude modulation of the r.f. carrier waves in accordance with the sound to be transmitted. Both types of modulation were explained in Chapter 13. Amplitude modulation is the process of modifying the amplitude of a radio-frequency carrier so that it corresponds with the strength of the audio (sound) signal. This is accomplished by converting the sound waves into equivalent electrical variations with a microphone or similar device, amplifying the resulting audio signal, and finally superimposing it on the anode or grid voltage of an r.f. power amplifier. The generation, amplification and frequency multiplication of the radio-frequency signal are done in exactly the same way as in c.w. transmitters, and the same types of transmitter arrangements are used. The only new components present are the microphone and audio amplifier (modulator), which accomplish the amplitude modulation of the r.f. carrier wave.

A basic block diagram of a typical amplitude-modulated radio-telephone transmitter is shown in Fig. 131. The electrical waveforms, when a pure audio tone (such as that from a tuning fork) strikes the microphone, are shown above each stage. The tone is converted by the microphone into an electrical signal, which is then amplified by a conventional audio amplifier. Power amplification to the required value of one-half the transmitter power (for anode modulation) takes place in the modulator stage, which—you will recall—is an ordinary audio power amplifier. The output of the modulator is applied through the modulation transformer either to the final r.f. power amplifier (high-level modulation) or to an intermediate r.f. amplifier stage (low-level modulation), as shown by the arrows in Fig. 131. The r.f. portion of the transmitter is conventional, consisting in this case of an oscillator, buffer amplifier, intermediate amplifier, final power stage, and aerial. A common power supply serves all stages of the transmitter.

MICROPHONES

There are many types of microphones operating on various principles, but their chief practical distinguishing marks are their output (voltage or current) and range of frequency response. Usually the better the frequency response, the lower is the output voltage (or current), and vice versa. The type of microphone used depends on its application. A microphone with a frequency response from 75 to 4,500 cycles per second (c/s) may be just the right thing for a radio-telephone, where speech intelligibility is of paramount importance, but it will never do for broadcast purposes. A frequency response of at least 30 to 10,000 c/s is required for amplitude-modulation broadcasting, and the whole audio range from about 20 to 15,000 c/s is necessary for frequency-modulation broadcasting or high-fidelity work. Three types of 'mikes', in ascending order of audio quality, are the carbon microphone, the crystal microphone, and the dynamic or moving-coil microphone.

Carbon microphone. A typical carbon microphone, as used in a telephone handset, is illustrated in Fig. 132. A loose pile of carbon granules is contained

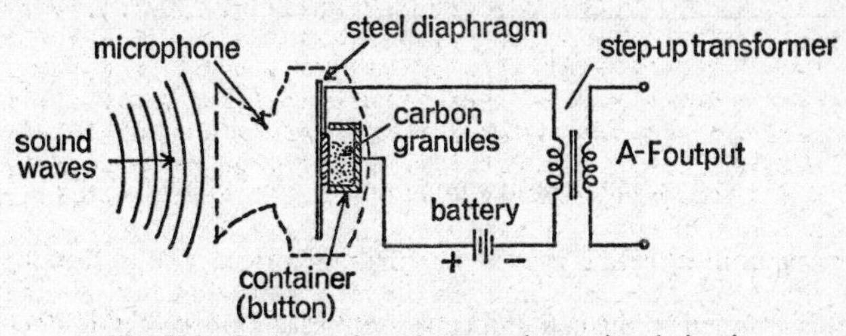

Fig. 132. Carbon microphone and associated circuit

in an insulated cup, called a button, which is in contact with a thin steel diaphragm. When sound waves strike the diaphragm, the resulting vibrations vary the pressure on the button and, thus, on the pile of carbon granules. These variations in pressure change the electrical resistance of the loosely piled granules in accordance with the vibrations of the diaphragm and, hence, in accordance with the pressure of the sound waves. The pile of carbon granules is in series with a battery and the primary of a step-up microphone transformer. Consequently, the changing resistance of the carbon granules produce corresponding changes in the current through the circuit, resulting in a pulsating direct current. The current pulsations in the primary of the transformer induce an alternating voltage in the secondary of the transformer, which is coupled to a suitable audio amplifier. The microphone transformer actually serves two purposes: it matches the relatively low resistance (a few hundred ohms) of the microphone to the high impedance of the grid circuit of the first audio amplifier tube, and it also steps up the voltage. With a battery from 1·5 to 6 V and microphone currents between 20 and 100 mA, a good carbon mike will develop up to 25 V peak in the secondary of the transformer.

However, in spite of its high output, the carbon microphone suffers from a number of disadvantages. Because of random changes in the resistance of individual carbon granules, there is a constant background noise or hiss, which drowns out weak sounds. The carbon granules may stick together, or pack, resulting in a loss in sensitivity and serious distortion. Finally, and most serious, the frequency response of a carbon microphone is limited to a few thousand cycles, which makes it unsuitable for anything but speech communication. There are various tricks to extend the response, such as stretching the diaphragm, but, owing to the superiority of other types of microphones, the effort is rarely worth while.

Crystal microphone. The operation of a crystal microphone is based on the piezoelectric effect, which we discussed in Chapter 11 (oscillators). Crystalline materials such as quartz and Rochelle salt generate a voltage when mechanical stress is applied to one of their faces. Rochelle salt is used for crystal microphones, since it has greater voltage output than other piezoelectric materials. A basic crystal microphone is shown in Fig. 133.

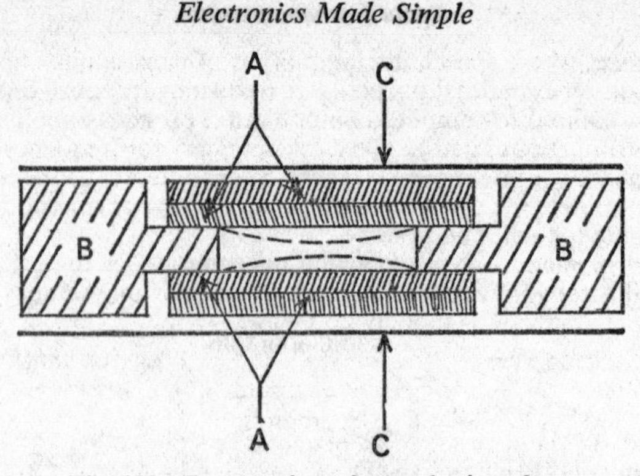

Fig. 133. Construction of crystal microphone

The basic unit, or cell, is made by cementing a metal foil to the surfaces of two thin crystal slabs and clamping the two slabs together. A sound diaphragm is coupled to the cell and causes a twisting of the crystal faces when sound waves strike the diaphragm. This in turn generates a voltage between the metal foil on the crystal face, which is proportional to the sound pressure.

The unit shown in Fig. 133 is known as a sound cell or grill type microphone, since it consists of a number of basic crystal units or cells connected together for greater sensitivity. This superior type of crystal microphone dispenses with the diaphragm, since sound strikes the crystal plate directly. The crystal cells (A) are mounted in a bakelite framework (B) and the assembly is covered with a flexible, airtight cover (C), which permits the cells to vibrate freely when sound strikes them.

Crystal microphones have a high output impedance and can therefore be connected directly into the grid circuit of the first audio amplifier valve. Because their output voltage is low (a few millivolts), crystal microphones

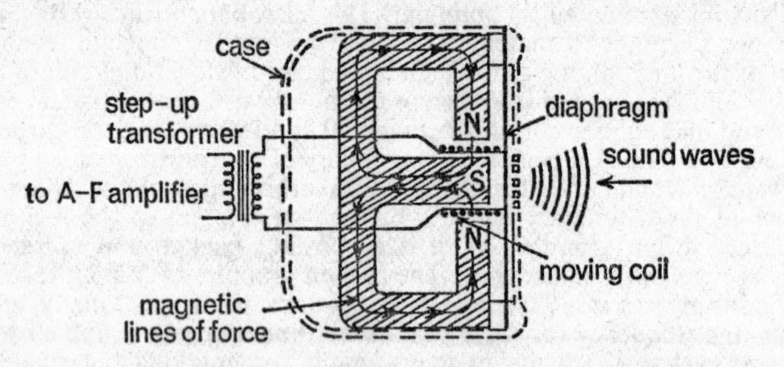

Fig. 134. Dynamic microphone with matching transformer

require several stages of high-gain audio amplification. However, their frequency response is relatively uniform from about 50 to 10,000 c/s, which is sufficient for good quality reproduction of speech and music.

Dynamic microphone. Better quality of reproduction is obtained from the dynamic or moving-coil microphone, illustrated in Fig. 134. Its operation is based on the fundamental electric principle that a voltage is induced in a wire or coil which moves across a magnetic field.

As shown in the illustration, a coil of fine wire is suspended in the field of a strong permanent magnet and one end is fastened rigidly to the back of a diaphragm. When sound waves strike the diaphragm, it vibrates, and the coil moves back and forth with it in unison. The motion of the coil cuts across the lines of force of the magnet and, hence, induces a voltage in the coil that is an electrical representation of the original sound.

The dynamic microphone is practically ideal for the most demanding purposes. It requires no external voltage, is light in weight, rugged, and practically immune to the effects of mechanical vibration, temperature, and moisture. It is highly sensitive and has a fair voltage output, a little lower than that of the crystal mike. Most important, the dynamic microphone has an essentially uniform output over the entire audio-frequency range from about 30 to 15,000 c/s. The output impedance of the moving coil, however, is only about 30 to 500 ohms maximum and, hence, a step-up transformer is required to match the microphone to the input stage of an audio amplifier.

One of the highest quality microphones commercially available is the *ribbon microphone*. Its operation is similar to that of the moving-coil type, but the coil is replaced by a single corrugated strip of aluminium foil which also acts as the diaphragm.

FREQUENCY-MODULATED TRANSMITTERS

As described in Chapter 13, in frequency modulation the amplitude of an r.f. carrier remains constant, but its frequency is continuously varied in accordance with the instantaneous amplitude (strength) of a sound signal. Thus, the original sound waves are converted into a frequency deviation of the r.f. carrier that is proportional to the intensity of the sound waves. The number of carrier frequency swings per second is equal to the frequency of the sound. We have already become acquainted with the reactance valve modulator (in Chapter 13), which changes the amplitude variations of an audio signal into corresponding phase or frequency variations. To construct an f.m. transmitter on paper, therefore, we need only fit together the pieces.

Fig. 135 shows the block diagram of a typical f.m. transmitter. The upper row of the diagram consists of a microphone, an audio amplifier, a reactance modulator (for converting the audio amplitude changes into frequency changes), a 10 Mc/s transmitter-oscillator, a series of frequency multipliers that multiply the oscillator frequency 10 times to the required operating frequency of 100 Mc/s, a final r.f. power amplifier and an aerial. Since we have already explained the operation of the reactance modulator, there is nothing unfamiliar in the upper row of Fig. 135.

What is the complex circuitry indicated by the blocks of the lower row in Fig. 135 for? The basic purpose of these circuits is to stabilize the centre frequency of the transmitter. Since the frequency of the f.m. transmitter is constantly being deviated, it is obviously not possible to use a highly stable, crystal-controlled transmitter oscillator. Yet broadcasting regulations require that the centre frequency of the carrier be held constant within very close limits. In this example, a separate crystal oscillator generates a 9-mc signal. A series of doublers multiply the oscillator frequency to 99 Mc/s, that is, 1 Mc/s below the carrier centre frequency. The output of these multipliers is applied to a mixer stage together with a portion of the 100-Mc/s output of the frequency multipliers in the upper (main transmitter) row. These two signals 'beat' together in the mixer to generate a 1 Mc/s difference or intermediate frequency (i.f.) in a manner which we shall explain in connexion with the superheterodyne receiver. The intermediate-frequency signal is boosted in

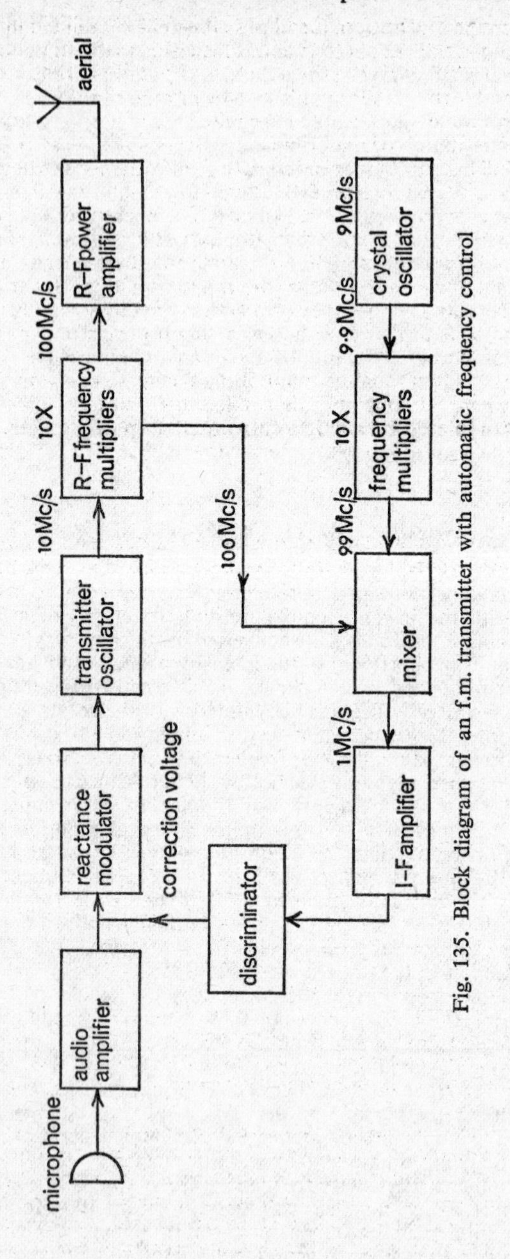

Fig. 135. Block diagram of an f.m. transmitter with automatic frequency control

amplitude by an i.f. amplifier tuned to 1 Mc/s. The 1 Mc/s output of the i.f. amplifier is finally applied to a phase-shift discriminator circuit, identical with that discussed in Chapter 13. The discriminator produces a fixed direct-voltage output as long as the frequency of the input signal to which it is tuned remains constant. As soon as the input frequency differs from the resonant value, however, the output voltage of the discriminator changes. This output voltage is fed to the grid of the reactance valve and serves as a reference potential for establishing the proper operating frequency of the transmitter.

If the centre frequency of the transmitter oscillator should momentarily rise, the mixer output frequency will also rise, since the difference between the transmitter and crystal oscillator frequency has increased. The frequency at the input of the discriminator, consequently, will increase and its direct-voltage output will change. In effect, a correction voltage is now being applied to the grid of the reactance valve modulator. This correction voltage changes the phase shift of the reactance valve in a direction that will automatically restore the correct transmitter frequency. The system will, of course, work similarly for a momentary lowering of the transmitter centre frequency. The same automatic frequency control (a.f.c.) system is used in television receivers.

AERIALS

We have investigated the behaviour of electrons within the atom, inside evacuated valve envelopes, in conductors and circuits, and we have seen them oscillating rapidly back and forth between the capacitor and inductor coil of a tank circuit. But so far the 'dance of electrons' has always been confined within the boundaries of a conductor, valve, or circuit and we have not explained the central point of radio, the radiation of waves into space. How indeed do the electrons oscillating at a high frequency in the anode-tank circuit of a transmitter set up electromagnetic (radio) waves to be radiated throughout space? Part of the answer is childishly simple, they do so automatically without any outside help: *accelerated* electric charges (i.e. electrons) always radiate energy in the form of electromagnetic waves. Note the essential word 'accelerated'. A charged glass rod or capacitor will not radiate; the charges must be accelerated, they cannot be static. Without going into mathematical theory, the fact is that the simple L–C tank circuit with its accelerated electrons (i.e. the electron oscillations) radiates energy in the form of electromagnetic waves. As a matter of fact, all the conductors in a high-frequency oscillator or amplifier radiate energy to some extent. The radiation from a tank or oscillating circuit, however, is extremely weak. To radiate or receive electromagnetic waves effectively we need an aerial. This is simply a system of conductors, which couples or matches the transmitter or receiver to space. A transmitting aerial, connected to the transmitter by a transmission line or lead-in, forces radiation into space by 'grasping it with long fingers', as one radio engineer expressed it. Similarly, a receiving aerial, connected to a radio receiver, grasps or intercepts a portion of the electromagnetic energy travelling through space.

It is the size of the aerial conductors that makes them effective radiators and absorbers of electromagnetic waves. The aerial must have the proper size to match, in effect, the impedance of the transmitter or receiver to that of space. (Space actually has a certain impedance, or opposition to the passage of electromagnetic waves.) You can visualize the effect of aerial size by the following simple analogy. If you stir up the water in a pond with your hands, small water wavelets will be produced. Using a paddle will create far larger waves with the same expenditure of energy.

BASIC ACTION OF AN AERIAL

Radiation of electromagnetic waves from an aerial is a complex process, which can be completely understood only by means of the mathematical theory. But you can form a fairly accurate picture of the action by going back to the description of tank-circuit oscillations in Chapter 11. You will recall that the electrons in a tank circuit rapidly flow from one plate of a capacitor to the other through the tank coil, thus charging and discharging the capacitor with alternating polarity. While the electrons rush from one plate to the other, they build up an intense *magnetic* field about the coil. When they have fully charged the capacitor the electron current stops, but now an intense *electric* field exists between the plates of the capacitor. Thus, the sequence of charge and discharge results in accelerated electron motion, or an oscillating current, with the energy alternately (each quarter-cycle) being stored in the electric field of the capacitor and in the magnetic field of the inductance coil.

Let us now attach a wire to each side of a tank circuit and observe what happens in this simple Hertz aerial during alternate quarter-cycles. The arrangement is illustrated in Fig. 136 and is similar to one used by Hertz during his early experiments.

Fig. 136a shows the original tank circuit at the moment of maximum current

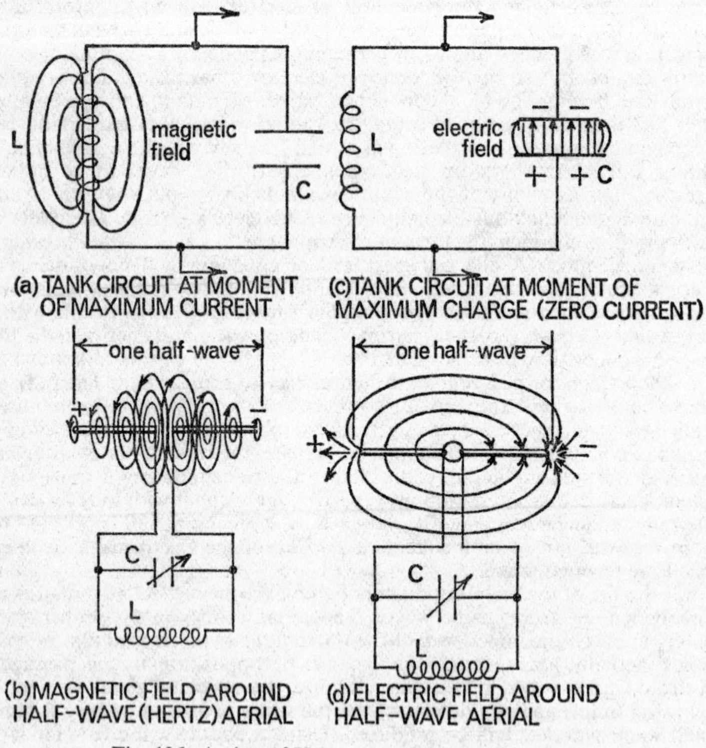

Fig. 136. Action of Hertz type of transmitting aerial

flow, when an intense magnetic field has built up around the coil. In (b), which portrays the same instant, two aerial wires have been attached to the tank circuit. The electron current now rushes to the end of each wire and back to the tank circuit, setting up a magnetic field along the whole length and perpendicular to the wires. As we shall see later, the length of the wires has to be just right so that the current can complete its round trip to the ends of the wires and back to the tank in just one half-cycle and is thus in phase with the current at the tank. Note that the magnetic field is strongest near the centre point of the aerial, where it is fed with energy, as indicated by the concentrated lines of force. This field extends far into space, although it grows progressively weaker with increasing distance from the aerial.

Fig. 136c portrays the tank circuit a quarter-cycle later, when the current has stopped, but the charge on the capacitor is a maximum and consequently a strong electric field exists between its plates. The action at the same moment in the Hertz aerial is illustrated in (d). The current has momentarily stopped

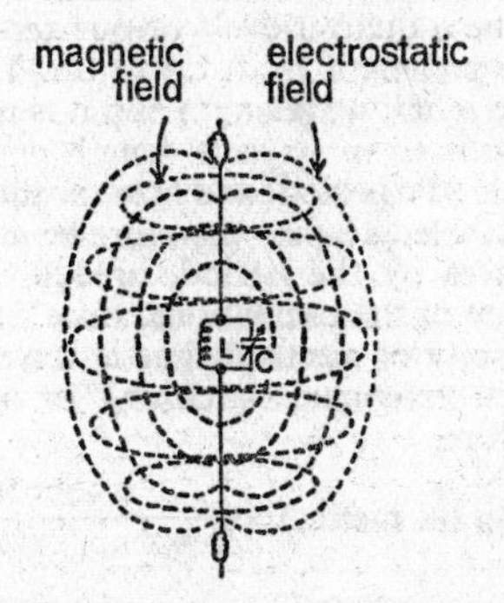

Fig. 137. Complex electromagnetic field surrounding the Hertz aerial

after reaching the ends of the aerial wires, thus charging the ends of the conductors with opposite polarity. The charged conductors are now surrounded by an electric field, with lines of force extending from the positive end to the negatively charged end parallel to the wires. This field, too, projects far into the space surrounding the aerial.

During the following quarter-cycle, the current rushes back from the ends of the wires to the centre, setting up a magnetic field of opposite polarity (in respect to Fig. 136b). After reaching the centre of the aerial, the current again dies down momentarily and the wires are charged up with opposite polarity. An electric field parallel to the wires, but with opposite polarity from that shown in (d), now surrounds the aerial conductors. The same sequence is repeated during the next cycle of the current oscillations.

It is evident that the aerial wires are surrounded alternately by electric and magnetic fields which are in step with the field alternations (every quarter-cycle) in the tank circuit. Since one field is building up, while the other is dying down, both the electric and magnetic fields effectively surround the aerial at all times. The resulting complex electromagnetic field is illustrated in Fig. 137. You have to imagine that the outermost field lines detach themselves from the aerial and extend far out into space as travelling electromagnetic waves.

Aerial length and resonance. It is correct to consider the conductors of an aerial as being in resonance with, or tuned to, the frequency of the tank circuit oscillations. You will recall that the oscillating current flowing through the tank circuit coil reverses in polarity or phase every half-cycle. (It takes a quarter-cycle to charge the capacitor and another quarter-cycle to discharge it.) For the oscillations in the aerial wires to be in step or in phase with the tank circuit oscillations, it must take just as long for the current to rush to the tips of the conductors and back as it takes to charge and discharge the capacitor. Since it takes a quarter-cycle to charge the capacitor and another quarter-cycle to discharge it, it should also take a quarter-cycle for the current to reach the ends of the two aerial wires and another quarter-cycle for it to return to the centre, so that the timing is just right. Furthermore, since the current flows in the opposite direction when returning from the aerial tips than when flowing out, its polarity or phase on the return trip has been automatically reversed. The returning current wave is thus in phase with the next half-cycle of the tank current oscillation, provided that the length is right. Under these conditions the aerial is in resonance with the frequency of the tank-circuit oscillations.

For the current to travel out to the ends of the aerial in a quarter-cycle, each aerial conductor must be a quarter-cycle or quarter-wave long, or the entire aerial must be a half-wavelength from tip to tip. The electrical length of a Hertz aerial is therefore a half wavelength and it is referred to as a half-wave aerial. What this amounts to in physical length depends, of course, on the time allotted to one cycle (the period) and how far the waves travel during this time (the velocity). A wavelength may therefore be determined by multiplying the velocity of the waves by the period, or equivalently, by dividing the velocity by the frequency of the oscillations, since frequency is the reciprocal of the period. The velocity of electromagnetic waves is the same as that of light (which is an electromagnetic radiation), or about 3×10^8 metres per second. We have, therefore,

$$\text{wavelength (in metres)} = \frac{3 \times 10^8}{\text{frequency (cycles per sec)}}$$

or more conveniently, converting to megacycles per second,

$$\text{wavelength (metres)} = \frac{300}{\text{freq. (Mc/s)}}$$

The wavelength of a 1,500 kc/s (1·5 Mc/s) broadcast station, for example, is 300/1·5 = 200 metres (656 ft). A Hertz aerial for this transmitter should be one half-wavelength, 100 m or about 328 ft long. Each aerial wire must therefore be one quarter-wavelength, or 100 m (164 ft) long. Actually, electromagnetic waves do not travel quite as fast in a conductor as in free space, so that a small correction must be applied. A convenient formula for the length of a half-wave or Hertz aerials, correct up to about 30 Mc/s, is:

$$\text{length of half-wave aerial (metres)} = \frac{143}{\text{freq. (Mc/s)}}$$

Substituting in the latter formula, we see that the half-wave aerial for the 1,500 kc/s broadcast station should be only

$$\frac{143}{1{\cdot}5} = 95 \text{ m (312 ft) long, instead of 100 m (328 ft)}$$

Each conductor must therefore be 47·5 m (156 ft) long.

Polarization of electromagnetic waves. We have stated that the electric field of a Hertz aerial is always parallel to the length of the aerial, while the magnetic field is at right angles to it. The magnetic and electric fields of an electro-

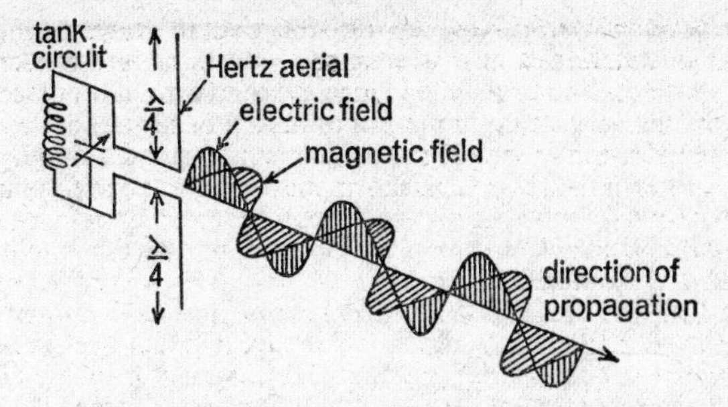

Fig. 138. Vertical polarization of an electromagnetic wave

magnetic wave, thus, always travel at right angles with respect to each other. The orientation of the electric field with respect to earth is known as the polarization of the wave. If the plane of the electric field is horizontal with respect to earth, the wave is said to be horizontally polarized; if the electric field is oriented vertically, the wave is said to be vertically polarized. Fig. 138 is a graphic presentation of an electromagnetic wave emanating from a half-wave aerial. Since the intensity of the fields varies sinusoidally with the amplitude of the current oscillations, the fields are represented by travelling sine-waves at right angles to each other. As you can see the aerial conductors are placed vertically with respect to earth, and hence the parallel electric field is also vertically oriented. The wave is therefore vertically polarized. If the aerial were turned to a horizontal position, the electric field would turn parallel with it, and the wave would be horizontally polarized. Evidently, therefore, the wave polarization is primarily a function of the aerial orientation. However, the plane of polarization of an electromagnetic wave may actually be turned somewhat from its original direction during its travels through space.

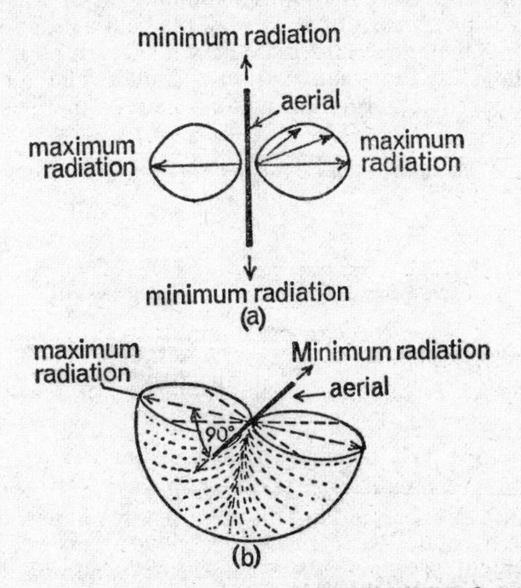

Fig. 139. Radiation pattern of a half-wave aerial: (a) in plane of paper, (b) doughnut pattern in space

Aerial radiation pattern. Note in Fig. 136 that the lines of force of the magnetic field are strongest near the centre of the aerial and weakest near the tips. Maximum radiation of the magnetic field therefore takes place broadside to the plane of the aerial, and minimum radiation occurs from the aerial tips. The effectiveness of the spread of the magnetic field is measured by the radiation pattern of the aerial. Fig. 139a shows the radiation pattern of a half-wave

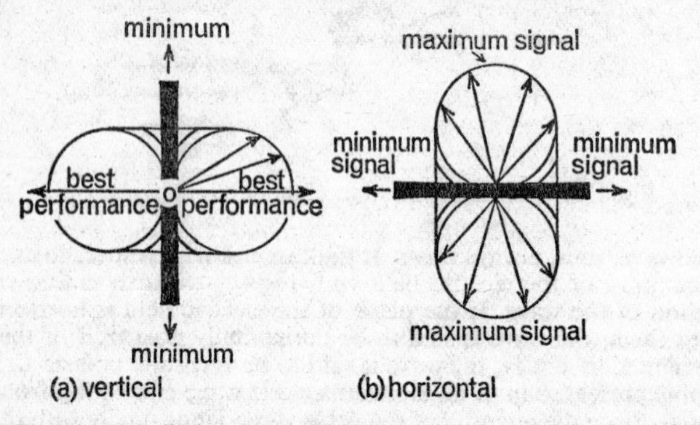

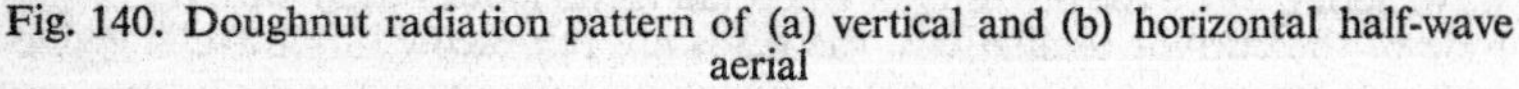
Fig. 140. Doughnut radiation pattern of (a) vertical and (b) horizontal half-wave aerial

aerial in one plane (the plane of the paper). As you can see, it is an 8-shaped pattern, extending from the front and rear of the aerial. The length of a line drawn from the centre to the periphery of the pattern is a measure of the amount of radiation in the direction of the line. The longest line that can be drawn is at right angles (broadside) to the aerial, representing maximum radiation. No line can be drawn along the length of the aerial, where the pattern is tangent to it, and hence the radiation is a minimum or zero in this direction.

The actual free-space radiation pattern of the half-wave aerial extends in all directions around the aerial and is doughnut-shaped, as illustrated in Fig. 139b. Only half of the doughnut pattern is shown. The complete radiation pattern is obtained by rotating the figure-8 pattern of Fig. 139a through a

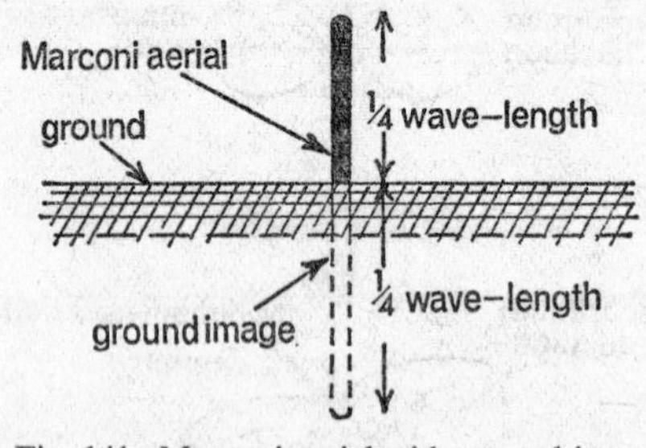

Fig. 141. Marconi aerial with ground image

circle around the axis of the aerial. This pattern may be considered attached to the aerial, since it rotates with it when the aerial is turned. Thus, when the aerial is vertical with respect to the earth, the doughnut-pattern will be horizontal, as shown in Fig. 140a, and maximum radiation takes place in all horizontal directions around the aerial. But if the aerial is turned horizontally the doughnut is vertical, and maximum radiation occurs in a vertical plane perpendicular to the aerial (Fig. 140b).

Marconi or grounded aerial. At low frequencies the length of a half-wave (Hertz) aerial becomes excessively large, as is apparent from our formula for aerial length. Marconi found in his experiments that the length of a Hertz aerial could be cut in half by earthing one end. The resulting Marconi aerial, illustrated schematically in Fig. 141, is a vertical earthed mast equal in length to one quarter-wavelength of the signal it is designed to radiate or receive. The theory of the Marconi aerial is similar to that of the Hertz half-wave aerial, but with the earth serving to provide a 'mirror image' one quarter wave in length (see Fig. 141). Since the aerial is vertical, the radiation emitted from it is vertically polarized.

Other transmitting and receiving aerials. Although little has been said about receiving aerials, in learning about the radiating properties of a transmitting aerial, we have also found out the criteria for a good receiving aerial. In general, a good radiator is a good absorber and, hence, a good transmitting aerial is an equally good receiving aerial. The action of a receiving aerial is, of course, the reverse of that occurring in a transmitting system. A receiving

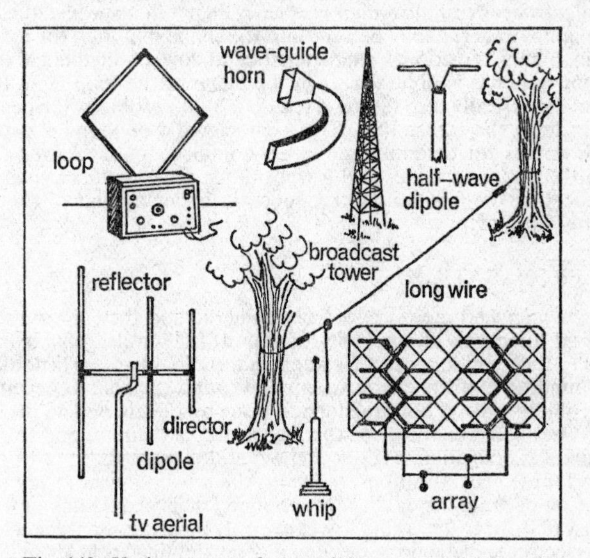

Fig. 142. Various types of transmitting and receiving aerials

aerial intercepts or absorbs a tiny portion of the electromagnetic energy travelling past it, and the resulting electron current along the conductor induces a small voltage of the same waveform as the radio waves. The voltage is passed on to the receiver, where it is amplified and demodulated. In principle, the design and shapes of receiving aerials are the same as those of transmitting aerials, but in practice the receiving aerial often consists only of a convenient length or loop of wire because of space restrictions. In areas close to the transmitter, sufficient voltage pickup is obtained by these makeshifts, but farther away from the transmitter, in the 'fringe areas' of reception, the design of receiving aerials becomes almost as elaborate as that of the transmitting aerial.

There are many types of transmitting and receiving aerials in additional to the half-wave (Hertz) and the quarter-wave (Marconi) systems we have discussed. The size and shape depends on many factors, such as the frequency of the signal, the required 'directivity' of the radiation pattern, the use to which the aerial is to be put, etc. Fig. 142 illustrates a variety of typical transmitting and receiving aerials.

The simplest type of aerial is a long wire, insulated from the ground and supported by trees or poles. The wire is tapped near one end and connected by a lead-in wire to the transmitter or receiver. Vertical quarter-wave grounded (Marconi) aerials may vary in size from a large broadcasting tower to a small mobile 'whip' mounted on a vehicle. Half-wave (Hertz) aerials, called dipoles, also come in many varieties from the simple two-conductor, centre-fed type we have studied, to sophisticated structures. Television aerials may consist of a 'folded' dipole, a director, and a reflector, which serve to improve the directivity and, hence, pickup from a particular station. Sometimes 'arrays' of half-wave aerials are stacked vertically or horizontally to obtain a specific radiation pattern and improve the 'gain' of the combination over that of a single conductor. An array may take the form of a metal horn, fed by a waveguide, to beam an ultra-high-frequency signal in a specific direction like a searchlight. An aerial may be bent into the shape of a loop (or looped coil) to obtain highly directional characteristics at low or medium frequencies. Maximum response is obtained when the plane of the loop faces the transmitter, and practically zero response (called a 'null') is obtained when the loop is broadside to the transmitter. This characteristic of loops is exploited in direction finders for determining the horizontal compass bearing (azimuth) between the direction finder and a transmitter. Two or more such bearings permit locating the position of the direction finder or that of an unknown transmitter on a map.

WAVE PROPAGATION

We have launched radio waves from an aerial and they are travelling outward in all directions at a velocity of over 186,000 miles per second. What happens to them out in space? We know that radio waves are little affected by the surrounding atmosphere, rain, snow, etc., and are able to penetrate non-metallic objects easily, but are stopped dead by metals or fine meshed wire screens. They seem to travel best in free space, a vacuum, and do not need a medium of propagation, as do sound waves, for example.

Beyond these generalizations, however, there is little we can say about the propagation of radio waves without specifying their wavelength or frequency. You must remember that radio waves are only a part of the tremendous spectrum of electromagnetic wave motion, a spectrum that includes such familiar radiations as heat and light, as well as such mysterious ones as cosmic radiation and X-rays. The electromagnetic spectrum roughly extends from the slow oscillations of electrical power generators (50 c/s), through the audio frequencies (20 to 15,000 c/s), and radio frequencies (about 15 kc/s to 300,000 Mc/s), the infrared or radiant heat region (from about 400,000 Mc/s to about 400 million Mc/s), the visible light region (400–800 million Mc/s, approx.), the ultraviolet region (to several thousand million Mc/s), and X-rays, gamma rays, and cosmic rays, oscillating at frequencies of billions of megahertz, or equivalently, wavelengths of fractions of an Ångstrom (1 Ångstrom = 10^{-8} cm). All these radiations are electromagnetic waves, travelling at 186,000 miles per second, different though their effects may be. In general, as the frequency goes up, the radiations tend to travel more nearly in straight lines, as light does, the effect already being noticeable at the higher radio (microwave)

TABLE 1

THE RADIO SPECTRUM

Frequency Band	Frequency Range	Wavelength Range	Typical Uses
Very low frequency (v.l.f.)	10–30 kc/s	30,000–10,000 m	Long-distance point-to-point communication
Low frequency (l.f.)	30–300 kc/s	10,000–1,000 m	Marine, navigational aids
Medium frequency (m.f.)	300–3,000 kc/s	1,000–100 m	Broadcasting, marine
High frequency (h.f.)	3–30 Mc/s	100–10 m	Communication of all types
Very high frequency (v.h.f.)	30–300 Mc/s	10–1 m	Television, f.m. broadcasting, radar, air navigation, short-wave broadcasting
Ultra-high frequency (u.h.f.)	300–3,000 Mc/s	1 m–10 cm	Radar, microwave relays, short-distance communication
Super-high frequency (s.h.f.)	3,000–30,000 Mc/s	10–1 cm	Radar, radio relay, navigation, experimental
Extremely high frequency (e.h.f.)	30,000–300,000 Mc/s	1–0·1 cm	Experimental

frequencies. The quasi-optical behaviour of these high-frequency radio waves makes it relatively easy to reflect, refract, and focus them in beams by appropriate microwave 'mirrors' and 'lenses'.

Since we are going to talk about radio waves from now on, let us further subdivide the radio portion of the electromagnetic spectrum into its accepted sub-classifications. (See Table 1.)

TYPES OF WAVE PROPAGATION

Depending primarily on its frequency, a radio wave may travel from the transmitting to the receiving aerial in a number of ways. The most important of these are known as the ground or surface wave, the ground-reflected wave, the direct or line-of-sight wave, and the sky wave. Fig. 143 illustrates these main radio wave paths.

Note that the ground wave does not reach the receiving aerial in the location shown since it is propagated for a short distance only. To receive it the receiving aerial would have to be moved much closer to the transmitter. You

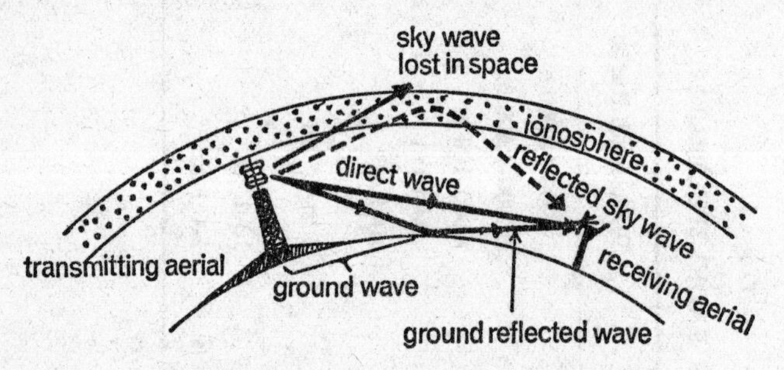

Fig. 143. Main types of radio wave propagation

can also see, in Fig. 143, that a portion of the sky wave travelling at a high angle is lost into outer space, while the portion travelling at a lower angle with respect to the horizontal is reflected back to the receiving aerial by the electrically charged layers of the ionosphere surrounding the earth.

Ground (surface) wave. In ground-wave propagation, the electromagnetic waves are actually guided by the earth travel just above its curved surface from transmitter to receiver. Because the waves hug the earth so closely, they are profoundly influenced by the electrical characteristics of the ground over which they travel. Moreover, since absorption of the waves increases considerably with frequency, the ground wave is generally useful only at low frequencies. Where it is present, however, reception is extremely reliable and not subject to the seasons or atmospheric conditions. Below 500 kc/s, reliable communication can be obtained over distances up to 1,000 miles by the ground wave alone. Amplitude-modulated radio broadcasts in the medium-frequency band are transmitted primarily via the ground wave. But at the higher frequencies used by frequency modulation and television, the ground wave is so reduced in strength by absorption as to virtually become useless beyond an area of a few miles around the transmitter.

Ground-reflected wave. As shown in Fig. 143, the ground-reflected wave travels from the transmitting aerial to the earth (which may be a mountain, for instance) and is reflected to the receiving aerial. Since it is not subject to continuous absorption by the earth, this wave travels considerably farther than the ground wave. The reflected wave exists even at very high frequencies and provides, together with the direct wave, the main means of transmitting f.m. radio and television programmes. This type of wave is reversed in phase (by 180°) at the point of reflection, which occasionally leads to undesirable results. If, for example, the ground-reflected and direct waves have travelled over the same length of path, they will arrive at the receiving aerial 180° out of phase with each other, and almost completely cancel out. In contrast, if the path lengths of the two waves differs by one half-wavelength (180°), the waves will reinforce each other, resulting in increased signal strength. By raising the height of the receiving aerial, the respective path lengths of the waves can be altered, and alternate points of maximum (reinforcement) and minimum (cancellation) signal strengths are encountered; this phenomenon is called selective fading. A similar phenomenon takes place by the interaction of the sky and ground waves.

Direct (*line-of-sight*) *wave*. As its name implies, this wave travels directly in an almost straight line from transmitter to receiver. The path is not completely straight, because of a certain amount of refraction or bending occurring in the earth's atmosphere. For this reason the direct wave travels somewhat farther than the optical line-of-sight path. We therefore speak of a 'radio horizon' over which these waves may be 'seen', to distinguish it from the optical horizon. This is illustrated in Fig. 144. Note that the width of the horizon depends on the relative heights of the transmitting and receiving aerials.

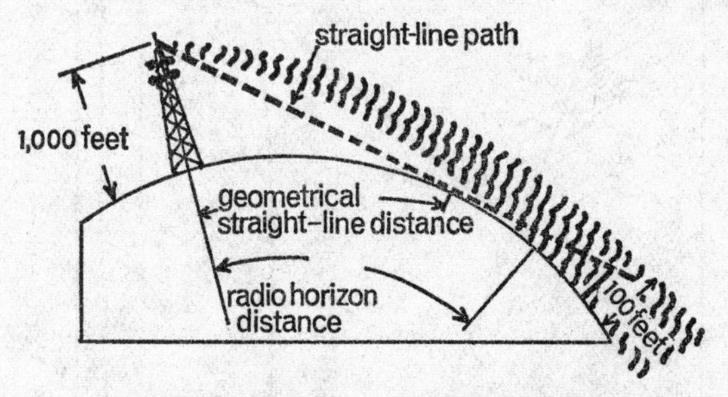

Fig. 144. Line-of-sight and radio horizons

The direct wave becomes increasingly important as the frequency goes up and the radio waves tend to travel more and more in straight lines. Direct line-of-sight transmission commences with the v.h.f. band above 30 Mc/s, approximately. F.M. and television signals are transmitted in this way, with contributions from the ground-reflected wave. U.H.F. transmission takes place exclusively by the direct wave. Thus, radar, microwave relays, air navigation aids, and many other services in the v.h.f. and u.h.f. bands rely solely on direct line-of-sight transmission.

The sky wave. Transmission by means of sky waves is an elusive and tricky business. It is subject to fading and erratic changes with the seasons, night and day, and atmospheric conditions. Yet it is the primary means of around-the-world short-wave communication and, hence, quite important. Let us see why the sky waves are so unreliable.

The fact that makes long-distance transmission via sky waves possible is the existence of a series of electrically ionized layers in the earth's upper atmosphere, called collectively the ionosphere. These act as a 'radio mirror' to

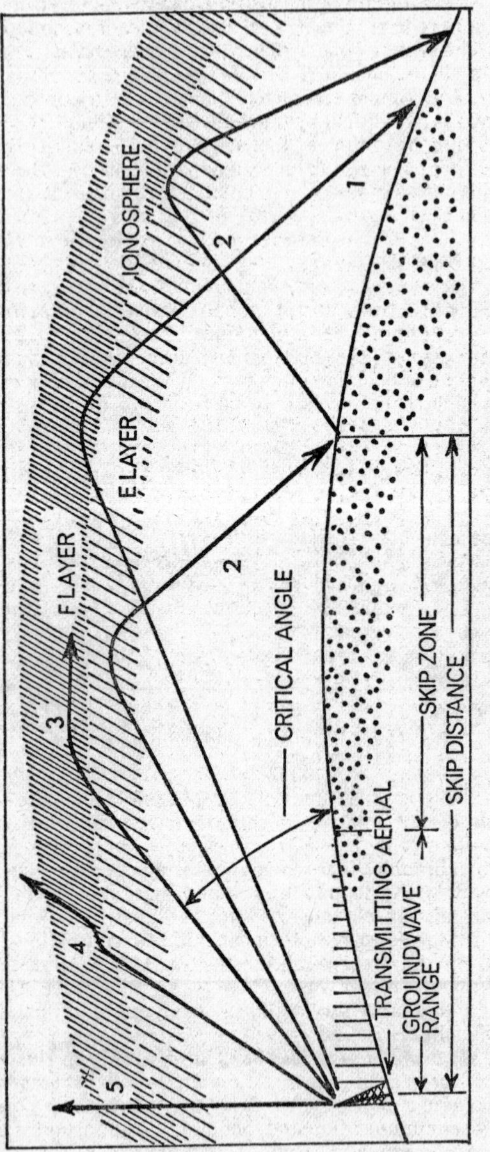

Fig. 145. Various sky-wave transmission paths

bounce the sky waves back to earth. (See Fig. 143.) The layers of ionized particles are created by various strong radiations from the sun and are therefore subject to its disturbances, as well as atmospheric conditions.

Although there are at least four distinct ionized layers in the ionosphere, the most important of these are the *E* and *F* layers. The *E* layer, sometimes called Kennelly–Heaviside layer after its discoverers, extends from about 55 to 90 miles above the earth's surface, its height varying somewhat with the season of the year. The *E* layer makes possible transmission of signals above 20 Mc/s (approximately) over distances of up to 1,500 miles. During the night the *E* layer disappears almost completely, making it useless for high-frequency (short-wave) radio communication. The highly ionized *F* or Appleton layer exists at heights from about 90 to 250 miles above the earth's surface and is the most useful for long-distance transmission via sky waves. Depending on season and time of day, frequencies from 2 to 30 Mc/s may be transmitted around the world through the *F* layer.

The actual mechanism of sky-wave transmission is illustrated in Fig. 145. Here the transmitting aerial is seen to radiate sky waves over a wide range of vertical angles with respect to the earth. Sky wave 1, radiated at a small vertical angle, is reflected from the *F* layer and returns to earth at a great distance from the transmitter. Sky wave 2 leaves the transmitting aerial at a somewhat greater vertical angle and is reflected back to earth by the lower *E* layer. Since it enters the ionosphere sooner, it is reflected back more quickly, at a shorter distance from the transmitter.

The distance between the transmitting aerial and the point where the sky wave is first received after returning to earth is called the skip distance. Near the transmitter, the signal is heard, of course, via the ground wave. As you leave the ground-wave range, however, you encounter a zone of silence, called the skip zone, where no signal at all can be picked up. If the sky-wave signal is sufficiently strong after returning to earth it may be reflected by the earth and again by the ionosphere, thus returning to earth in a series of hops, separated by the skip distance. The signal may be transmitted in this way one or more times around the world. Since it takes about 1/7th second for a radio signal to travel round the earth, you would then hear a series of echoes each spaced 1/7th of a second, depending on how many trips around the earth the signal completes.

As the vertical angle of the sky wave with respect to earth is increased, the ionospheric layers are no longer capable of reflecting the energy back to earth, as shown by sky waves 3, 4, and 5. The angle above which the sky wave no longer returns to earth, but travels outward into space, is known as the critical angle of radiation. Sky waves at or above this angle may be refracted (bent) by the ionosphere but they are not reflected back to earth. The critical angle depends primarily on the density of ionization and on the frequency of the signal. As the frequency increases, the ionosphere becomes progressively less effective and lower and lower angles of radiation must be used to reflect the signal back to earth. At frequencies above about 30 Mc/s the signal is no longer returned even for the smallest vertical angle, and the ionosphere has little effect upon it. The exact frequency at which this occurs depends on the density of ionization, which is affected, in turn, by sun-spot cycles, daily and seasonal variations.

If you realize that the actual number of ionized layers, their heights and density of ionization vary from hour to hour, day to day, month to month, season to season, and from year to year, you will no longer wonder why sky-wave transmission is less than reliable and why sudden 'fadeouts' and disappearances of signals occur.

Scatter propagation. Though we have stated that signals in the v.h.f. and

u.h.f. bands (above 30 Mc/s) penetrate the ionosphere and can be received only via the direct wave over a line-of-sight distance, exceptions to this have been reported many times in recent years. Continuing research after the Second World War revealed that consistently strong reception of v.h.f. and u.h.f. signals far beyond the radio horizon (trans-horizon) is possible. Although not much is known about this phenomenon, it is generally explained by the presence of minute disturbances in the electrical properties of the atmosphere. These disturbances result in a forward scattering of high-frequency radio waves, and since disturbances are always present, consistent trans-horizon reception is possible at all times. A new form of high-power signal transmission, based on this scatter propagation, is now in use for long-distance v.h.f. and u.h.f. communication.

RADIO RECEIVERS

We have followed the radio signal from the transmitting aerial through the radio path to a receiving aerial somewhere within radio range of the transmitter. The receiver with its aerial must intercept a portion of the electromagnetic energy and recover the information contained in it. Regardless of the type of signal or modulation, every radio receiver must perform the following six functions, illustrated in Fig. 146:

1. The receiving aerial must intercept a portion of the passing radio waves.
2. Tuned (resonant) *L–C* circuits must select the desired signal from the mass of radio signals intercepted by the aerial.
3. The weak signals from the aerial must usually be amplified by one or more tuned radio-frequency amplifiers, before they can be demodulated.
4. The original intelligence (modulation) imposed on the r.f. carrier wave must be recovered in the form of an audio-frequency signal by a detector or demodulator circuit. For frequency-modulated waves, limiting of the carrier amplitude variations is usually also necessary. In the case of c.w. (radiotelegraphy) reception, an additional beat-frequency oscillator is required to make the carrier-wave interruptions audible.
5. The audio signal from the detector must be strengthened by one or more stages of audio-frequency amplification to a level sufficient to operate a set of headphones or a loudspeaker.
6. The amplified audio-frequency signal must be converted into corresponding sound waves by an electromechanical reproducer, such as a loudspeaker or headphones.

AMPLITUDE-MODULATED (A.M.) RECEIVERS

Although the basic functions of any radio receiver are the same, certain differences in the receiving circuits make it convenient to consider receivers for different types of modulation separately. Let us take the standard a.m. receiver first; we can then point out significant differences in f.m. and c.w. reception.

THE CRYSTAL RECEIVER

Though it is more a toy than a well-functioning radio receiver, the crystal receiver (Fig. 147) illustrates the basic principles of radio reception and deserves brief consideration.

The signal intercepted by the aerial of the receiver (Fig. 147a) is coupled to

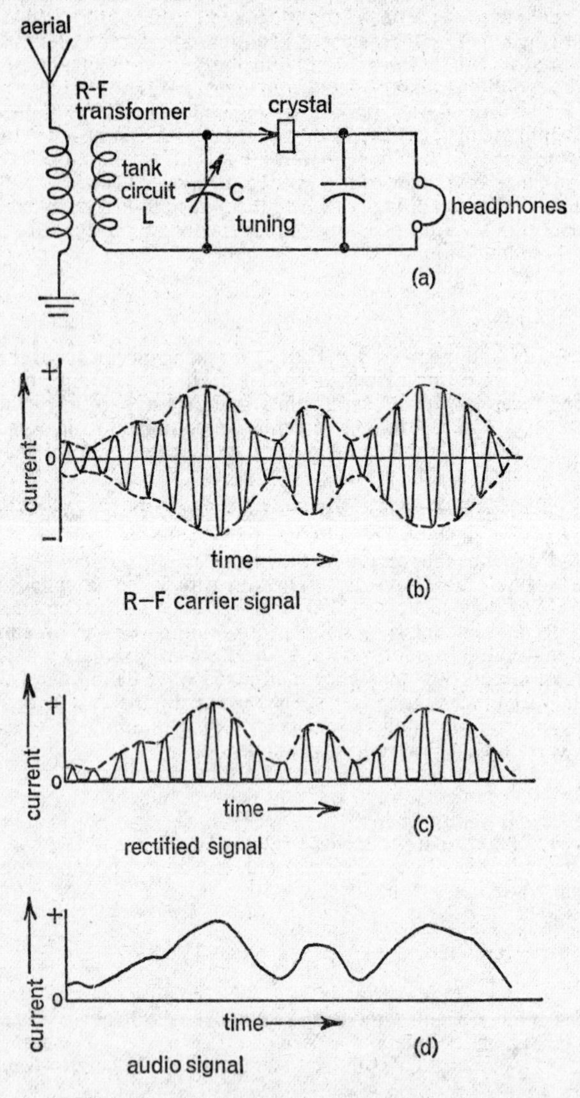

Fig. 147. Circuit and action of crystal receiver

the *L–C* tank circuit through a step-up radio-frequency transformer, which provides some amplification. Signal selection takes place in the *L–C* tank circuit, which is tuned to the frequency of the desired radio-frequency carrier wave (Fig. 147b). The signal is rectified by the crystal (Fig. 147c) which permits current to pass in one direction much more readily than in the other. The crystal is usually some mineral, such as galena, silicon, or carborundum. An elementary filter, consisting of a by-pass capacitor, removes the remaining r.f. current pulses and keeps them out of the headphones. The resulting audio-frequency variations (Fig. 147d) are translated by the headphones into sound waves corresponding to the original information at the transmitter. Since a crystal receiver is not sufficiently selective to discriminate well between adjacent r.f. carrier signals and does not provide sufficient amplification to operate a loudspeaker, it is rarely used today.

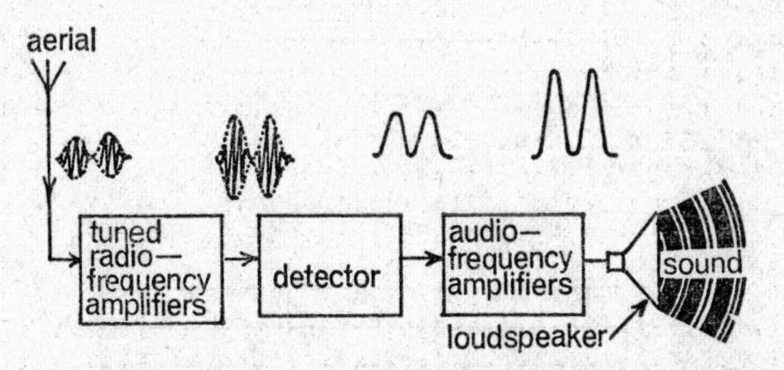

Fig. 148. Block diagram of a t.r.f. receiver and its waveforms

TUNED RADIO-FREQUENCY RECEIVER

The simplest type of a practical receiver for amplitude-modulated radio waves is the tuned radio-frequency (t.r.f.) receiver. It consists of several stages of radio-frequency amplification with tuned (*L–C*) tank circuits at the input of each stage, an a.m. detector, and one or more audio-frequency amplifier stages driving a loudspeaker. A block diagram of a typical t.r.f. receiver with the waveforms of each stage is shown in Fig. 148.

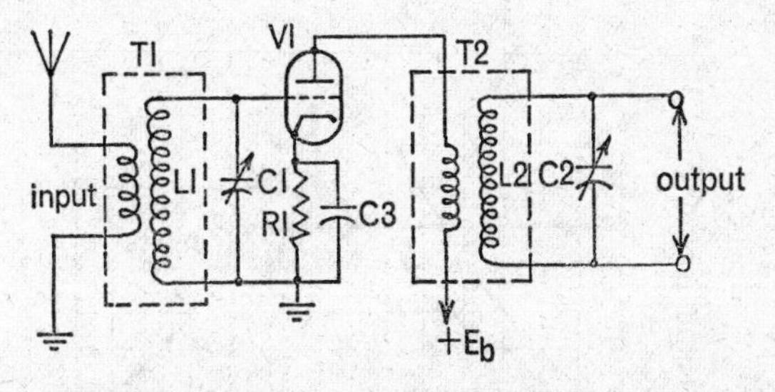

Fig. 149. Tuned r.f. input stage of a t.r.f. receiver

The signal intercepted by the receiving aerial, portrayed as an r.f. carrier modulated by two cycles of an audio tone, is selected and amplified by several stages of tuned radio-frequency amplification. The amplified r.f. signal is demodulated in the detector stage, so that only audio-frequency variations of

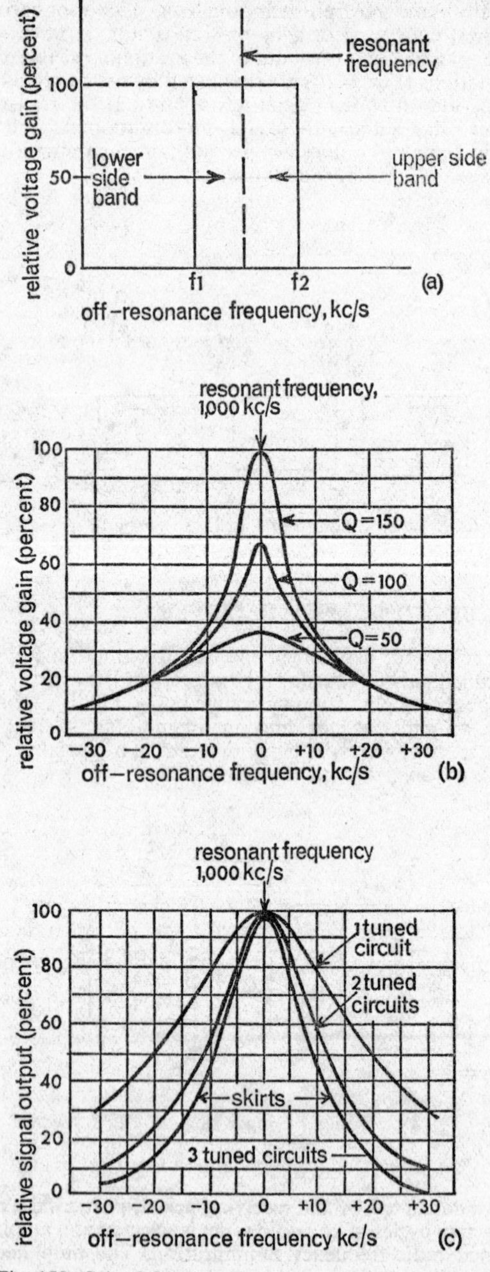

Fig. 150. Selectivity characteristics of a t.r.f. receiver

the carrier remain. The detector may be of the grid-leak, or infinite-impedance type discussed in Chapter 13. (Diode detectors are not generally used in the low-gain t.r.f. receiver because of their insensitivity.) The a.f. signal output from the detector is amplified by one or more stages of audio amplification until it is sufficiently strong to drive a loudspeaker. The loudspeaker (or head-phones) converts the audio variations into sound waves.

Selectivity of tuned r.f. amplifiers. Fig. 149 illustrates a triode input stage of a typical t.r.f. receiver. A triode is shown for convenience, though the higher-gain pentodes are usually used. (See also Fig. 100.)

In Fig. 149, the signals intercepted by the aerial are coupled to the tuned-grid tank circuit, L_1–C_1, through r.f. transformer T_1. Since the tank circuit is resonant at the desired carrier frequency, it strengthens the r.f. carrier currents while partially suppressing all other frequencies. Equivalently, the impedance of the L–C tank circuit is high at the desired r.f. carrier frequency, thus building up a large voltage at the grid of the triode, while the impedance is low for other frequencies, which are therefore by-passed to ground.

The signal selected by the L_1–C_1 tank is amplified in the anode circuit and is coupled to the input tank circuit, L_2–C_2, of the next stage through r.f. transformer T_2. Tank circuit L_2–C_2 is tuned to the same frequency as L_1–C_1 and discriminates further against unwanted, adjacent carrier signal frequencies. The stage is operated Class A for minimum distortion. It obtains grid bias through cathode resistor R_1, which is by-passed for radio frequencies by capacitor C_3.

So much for the physical action of the t.r.f. amplifier. You may wonder why a tuned tank circuit is needed at the input of each stage, since one should be enough to select the desired frequency. The reason for this is to obtain sufficient selectivity; that is, the ability to select one frequency and suppress all others. One tank circuit just cannot do the job. You will recall from our discussion in Chapter 12 that it is the sharpness of the resonance curve, which makes a tuned circuit selective. The sharpness of resonance, in turn, is determined by the Q-factor of the tank circuit, which is the ratio of the coil reactance to the resistance of the tank circuit at the resonant frequency. The higher the Q of the tuned circuit, the sharper is its resonance curve and, hence, the more peaked is the response to the desired frequency. (See Fig. 150b.) A sharply peaked response means, of course, that only the selected resonant frequency is amplified and all others are rejected. But there is a limit to the usable sharpness of the resonance curve and, hence, the selectivity, since not only the carrier frequency of the desired signal must be amplified, but a whole band of frequencies—the upper and lower sidebands—that contain the audio modulation. If the response of the tuned circuit is too sharp, a portion of the sidebands containing the audio modulation will be suppressed, and we shall gain selectivity at the price of audio quality (fidelity). Thus, a compromise is necessary.

Ideally, the resonance curve of the tank circuit in an r.f. amplifier should be rectangular, flat at the top with vertical sides, or skirts, as they are called. Such a rectangular selectivity characteristic, illustrated in Fig. 150a, would pass the carrier with its upper and lower sidebands, contained between frequencies f_1 and f_2, and nothing else. Unfortunately, actual tank circuits do not work that way. The response of a real tank circuit, tuned to a carrier frequency of 1,000 kc/s, is illustrated in Fig. 150b for various Q-factors. An arbitrary voltage gain of 100% has been assumed for a Q of 150. Assume further that the desired carrier wave contains upper and lower sidebands of 10 kc/s each and, hence, extends from 990 to 1,010 kc/s. This means that audio frequencies up to 10,000 c/s (10 kc/s) are contained in the transmitted signal.

As shown in Fig. 150b, the selectivity (resonance) curve for $Q=50$ is very broad and discriminates little against unwanted frequencies. The carrier frequency and its two sidebands are easily passed, but so are adjacent channels as much as 30 kc/s away. The curve for $Q=100$ discriminates much more sharply against unwanted, adjacent carrier frequencies, but also cuts down the 10 kc/s sidebands considerably. Finally, the curve for $Q=150$ has a very sharp 'nose' and, hence, provides excellent selectivity against unwanted adjacent channels. The response to an adjacent carrier signal, say, 30 kc/s distant on either side of the resonant frequency (i.e. either at 970 kc/s or 1,030 kc/s) is only about 10% of the resonant response and, hence, the undesired signal will be amplified only one-tenth as much as the desired carrier. This is fine, but what happens to the sidebands of the desired carrier? As is apparent from Fig. 150b, the upper and lower limits of the 10 kc/s sidebands, representing a 10,000 c/s audio tone, are amplified only about 35% of the carrier frequency. This means that the 10,000 c/s audio note would be amplified, roughly, only one-third as much as the carrier and, hence, would sound only about a third as loud as a low-frequency (bass) tone, close to the carrier frequency. (A 100 c/s audio tone, for example, would produce side frequencies on a 1,000 kc/s carrier of 999·9 kc/s and 1,000·1 kc/s; these would be amplified just as much as the carrier signal.) Clearly, we have now attained excellent selectivity with very poor audio fidelity.

The problem of selectivity versus fidelity can be partially solved by adding a number of tuned r.f. stages in cascade. As illustrated in Fig. 150c, each added tuned circuit affects chiefly the sides or skirts of the resonance curve, but does not change much the sharpness (nose) of the curve. As tuned stages are added, therefore, the selectivity curve comes progressively closer to the ideal rectangular curve. By choosing the Q of each tuned circuit properly, the nose of the resonance curve can be made sufficiently broad to pass the two sidebands, while the skirts of the curve may be made sharp by adding tuned tank circuits, so that good selectivity against adjacent interfering signals is attained. Thus, for three tuned circuits, for example, the 10 kc/s sidebands are down to about 50% of the carrier response (Fig. 150c), which can be compensated for by tone controls in the audio amplifier. An interfering adjacent carrier 30 kc/s away from the desired carrier frequency, however, is down to about 2% of the desired resonant response and, hence, is almost completely rejected. By adding still more stages with a lower Q each, equal selectivity with even better audio fidelity could be achieved. Sometimes special band-pass filters are used in place of the conventional tank circuits, to approach more nearly the desired flat-topped, rectangular curve of Fig. 150a.

Sensitivity. The sensitivity of any receiver is the amount of r.f. input voltage needed to produce a specified amount of audio output power. It would seem that any amount of sensitivity could easily be obtained by simply adding more r.f. amplifier stages. But this is not so, because of the presence of noise. Noise is any undesired electrical disturbance within the desired frequency band. Noise is already picked up by the aerial in the form of atmospheric disturbances, called static, and man-made electrical interference, produced by a variety of electrical and electromechanical devices. More noise is added chiefly by the first stage of the receiver in the form of power-supply hum (the ripple), random fluctuations in the emission of valves (called shot effect) and random thermal motions of free electrons in conductors, resistors, etc. (called thermal agitation). These external and internal receiver noises are amplified by all the stages of the receiver along with the desired signal and eventually tend to drown out and mask the signal.

It is evident, therefore, that the usable sensitivity of a receiver is not determined by the absolute value of the signal strength at the receiver input, nor by the number of stages of amplification, but rather by the ratio of the signal strength to all noise present at the input of the receiver. It is this signal-to-noise ratio at the input of a receiver, therefore, that limits its maximum usable sensitivity. Little can be done to improve the signal-to-noise ratio for a given receiver circuit or system of modulation. There is, obviously, no point in amplifying the signal with the noise beyond the point, where the noise itself becomes objectionable, since this will only make the signal (plus noise) louder, but not more intelligible. However, since noise is more or less uniformly distributed over the entire frequency spectrum, the noise pickup can be reduced by limiting the bandwidth and hence audio fidelity passed by the receiver. In communication-type receivers, where audio quality is of secondary importance, the bandwidth is generally cut down to the bare minimum for acceptable intelligibility. It now becomes evident, why frequency modulation—which does not respond to amplitude noises—is such a boon compared with a.m. reception. The signal-to-noise ratio and, hence, sensitivity of f.m. receivers is far greater than that of a.m. receivers.

The number of stages that can be used in a t.r.f. receiver is also limited by possible instability. Since the entire amplification takes place at the same frequency, the slightest amount of coupling between input and output stages (through the power supply, for example) leads to large amounts of regenerative feedback. If sufficiently amplified, this regenerative (positive) feedback results in instability and oscillations. In general, a t.r.f. receiver performs well only on a single low- or medium-frequency band, such as the broadcast band. It is not suitable for use at high frequencies or as a multiband receiver, since the amplification and selectivity of r.f. stages fall off rapidly at higher frequencies. We shall see in the following paragraphs how the superheterodyne receiver overcomes some of these limitations.

SUPERHETERODYNE RECEIVERS

The superheterodyne receiver, invented during the First World War by Major Armstrong, achieves its unique advantages over the t.r.f. receiver by converting all incoming carrier frequencies to a fixed, lower value (the intermediate frequency). At this fixed intermediate frequency, the amplifier circuits can operate with maximum stability, selectivity, and sensitivity and are not subject to the variable amplification and instability of the t.r.f. circuit. The conversion of the received signal frequency to the intermediate lower frequency (i.f.) is achieved by heterodyning or beating the carrier frequency against a locally generated frequency.

A block diagram of the basic superheterodyne circuit is shown in Fig. 151. As in the t.r.f. receiver, the modulated r.f. carrier signals intercepted by the aerial are coupled to a tuned r.f. stage, where the initial signal selection and amplification takes place. Because of the inherently high gain (sensitivity) and selectivity of the superheterodyne circuit, the input r.f. stage is sometimes untuned (i.e. without an *L–C* tank circuit) and, occasionally, it is omitted altogether.

The amplified r.f. signal is coupled to the input of a mixer stage, where it is combined with the output of a local oscillator. The two frequencies beat together in the mixer, as we shall see presently, and generate an intermediate frequency equal to the difference between the r.f. signal and local oscillator frequency. The frequency of the local oscillator may be either above or below the r.f. signal frequency by the amount of the desired intermediate frequency. (In practice, it is usually above the signal frequency.) Note that the intermediate-frequency (i.f.) signal at the output of the mixer, though lower in frequency, contains the same modulation as the original r.f. carrier signal. (See waveform in Fig. 151.)

The i.f. signal from the mixer is then amplified by several stages of fixed-tuned i.f. amplification (see Fig. 101) and is coupled to the input of a detector stage, where it is demodulated. Since the circuit has plenty of gain, a diode detector is usually used because of its low distortion and excellent audio fidelity. The audio signal from the detector is amplified by one or more stages of audio amplification until it is sufficiently strong to drive a loudspeaker. The loudspeaker converts the audio signal into sound waves corresponding to the original sound at the transmitter. (We shall have more to say about loudspeakers in the next chapter.)

Frequency conversion (*heterodyning*). If you have ever listened to the throbbing, pulsating sound emanating from two tuning forks (or organ pipes) differing slightly in frequency, you know what beats or heterodyning sound like. The pulsations or beats, in this case, are caused by alternate destructive interference and reinforcement between the two tones. Since the notes differ only slightly in frequency, their waves will be alternately in phase and out of phase with each other. Whenever they are in phase, the notes reinforce each other and the sound is strengthened; whenever they are out of phase, the sound waves tend to cancel each other, and the sound is weakened. The in-phase and

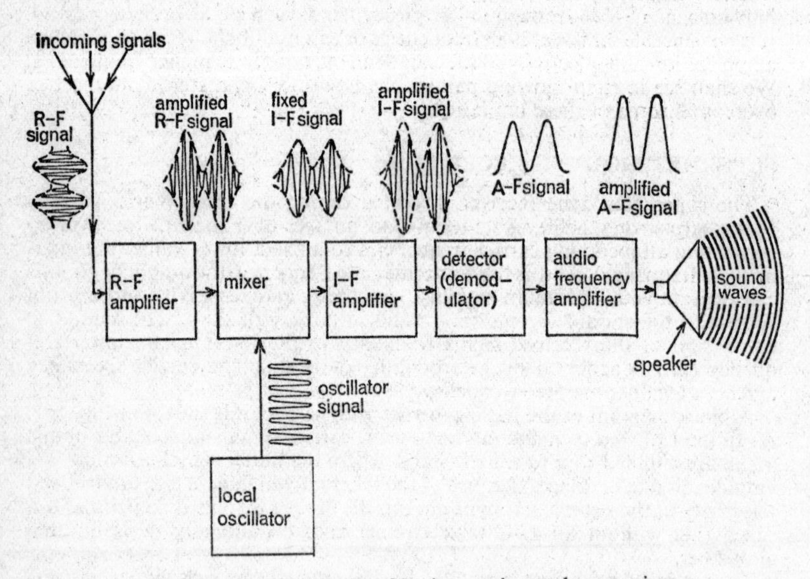

Fig. 151. Block diagram of basic superheterodyne receiver

out-of-phase conditions take place at the difference between the frequencies of the two tones. Thus, a beat frequency tone is heard equal to the difference between the two frequencies.

The same principle of beats is utilized to convert a high-frequency wave into a lower-frequency one. It does not matter whether the high-frequency signal carries amplitude or frequency modulation. The process of receiving a modulated r.f. signal by the heterodyne (beat) principle is illustrated in Fig. 152.

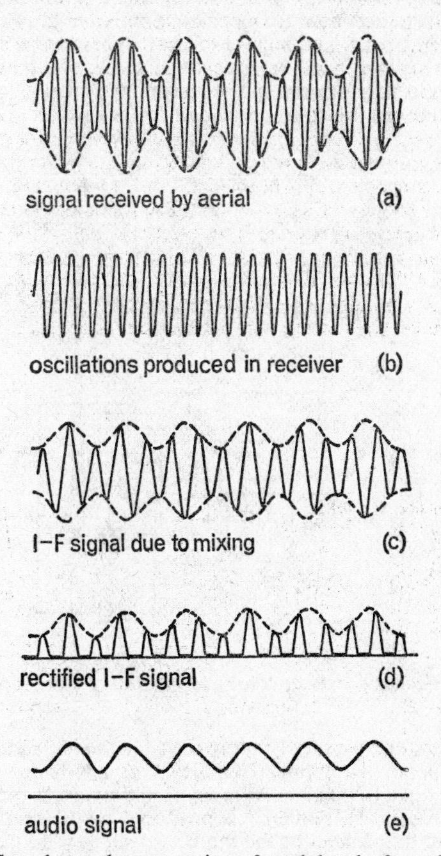

Fig. 152. Superheterodyne reception of modulated r.f. carrier signal

Part (a) of Fig. 152 shows the r.f. carrier signal, at a frequency of perhaps 1,500 kc/s, at the aerial. A local oscillator in the receiver generates unmodulated oscillations at a frequency of, say, 1,955 kc/s, as shown in (b). These heterodyne or beat together in the mixer stage to produce an intermediate or difference frequency of 1,955–1,550, or 455 kc/s (c). (A local oscillator frequency of 1,045 kc/s would have produced the same intermediate frequency.) Although the signal frequency has now been lowered to 455 kc/s, the same modulation is imposed upon the i.f. carrier as on the original r.f. signal, as shown by the peak amplitudes of the i.f. signal in Fig. 152. The i.f. signal is then amplified by several stages, rectified (d) and filtered in the detector, and the resulting audio signal (e) is amplified to drive a speaker. It is of interest to note that the process of heterodyning is essentially the same, except for the different waveforms involved, for the reception of amplitude-modulated, frequency-modulated, or continuous waves.

Frequency-converter circuits. As we have seen in the case of non-linear (harmonic) distortion and amplitude modulation, a non-linear characteristic is required to produce new frequencies. Similarly, beat or difference frequencies between the r.f. and local-oscillator signal can be produced only by mixing the two signals in a non-linear device; i.e. one that does not follow Ohm's law. Again valves come to the rescue by providing the required non-linearity in the curved (bottom) portion of their characteristic. To obtain beat

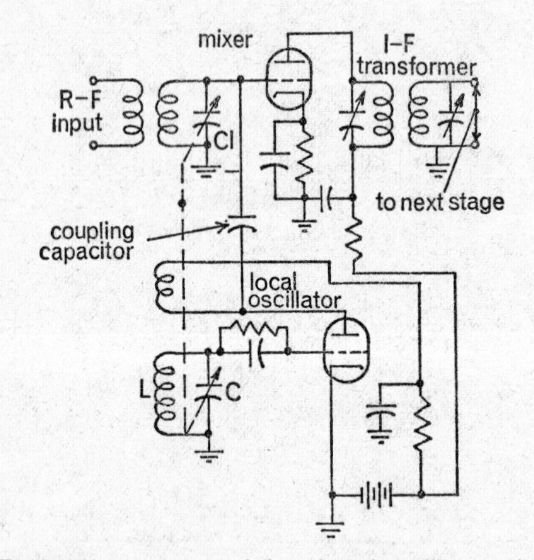

Fig. 153. Frequency converter consisting of triode oscillator and triode mixer

frequencies, therefore, we simply apply the r.f. and local oscillator signals to a mixer valve, operated non-linearly, so that its anode current varies at the desired intermediate frequency. As is so often the case, however, not only the desired difference (beat) frequency is produced by the non-linear mixer, but also the original frequencies at the input, as well as a frequency equal to the sum of the originally applied frequencies. These undesired frequencies must be suppressed by tuning the output of the mixer to the intermediate (difference) frequency only.

A typical frequency-converter circuit using a triode mixer and a triode local oscillator is shown in Fig. 153. Ordinary triodes and pentodes make excellent mixers, especially at high frequencies, and they are still used to a considerable extent.

The mixer valve functions as an ordinary anode detector biased approximately to cut-off; i.e. in the non-linear portion of its characteristic. (The use of this simple detector circuit in the early days led to the name of first detector to distinguish it from the later demodulator stage, often called second detector.) The input tank circuit of the mixer is tuned to the frequency of the incoming signal, while the oscillator grid tank circuit is tuned above (or below) the r.f. signal frequency by an amount equal to the desired beat or intermediate frequency. In a.m. superheterodynes, this is between 455 and 470 kc/s. The i.f.

transformer in the anode circuit of the mixer is then tuned to the desired intermediate (difference) frequency. The oscillator and mixer tuning capacitors are 'ganged together', as shown, to permit tuning by a single knob.

A conventional tickler feedback oscillator is shown in Fig. 153. Any of the other types of oscillators discussed in Chapter 11 could be used equally well as local oscillator, since little power is required. Oscillations from the anode circuit of the local oscillator are coupled to the grid of the mixer, in this case, by means of a coupling capacitor, a method known as grid injection. Any

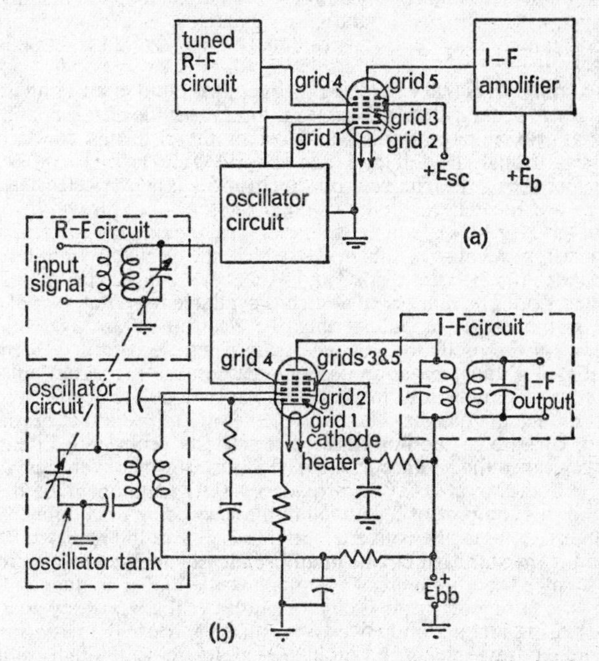

Fig. 154. Pentagrid converter: (a) block diagram; (b) circuit

convenient means can be used, however, to couple the local oscillator voltage to the mixer valve. The oscillator voltage may be inductively coupled (by means of a coil) to the cathode circuit of the mixer, a popular method known as cathode injection, or the oscillator grid tank can be inductively coupled to the mixer input tank circuit.

Separate triode or pentode mixers and oscillator perform very well and have an excellent signal-to-noise ratio, which makes them suitable for use at high frequencies. Triode and pentode converters have the disadvantage, however, of producing undesirable coupling between the mixer and oscillator portions, which may result in the oscillator locking on to some strong interfering signal. Special valves with five grids, called pentagrid tubes, have been developed to provide the necessary isolation between the oscillator and mixer portions of the converter. In the pentagrid mixer, isolation is achieved by means of two independent control grids, one for the r.f. signal and one for the local-oscillator signal. A separate triode or pentode is used in the local-oscillator circuit. A more elegant solution is the pentagrid converter, which combines the functions of the local oscillator and the mixer within a single pentagrid tube; coupling between the two circuits is accomplished by means of the common electron stream. (This is similar to the electron-coupled oscillator discussed in Chapter 11.) A block diagram and actual circuit of a pentagrid converter are shown in Fig. 154.

The cathode, grid 1 and grid 2 of the pentagrid converter are connected to an ordinary triode tickler feedback oscillator. Grid 1 functions as oscillator control grid, and grid 2 as oscillator anode. The local-oscillator frequency is controlled by the variable capacitor in the oscillator tank circuit. This capacitor is ganged with the input signal tuning capacitor in such a way that it maintains a constant frequency difference equal to the desired intermediate frequency.

The incoming r.f. signal is selected by the tuned input circuit and is applied to grid 4 of the pentagrid valve, so that it modulates the electron stream. The action is approximately as follows. Most of the electrons emitted by the cathode pass through the oscillator 'anode' (grid 2) and go on to the screen grid (grid 3), which accelerates the electron stream and acts as an electrostatic shield. Again, most of the electrons pass through the screen and travel towards the r.f. signal grid (grid 4), which is biased negatively through the self-bias cathode resistor. Because of this negative bias, the electrons are retarded and form a space charge between grids 3 and 4 that constitutes a virtual cathode for the mixer section. The number of electrons available from this virtual cathode depends on the oscillator pulses; thus the electron stream available to the mixer varies at the oscillator frequency. Since the electron flow between the r.f. signal grid 4 and the anode depends also on the r.f. signal voltage, the electron current actually arriving at the anode is modulated by both the oscillator and r.f. signal voltages. The two voltages are mixed in the output of the valve and, because of the non-linear characteristic, sum and difference frequencies appear in the anode current. The i.f. transformer in the anode circuit is tuned to the correct difference-frequency (i.f.) component of the anode current and this component is coupled to the succeeding i.f. amplifier.

Although the pentagrid converter performs well in the medium-frequency range, and is the standard circuit in all broadcast superheterodyne receivers, its performance becomes increasingly poor at the higher frequencies. This is caused by the falling off in the oscillator output as the frequency goes up and by an increasing interaction between the mixer and oscillator portions. For this reason, separate mixers and oscillators are preferred at high frequencies.

Automatic volume control (*a.v.c.*). In a modern radio receiver, a gain or volume control is provided in the output of the detector to permit varying the volume of the incoming signal. Once the volume control has been set, the output of the receiver should remain constant regardless of variations in signal strength, fading, etc. These signal variations can overload the r.f., i.f., or detector stages and lead to distortion in the signal, in addition to volume variations. The automatic volume control (a.v.c.) circuit overcomes these deficiencies and maintains approximately constant loudspeaker volume.

Automatic volume control can be achieved by very simple means. The r.f. and i.f. stages of a superheterodyne receiver utilize variable-mu (remote-cut-off) pentodes, whose gain can be controlled by varying the grid bias. If we make the grid bias of these r.f. and i.f. stages more negative when the signal strength increases, thus cutting down the gain, and less negative when the signal strength is fading, we will attain more or less constant output volume, regardless of signal strength. All that is needed, therefore, is a negative bias voltage whose value is controlled by the signal strength. As we have seen in Chapter 13, the load resistor of a diode detector is an excellent source of this variable voltage, since the rectified signal voltage will increase and decrease with variations in the signal strength. A typical diode detector with an automatic volume control circuit is shown in Fig. 155. Manual volume control is also provided.

As in the conventional diode detector, the signal is rectified by the diode and the rectified current flows through the load resistor, R, causing a voltage drop with the polarities indicated. The load resistor is in the form of a potentiometer, so that a portion of the rectified voltage can be tapped off and fed to the first audio amplifier stage through capacitor C_o. This provides manual volume control.

The output from the negative end of the load resistor is fed through filter R_1–C_1 to the grids of the controlled r.f. and i.f. stages and provides variable bias in accordance with the signal strength. The filter R_1–C_1 removes the audio-frequency conponent of the signal, so that the a.v.c. voltage cannot follow the audio modulation of the signal, which would result in distortion. The a.v.c. thus follows only the slower variations in signal strength due to fading and other causes.

The a.v.c. diode is usually combined with the regular diode detector in the envelope of a single double-diode valve and sometimes the first audio amplifier is thrown in also by using a double-diode-triode valve. Since some a.v.c.

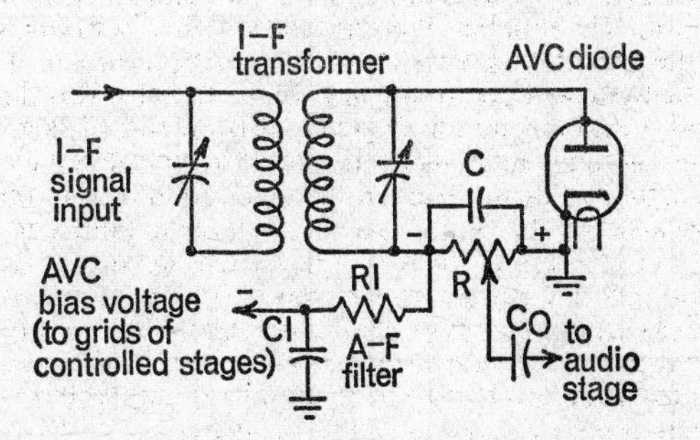

Fig. 155. Automatic volume control (a.v.c.) circuit

bias is developed even for weak input signals, where maximum amplification is desired a special delayed-a.v.c. circuit is occasionally employed, which prevents the application of the a.v.c. bias until the signal strength exceeds a certain predetermined value. This necessitates another diode.

Complete superheterodyne receiver. A complete superheterodyne receiver, typical of those found in almost every home, is shown in Fig. 156. This is a high-quality circuit because of the presence of push-pull audio amplification and an initial r.f. amplifier; both features are frequently omitted in cheap sets.

Although the circuit diagram at first glance looks rather forbidding, a closer inspection reveals only familiar circuits. The receiver is seen to consist of a stage of tuned r.f. amplification; a pentagrid mixer and local Hartley oscillator; one stage of i.f. amplification; a combined detector, a.v.c., and first audio amplifier stage (V4); a push-pull audio power amplifier, and a loudspeaker. The set is operated from the a.c. mains through a full-wave rectifier power supply with a two-stage capacitor-input filter circuit. The power supply provides the high-voltage d.c. for the anodes and screen grids as well as the low-voltage a.c. for the heaters.

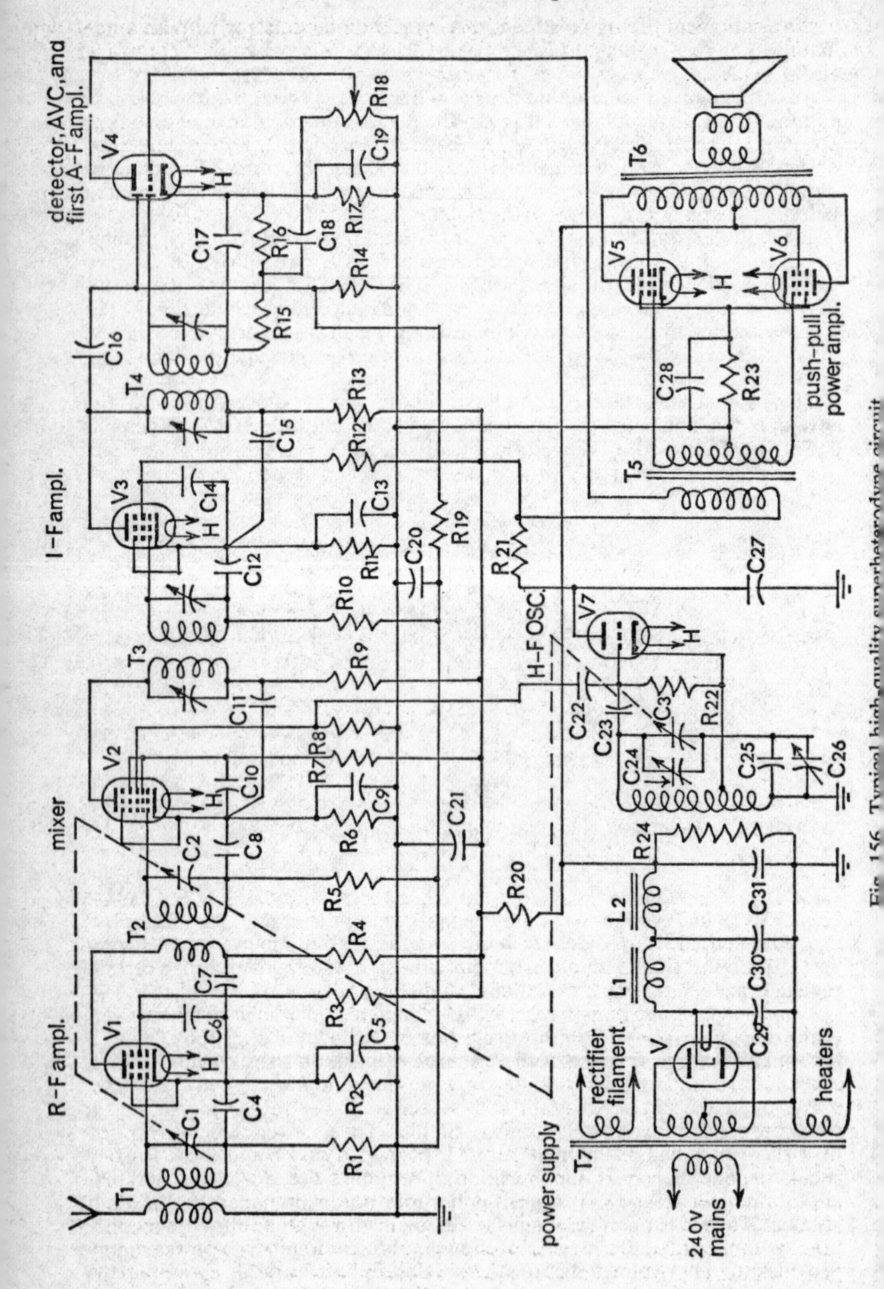

Fig. 156. Typical high-quality superheterodyne circuit

There are no unusual features in the receiver, but some points are worth noting. The tuning capacitors C_1, C_2, and C_3, of the r.f. amplifier, mixer, and local oscillator, respectively, are all ganged together to provide single-control tuning. The small variable capacitor C_{24}—called a trimmer—in parallel with oscillator tuning capacitor C_3 is factory-adjusted or aligned to assure that the oscillator tuning capacitor follows or tracks the r.f. signal tuning capacitors (C_1 and C_2) at exactly the right intermediate frequency, above the r.f. signal frequency. Since this trimmer capacitor affects the tracking chiefly at high frequencies, another preadjusted capacitor, called a series padder, is needed to ensure correct tracking at low frequencies. Padder capacitor C_{26}, in parallel with fixed capacitor C_{25}, fulfils this purpose. Note that the oscillator output signal is injected through capacitor C_{22} into grid 3 of the pentagrid mixer tube, V2.

Double-diode-triode V4 performs the functions of a diode detector, a.v.c., and first triode audio amplifier. The rectified signal voltage from the diode detector is developed across load R_{16} and C_{18}, and is coupled through manual volume control R_{18} to the grid of triode V4. Here it is amplified and then applied to the push-pull power amplifier, which drives the speaker. A separate diode in V4 provides the a.v.c. bias voltage across resistor R_{14}. This voltage is applied, through filter R_{19}–C_{20}, to the grids of the r.f. amplifier (V1), mixer (V2), and i.f. amplifier (V3).

FREQUENCY-MODULATED (F.M.) RECEIVERS

Receivers for frequency-modulated signals always use the superheterodyne circuit because it is the only circuit that provides sufficient gain to boost the amplitudes of weak signals up to the point where they can be maintained constant by the limiter. (The noise-free feature of f.m. is only attained when the carrier amplitudes are limited to a fixed value.) Fig. 157 shows a comparison between an a.m. and an f.m. superheterodyne receiver.

It is evident from Fig. 157 that the a.m. and f.m. superheterodyne circuits are very similar, except for the presence of the limiter and discriminator in the f.m. receiver in place of the conventional detector in the a.m. receiver. As was pointed out in Chapter 13, the discriminator is sometimes replaced by the similar ratio-detector circuit, and the limiter stage may then be omitted. In either case, the functions of the limiter-discriminator or the ratio detector are to maintain the amplitude of the carrier constant, while at the same time converting the carrier frequency variations into equivalent amplitude (audio) variations.

Another significant difference between a.m. and f.m. receivers exists in the i.f. amplifier. The bandwidth of a broadcast f.m. signal is 150 kc/s or more and, hence about ten times as much as that of the a.m. signal. The i.f. amplifier frequency response must be sufficiently wide to pass this 150 kc/s band. Furthermore, since f.m. operates in the 100 Mc/s frequency range, the i.f. frequency is chosen much higher than for a.m.; its value is usually 10·7 Mc/s.

C.W. SUPERHETERODYNE RECEIVER

The a.m. superheterodyne receiver cannot detect the continuous waves (c.w.) of radio telegraphy, because these waves are not modulated. You will recall that c.w. consists of an unmodulated carrier signal which is interrupted in accordance with a telegraphic (Morse) code. To make these continuous waves audible, they must be modulated locally in the receiver with a convenient audio tone. The modulated waves may then be detected by a conventional detector circuit and the interrupted audio tone may be listened to in a headset.

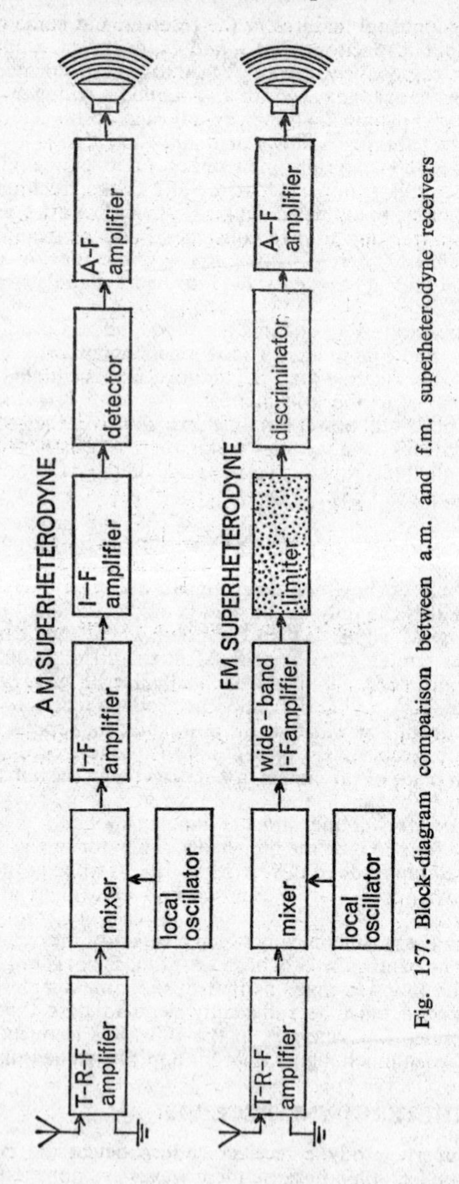

Fig. 157. Block-diagram comparison between a.m. and f.m. superheterodyne receivers

As already indicated in Fig. 146, the addition of a beat-frequency oscillator (b.f.o.) is all that is required to convert an a.m. superheterodyne receiver for the reception of c.w. signals. When the oscillator is switched on with the b.f.o. switch, its output heterodynes with any incoming c.w. signal to produce an audio-frequency beat tone. Any standard oscillator circuit may be used as beat-frequency oscillator. Fig. 158 illustrates a typical Hartley pentode b.f.o.

In operation, the main tuning capacitor, C_4, of the oscillator is preset to the intermediate frequency of the receiver. A parallel trimmer capacitor, C_5, then permits detuning the oscillator frequency over a band ranging from a few

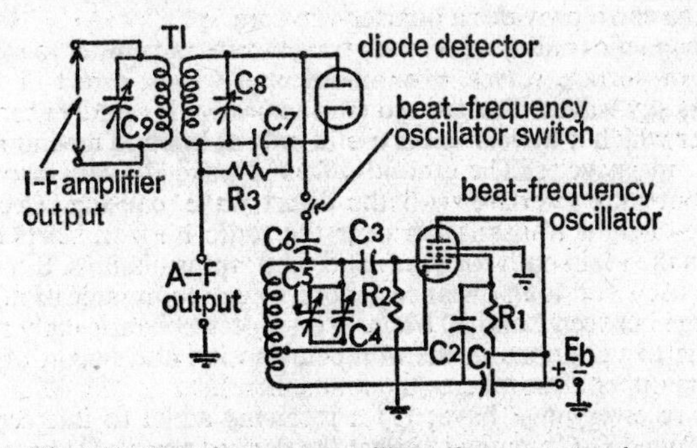

Fig. 158. Beat-frequency oscillator for c.w. reception

hundred to about 1,000 c/s. The output of the beat-frequency oscillator is injected, in this case, through C_6 and the b.f.o. switch into the secondary of the last i.f. transformer (the anode of the detector). Here the off-tune oscillations are mixed with the i.f. current, and the audio-frequency beat note results. Injection of the b.f.o. signal into an earlier i.f. stage is equally suitable.

SUMMARY

The basic elements of a radio communication system are: (1) a radio transmitter to generate radio-frequency waves; (2) a telegraph key or microphone to control the waves in accordance with the information to be conveyed; (3) a transmitting aerial to radiate the waves; (4) a receiving aerial to intercept the waves; (5) a radio receiver to select and amplify the desired signal, and recover (demodulate) the information contained in it; (6) a loudspeaker or headphones to convert the demodulated signal into sound, reproducing the original information.

In c.w. (radio telegraph) transmission, the r.f. carrier wave is periodically interrupted by a key in accordance with a telegraphic (Morse) code. The c.w. signal may be made audible at a distant receiver by converting it to a low-frequency (audio) tone.

Neutralization is used to balance out the feedback produced by the grid-to-anode interelectrode capacitance of a triode.

In amplitude- or frequency-modulated transmitters the required modulation of the carrier wave is obtained by microphones that change the sound waves into corresponding electrical variations.

Accelerated electric charges (electrons) in aerials radiate energy in the form of electromagnetic waves.

A Hertz aerial, or dipole, is a half-wavelength long and is resonant to the r.f. oscillations. The electric field of a half-wave dipole is parallel to the plane of the aerial; the orientation of the electric field with respect to earth is called the polarization of the radio wave.

The radiation pattern of an aerial is a measure of the spread of the magnetic field. The radiation pattern of a half-wave aerial in the aerial plane is a figure 8, with maximum radiation occurring broadside to the plane of the aerial, and minimum radiation taking place along the length of the aerial (from tip to tip). The complete free-space radiation pattern of a half-wave dipole is doughnut-shaped.

The Marconi or grounded aerial is a quarter-wave long and is earthed at one end. The earth provides a quarter-wave image.

Electromagnetic (radio) waves may travel from transmitter to receiver either as ground or surface waves, ground-reflected waves, direct or line-of-sight waves, or as sky waves. The ground (surface) wave is rapidly absorbed by the ground over which it travels and is useful only at low and medium frequencies up to a few megacycles. The ground-reflected wave is useful also at high frequencies, but may interfere with the direct wave, causing selective fading. Direct, line-of-sight transmission over the radio horizon starts at about 30 Mc/s and is the main path for v.h.f. and u.h.f. transmissions. Sky-wave transmission is used for long-distance (short wave) communication in the frequency range between 2 and 30 Mc/s. It is subject to erratic daily and seasonal changes due to variations in the number, density, and height of the ionized layers in the upper atmosphere (the ionosphere).

A radio receiver must have: (1) a receiving aerial to intercept the radio waves; (2) tuned *L–C* circuits to select the desired signal; (3) tuned r.f. amplifiers to amplify the weak r.f. signals; (4) a detector for demodulating the r.f. carrier and recovering the audio signal; (5) an audio-frequency amplifier; and (6) a reproducer.

A tuned radio-frequency (t.r.f.) receiver consists of several stages of tuned r.f. amplification, an a.m. detector, and one or more audio-frequency amplifier stages driving a loudspeaker (or headphones).

Selectivity and audio fidelity are interdependent. The sharper the resonance curve (Q) of the tuned circuit, the greater is its selectivity (to discriminate against unwanted signals), but also the poorer is its sideband (audio) response.

Sensitivity is a measure of the r.f. signal voltage needed to produce a specified amount of audio power. Usable sensitivity depends on the signal-to-noise ratio at the input of the receiver.

The superheterodyne receiver converts all incoming carrier frequencies to a fixed intermediate frequency and thus attains maximum sensitivity, stability, and selectivity. It comprises a tuned r.f. amplifier (usually), a frequency converter—consisting of mixer and local oscillator—an i.f. amplifier, a detector stage, an audio amplifier, and a loudspeaker.

Frequency conversion in a superheterodyne is achieved by beating or heterodyning the output of a local oscillator, tuned above or below the incoming r.f. signal by the difference or i.f. frequency, with the r.f. carrier signal in a non-linear mixer stage. The output of the mixer is tuned to the resulting difference frequency (i.f.).

A pentagrid converter combines the functions of a local oscillator and mixer within a single five-grid valve.

Automatic volume control (a.v.c.) to compensate for varying signal strength is obtained by rectifying the i.f. signal in a diode detector and using the variable bias voltage to control the gain of the r.f., i.f., and mixer stages.

A beat-frequency oscillator (b.f.o.) makes possible superheterodyne reception of c.w. signals by providing an audible beat note.

CHAPTER SIXTEEN

HIGH FIDELITY AND STEREO

High fidelity is the attempt to reproduce by electronic and electromechanical means and without distortion the full range of audible sound, from about 20 c/s (cycles per second) to about 15,000 c/s. A reproducing system is considered 'high fidelity' if it fulfils these conditions and thus comes as close as possible to sounding like the original.

High-quality reproduction of music is not new. The principles of designing high-quality amplifiers and radio receivers have been known for many years, and we have discussed them in earlier chapters. However, there has been a remarkable improvement in recent years in the weakest links of high-fidelity, which are the electromechanical components, such as loudspeakers, gramophone pickups and records, and tape recorders. These improvements have given new impetus to the quest for hi-fi and it might therefore be of interest to explore the subject further.

FUNDAMENTALS OF SOUND

We have repeatedly stated that the range of human hearing extends from about 20 to 15,000 c/s. This is an average figure, since many of us cannot hear sounds much below 50 c/s or above 12,000 c/s. As we get older, our range of hearing is likely to become more restricted, while many children, in contrast, can hear musical sounds up to 18,000 c/s or higher. It is no accident, of course, that the frequency range of musical sounds fits comfortably within the range of human hearing, since musical instruments have been adapted to our ears. Fig. 159 illustrates the frequency range of some musical instruments compared with the range of human hearing.

As you can see, the range of the pipe organ, for example, encompasses the entire range of human hearing, while the piano keyboard has a range of fundamental notes from 28 to 4,186 c/s. Note the stress on 'fundamental'. When you strike the key to sound an A-note, for example, the string will vibrate at a fundamental frequency of 440 c/s, which is the characteristic pitch to which all instruments of an orchestra are tuned. In addition to the fundamental, however, the string will also be excited to 'sympathetic' vibrations or overtones at multiples of the fundamental; that is, an overtone of 880 c/s (called the second harmonic), one of 1,320 c/s (the third harmonic), one of 1,760 c/s (the fourth harmonic), and so on. The fourth harmonic of the 4,186-c/s piano note thus reaches up to 16,744 c/s—the upper limit of human hearing. Although the intensity (loudness) of these harmonics is much less than that of the fundamentals, and varies from instrument to instrument, their presence determines the characteristic quality and timbre of different instruments. A high-fidelity system, therefore, must reproduce the full range of fundamentals and harmonies of all musical instruments.

Loudness. How loud a sound appears to us depends on the intensity of the sound, the sensitivity of our ears, our state of sobriety, interest, attention, and other psychological factors. In other words, the loudness of a sound is a subjective, psychological sensation, depending to some extent on the listener, while the intensity of a sound is an objective indication of its power, measurable by scientific instruments. The distinction is important, since loudness and

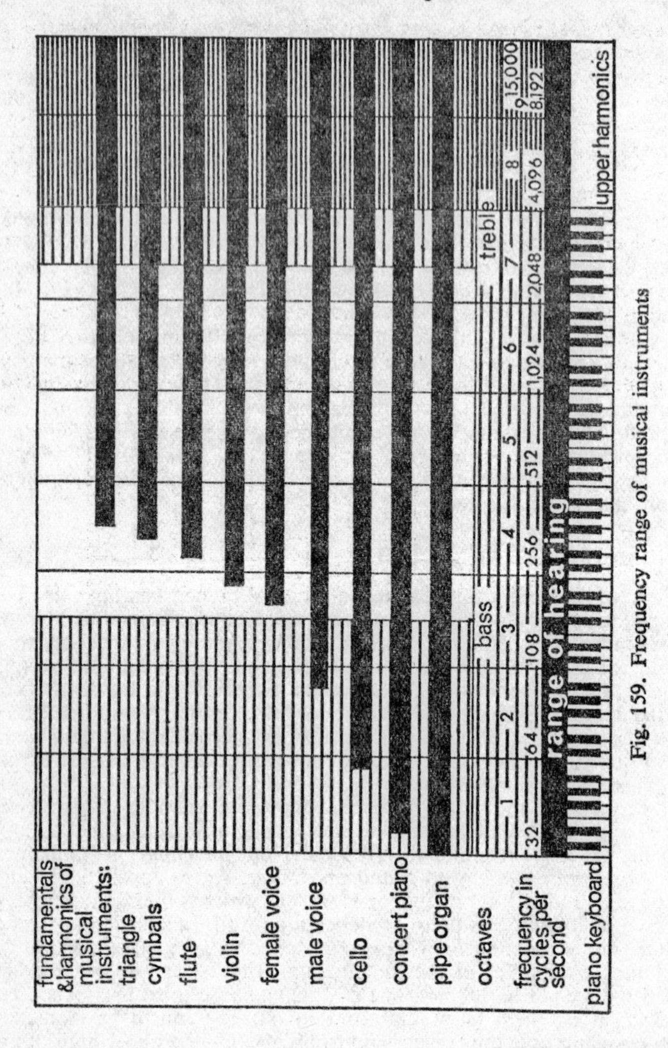

Fig. 159. Frequency range of musical instruments

intensity do not go hand in hand, as it might appear. The human ear is rather slow to respond to increased stimulation. Thus, a sound that is increased in intensity to twice its original value (twice the power) is not heard twice as loud, but seems to have hardly changed in loudness. To make it sound twice as loud, the intensity must be increased ten times. Moreover, to make the sound appear three times as loud as before, the intensity must be increased a hundred-fold over the original value, and to make it appear four times as loud, the sound intensity must be increased a thousand times over the original value. This is a little like lighting candles in the darkness: the first candle brings a flood of light into the dark room; the second candle has much less effect, and you may need ten candles to get the same sensation of increased brightness as was produced by the first.

A relationship between two quantities, where one goes up by jumps of 1, 10, 100, 1,000, etc., while the other (dependent) quantity increased by 1, 2, 3, 4, and so on, is called logarithmic. The response of the ear to increased stimulation is thus logarithmic in nature: to obtain equal increases in loudness of a

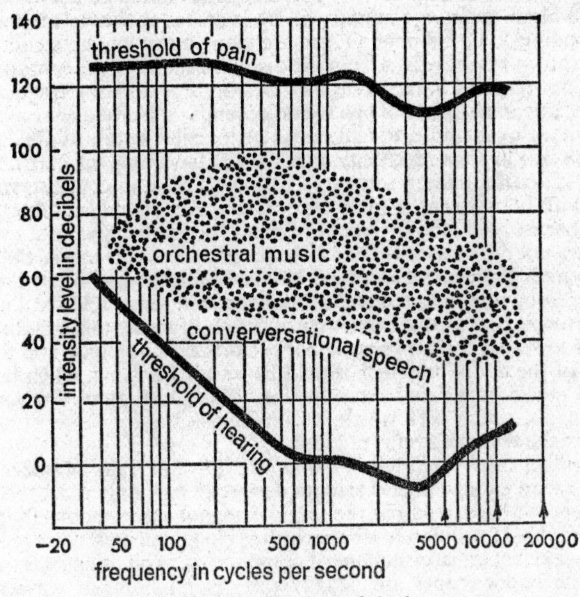

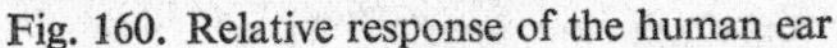

Fig. 160. Relative response of the human ear

sound, its intensity must go up by ratios of ten each time. It is convenient, for this reason, to measure the sound intensity in logarithmic units which go up in equal steps for tenfold increases in intensity. Such a unit is the decibel, which is used to express ratios of intensity and increases in loudness (as well as many other logarithmic quantities). Thus, when the ratio between two sound intensities is ten, they are said to differ by 10 decibels (dB), when it is a hundred, they differ by 20 dB, a ratio of 1,000 equals 30 dB, and so on. Moreover, each 10-dB intensity increase boosts the loudness by a factor of 1.

Fig. 160 illustrates the relative response of the ear at different frequencies and sound intensity levels. The two curves in Fig. 160 tell a big story. The lower curve, entitled 'threshold of hearing', is a measure of the minimum sound intensity required to barely move the eardrum and thus become audible. The upper curve, entitled 'threshold of pain', refers to a sound intensity so great that it causes a tingling or painful sensation in the ear. All sounds lie between these two loudness or intensity limits. The range encompassed by these limits is truly tremendous. You will note that the point where the lower (threshold) curve crosses a frequency of 1,000 c/s has been arbitrarily marked zero decibels. The intensity at which a 1,000 c/s tone becomes barely audible, thus, is our reference or zero-dB level. Since we are interested only in ratios of intensity or loudness, and not in absolute values, this is a logical point for the zero reference level. Note further that to boost the same 1,000 c/s note to the threshold of pain, the intensity of the sound must go up by about 120 decibels. An increase of 120 dB expresses an intensity ratio of a billion (10^{12}) to one between the loudest and softest sound, a fantastic figure. However, since the intensity has gone up by only 12 logarithmic (10-dB) steps, from 0 to 120 dB, the loudness of the loudest possible sound appears only about 12 times as great as that of the softest, barely audible sound.

The two curves of Fig. 160 are also known as loudness contours, because anywhere along a curve a sound will be heard at the same loudness as the corresponding 1,000 c/s note. (There are many more loudness contours than we have shown.) It is evident from the distorted shape of the contours that the ear does not respond in the same way to sounds of differing frequencies and, hence, the apparent loudness of a sound depends on its frequency as well as on its intensity. For example, to make a 40-c/s tone just barely audible requires a 60 dB increase in sound intensity over the 1,000-c/s tone (at 0 dB). This is an intensity ratio of a million to one. A 10,000-c/s tone, in contrast, will require only a 10-dB boost over the 1,000-c/s note to make it barely audible. Note that at the other extreme, the threshold of pain, the loudness variations with frequency are not nearly as great as for the lower curve. This means that both the low and high frequencies must be boosted in intensity, when the level of musical reproduction is reduced, to obtain the same relative loudness of different tones. Since most home reproduction of music takes place at a considerably lower level than the original performance, the bass and treble tone controls of the audio amplifier should be set for boosting, when listening to recorded music or a radio reproduction. Many hi-fi amplifiers have 'loudness' controls, which perform the required bass and treble boost automatically, when the volume level is reduced.

It is evident from the foregoing that a uniform or 'flat' amplifier response over the entire audio range is seldom desirable. Not only must the amplifier response be compensated for the level of reproduction, compensation must also be provided for different disc recording characteristics, poor room acoustics, and deficient electromechanical components, such as speakers and pick-ups. These requirements are usually met by providing a variety of tone equalization controls on the amplifier.

BASIC COMPONENTS

A complete high-fidelity system usually consists of a number of distinct, separately-bought components, which makes for flexibility in fitting various needs and various pockets. A minimum system must consist of at least one input source, such as a record player or radio tuner; an audio amplifier; and a loudspeaker in an enclosure (sometimes called a baffle). A more elaborate system, illustrated in Fig. 161, may contain many input sources, separate pre-amplifier (voltage) and main amplifier (power), and a multiple speaker system at the output. The reason why people bother with the mess and considerably higher cost of separate components instead of a radiogram in a single cabinet, is the better quality of each individual component, especially the amplifiers and loudspeakers, and also because of the absolute need for the speakers to be housed separately for high-quality sound reproduction.

Note that the hi-fi system of Fig. 161 not only incorporates a record player, an a.m.-f.m. radio tuner and a television set, but also a tape recorder. The tape recorder, which records sound on magnetized tape, is both an input and output device. When a recording is to be made, the audio signal from a

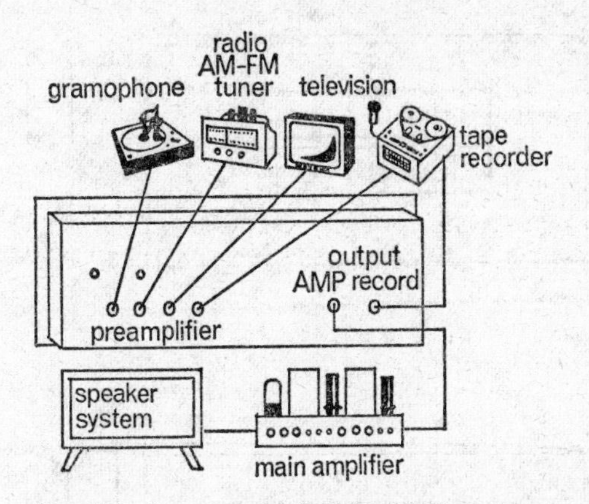

Fig. 161. Components of a high-fidelity system

microphone or other source (such as a radio tuner) is fed through the preamplifier and its 'record' output jack to the input of the tape recorder, where the signal is impressed on magnetic tape. When a tape recording is to be played back, the output of the tape recorder feeds to the input of the preamplifier, and from there to the main amplifier and speaker system, where the recording is reproduced.

Let us now consider separately the various components of a high-fidelity system, with the exception of the television receiver, which will be explained in detail in the next chapter.

THE RADIO TUNER

A radio tuner is a radio receiver minus audio amplifier. It therefore performs all the functions of the radio receiver in selecting (tuning), amplifying, and demodulating a signal, except that it does not amplify the audio signal from the detector to a point where it is capable of driving a loudspeaker. Since a hi-fi system has its own audio amplifiers, this latter function, obviously, is not required.

To qualify as a hi-fi component, a radio tuner (a.m. or f.m.) must incorporate special high-quality design features. Most important, the r.f., i.f., and detector circuits of the tuner must pass the entire audio modulation band impressed on the carrier. Despite this large bandwidth, the tuner must be sufficiently selective to separate two closely spaced stations, to prevent any interference between them. The sensitivity of a tuner should be in the order of a few microvolts to pull in weak signals. Moreover, a tuner should incorporate automatic frequency control (a.f.c.), to prevent the tuner frequency from drifting away from the selected station. The output circuit of the tuner should be a cathode follower (see Chapter 10), to provide isolation and low loading of the detector stage when a long output cable is connected. Finally, a hi-fi radio tuner should have complete band coverage and exceptionally low distortion, hum, and noise level.

Fig. 162 illustrates two general types of commercially available f.m. or a.m.-f.m. tuners. The basic tuner, shown in (a), consists of the set-up just discussed; it provides an audio signal from the detector, which is to be fed to the

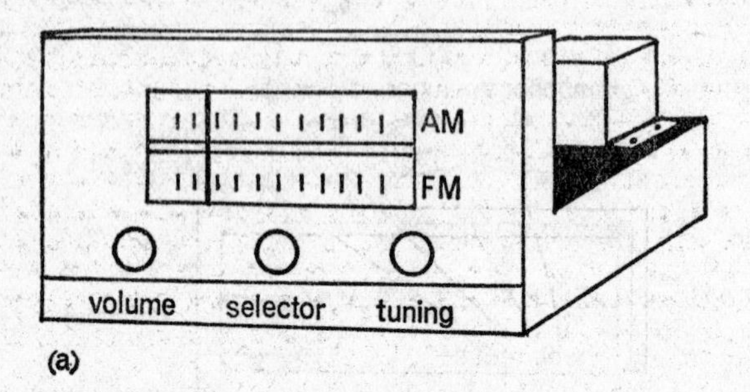

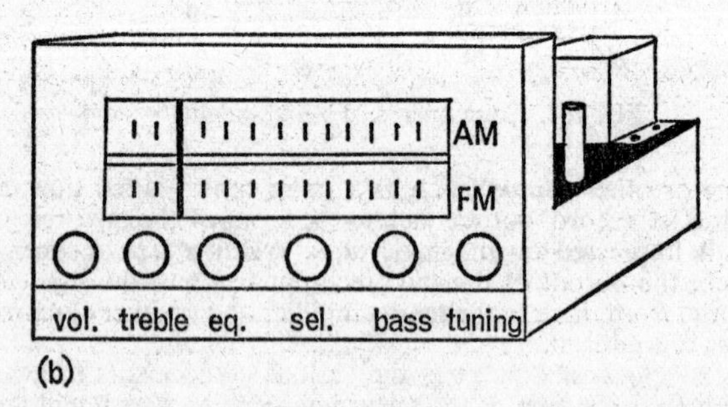

Fig. 162. Radio tuners: (a) Basic tuner, (b) Tuner-amplifier

preamplifier of the hi-fi system. A basic tuner is usually equipped with a volume control, a tuning control, and a control for selecting a.m. or f.m. wavebands. The type of tuner shown in (b), incorporates in addition an audio amplifier. As we shall see, such a preamplifier strengthens and equalizes the output signal from the gramophone pickup, so that it can be applied to a main amplifier. It also acts as a general control centre of the hi-fi system, permitting front-panel control of the main amplifier and various inputs, plugged into the back of the tuner.

PICKUPS AND RECORD PLAYERS

Any record player, whether it be a single-record turntable (transcription player) or a changer type, must have: a motor and turntable for rotating the records at the correct, uniform speed; a tone arm that moves over the record surface; and a pickup consisting of cartridge and stylus to pick off the sound vibrations from the record grooves and translate them into a corresponding audio signal. A record changer has, in addition, a complicated mechanism for playing a number of records automatically and shutting itself off after the last one.

Motor and turntable. The function of the motor is to rotate the turntable and disc at the same speed at which the music was recorded. This sounds simple, but it is surprising how many gramophone motors cannot do this job. Motors and turntables are subject to a number of mechanical distortions and speed variations, which are quite noticeable to the discerning ear. If the speed of the motor is not absolutely uniform, the turntable is uneven, the record has an off-centre hole, or is warped, you will hear unpleasant variations in the pitch of the sound. Pitch variations that take place at a relatively low rate (less than ten times per second) are described as low-frequency wow, while the term flutter is used for higher-frequency pitch variations. Furthermore, if the motor and drive mechanism are not perfectly machined and aligned, and the motor and turntable are not properly mounted, a low-pitched growling noise, called rumble, is transmitted to the turntable and superimposed on the music.

A really well-made motor drive mechanism and single turntable can get around all these deficiencies. The motor is usually of the induction type, used for a.c. only. Cheaper units employ a two-pole induction motor, while more expensive units use a four-pole motor, which provides more power and uniform speed. The ultimate in the hi-fi field is the hysteresis-synchronous motor, which provides a speed as constant as that of an electric clock, because it is geared to the frequency of the a.c. mains.

Although the speed of an induction motor is governed by the frequency of the mains supply, and is therefore constant, it is very much higher than

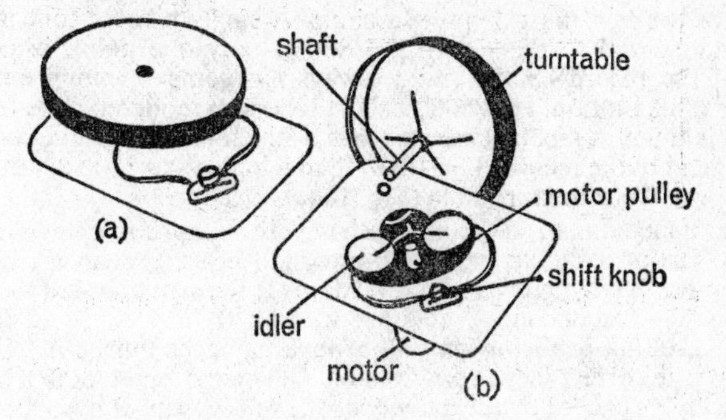

Fig. 163. Typical turntable and drive mechanism

the speed of rotation of any gramophone record. Moreover, the equipment is usually required to play records at 78, 45, $33\frac{1}{3}$ and $16\frac{2}{3}$ r.p.m. These problems are overcome in the most popular type of transcription (single) player by having a stack of four differently sized small pulleys on the motor shaft; a rubber 'idler' wheel is held against one of these pulleys (selected by the position of the speed-control or shift knob) and at the same time makes contact with—and drives—the inner surface of the turntable rim. (See Fig. 163.)

A heavy, die-cast turntable (Fig. 163a), well balanced and machined on a lathe, will give the stability and smoothness required for high fidelity. The rim of the casting is made very heavy to give a flywheel action to the turntable, which helps to smooth out minor speed variations in the motor and drive mechanism. The turntable must rest concentrically on its supporting shaft and must run absolutely true. You can check this by observing the lower rim of the turntable as it revolves. There should be no up-and-down or sideways motion. It is of course essential that the turntable should be non-magnetic and hence does not 'pull' a magnetic pickup cartridge.

Tone arm. The tone arm must move the stylus of the pickup across the record surface, following as closely as possible a straight line along a radius of the turntable circle. (See Fig. 164a.) This is necessary so that the tone arm always rides tangentially to the record grooves at the point of contact, regardless of its position along the record surface. If the stylus does not 'track' well in the grooves, it may exert a side pressure on the grooves which results in distortion and eventually damages the record. Of course, any arm mounted on a single pivot cannot track perfectly along a straight line, as in Fig. 164a, since

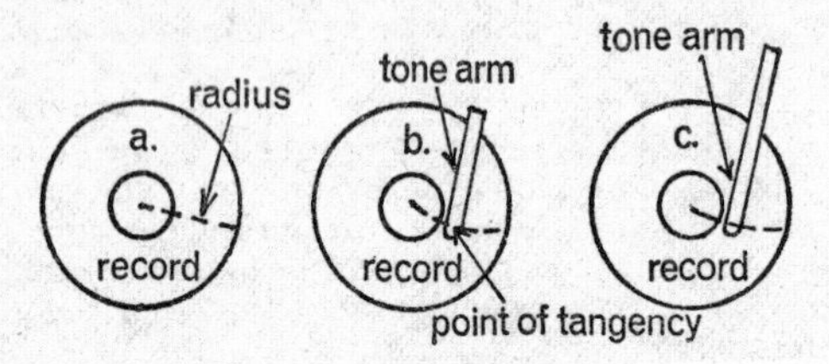

Fig. 164. Tracking error: (a) Ideal—no error, (b) Large error of short arm, (c) Small error of long tone arm

the end of the arm must describe a circle. A singly pivoted tone arm, therefore, is tangential to the record grooves at only one point, usually at the centre of the record's radius, and always has some 'tracking error' at the extremes of its motion. However, a short tone arm mounted close to the turntable, as is done in record changers, has a much larger tracking error at the start and end of the record (Fig. 164b) than a long, 9- to 11-inch arm mounted farther away from the turntable (Fig. 164c).

Besides its length, a tone arm must have other desirable characteristics for hi-fi use. The arm should have high-precision bearings with a minimum of friction in horizontal and vertical directions. If the arm does not move freely, the stylus may exert considerable pressure on the side walls of the record grooves, resulting in distortion and groove damage. Furthermore, the arm must have the correct inertia and weight. The inertia operates in a horizontal direction and is caused by the mass (weight) of the arm. It must be considerable to keep the arm from vibrating or resonating at the lower bass notes. The resonant frequency of the arm vibrations must be kept below audibility. Despite high inertia the arm must have the minimum vertical weight pressing down on the record. The pressure exerted on the stylus by the weight of the tone arm and pickup must be extremely low for hi-fi reproduction, usually below 5 grams for modern L.P. records and about 15 grams ($\frac{1}{2}$ ounce) for 78s. To serve these varying requirements, the stylus pressure (effective weight) of tone arms is usually adjustable through some spring or counter-balance arrangement. A typical modern tone arm is illustrated in Fig. 165. The screw adjusts the stylus pressure.

Pickups. The gramophone pickup, consisting of cartridge and stylus mounted at one end of the tone arm, constitutes the electronic starting-point of the hi-fi system. It must faithfully translate the mechanical (sound) modulation contained in the record grooves into corresponding audio variations. Not so long ago this job was done more or less satisfactorily by a Rochelle salt crystal cartridge. You will recall that Rochelle salt is one of the substances that shows the piezoelectric effect. When the stylus follows the lateral motion of the record groove modulation, the crystal faces become slightly deformed and generate a voltage proportional to the deformation and, hence, needle excursion. Crystal pickups have a high voltage output (1–2 V) and compensate, to some extent, automatically for the recording frequency characteristics. They therefore do not need a preamplifier, which makes them popular in portable record players. However, the non-uniform frequency response of crystal pickups (they have a large bass boost) cannot compensate for all types of recording characteristics used by different record manufacturers. This compensation, hence, is best left to the specially designed equalizer circuits in the preamplifier, and a pickup with a 'flat' frequency response over the audio band is preferred. Crystal pickups, moreover, function poorly in warm, damp climates and the crystal may be destroyed by high temperatures. In this respect cartridges containing an artificially made ceramic 'crystal' are much better.

The most expensive pickups for hi-fi use contain a magnetic cartridge to which a diamond stylus is permanently attached. Magnetic pickups depend on variations in a magnetic field to generate an electrical signal proportional to the stylus motion. They are, in general, characterized by a low output voltage and impedance, and excellent, uniform frequency response over the entire

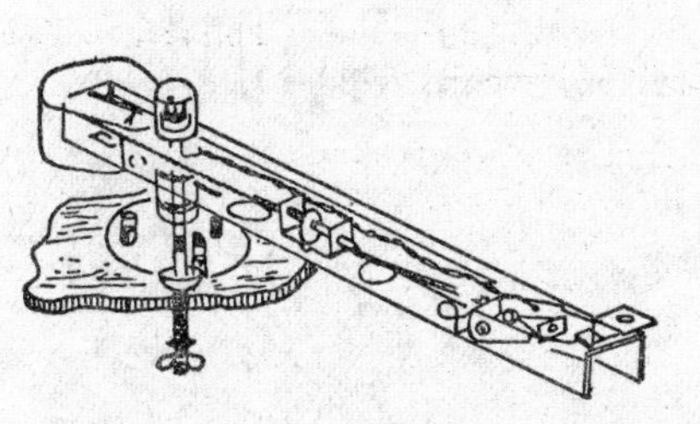

Fig. 165. Tone arm

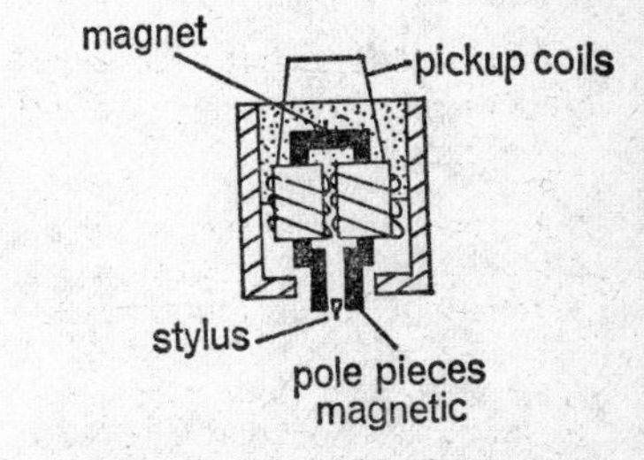

Fig. 166. Variable-reluctance cartridge construction

audio range. Because of their low output voltage and their lack of compensation for the recording characteristic, magnetic pickups always need a preamplifier that will boost the voltage and equalize the frequency response.

Two types of magnetic cartridge are commercially available. One is the variable-reluctance type, the other the moving-coil or dynamic pickup. The moving-coil (dynamic) pickup works like the dynamic microphone discussed in the last chapter and is similar to it in quality. An armature, consisting of a stylus and a small coil wound on a thin steel sleeve, moves in the airgap between the poles of a strong permanent magnet. The lateral motion of the stylus in the record groove moves the coil across the lines of force of the magnetic field, thus generating a voltage in the coil that is proportional to the velocity of the motion. Since the moving system is very light, substantially flat frequency response from about 10 to 30,000 c/s can be obtained. The quality of the dynamic pickup is high, but so is the price.

Almost equal quality at less expense is provided by the variable-reluctance cartridge, which operates on a slightly different principle. Here the stylus is mounted on a magnetic armature and moves freely in the airgap between the pole pieces of a strong permanent magnet. Two coils are fixed-mounted on the pole pieces, one on each side of the stylus and armature, as illustrated in Fig.

166. When the magnetic armature is deflected by the stylus motion towards either side, the resistance, or reluctance, of the magnetic circuit on that side is decreased and, hence, the magnetic field (flux) is increased. At the same time, the reluctance on the other side—away from the stylus—is increased and the magnetic flux on that side is decreased. As a result, the magnetic flux increases through one of the coils and decreases through the other, and voltages proportional to the difference in magnetic flux are induced in the coils. Since the coil windings are in series, a push-pull action results, with the total coil voltage being proportional to the lateral stylus motion. No voltage is generated by vertical (up and down) motions of the stylus, which considerably reduces surface noise. The output voltage of a variable-reluctance cartridge is a few millivolts and the frequency response is smooth from about 20 to 20,000 c/s.

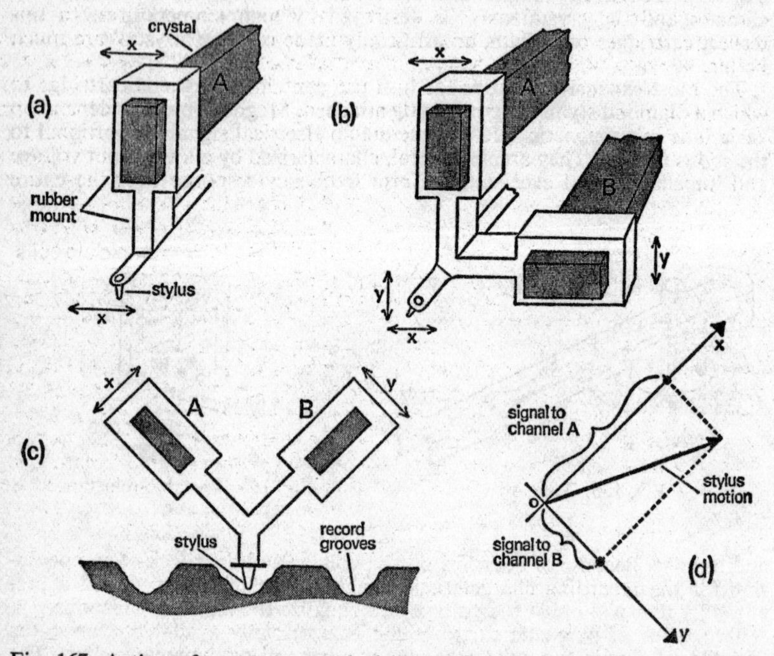

Fig. 167. Action of a stereo crystal pickup: (a) single crystal for monophonic reproduction; (b) two crystals at right-angles respond to stylus movements in any direction; (c) section through a stereo cartridge; (d) vector resolution of stylus movement

Stereophonic Pickups. A stereophonic recording on tape contains two physically separate 'channels' of sound, which are reproduced separately to create for the listener an illusion of 'solid' sound. A stereophonic gramophone record also contains two channels of sound, but instead of being physically separate both channels are recorded in the same groove. Thus a stereo pickup has the difficult task of converting a complex mechanical movement into two simultaneous electrical signals.

Fig. 167a shows a monophonic (single-channel) crystal cartridge. Horizontal movements of the stylus are transmitted to one end of the crystal: the other end is clamped. The voltage signal generated by the crystal is proportional to the amplitude of the horizontal motion (x) of the stylus; any vertical motion should have no effect. In Fig. 167b a second crystal (B) has been added at right-angles to the first. Vertical motion (y) does now have an effect: it generates a signal voltage in crystal B (but not in crystal A). In Fig. 167c the whole arrangement has been turned through 45° so that the stylus is again vertical. It should be plain that if the motion of the stylus has any component in the x direction a signal will appear in crystal A, and if the motion has any component in the y direction a signal will appear in crystal B. These two signals are amplified quite separately and fed to the channel-A loudspeaker and channel-B loudspeaker, respectively. Fig. 167d shows by way of example an almost horizontal excursion of the stylus: this has a relatively large component in the x direction and hence generates a large channel-A signal; it has only a small component in the y direction and so generates only a small channel-B signal. (See page 249.)

PREAMPLIFIERS

The voltage output of a magnetic pickup cartridge is only a few millivolts (thousandths of a volt), whereas the output of a piezo (crystal) cartridge is at least 50 millivolts and may be as high as 1–2 volts. Consequently, when magnetic pickups became popular, it was necessary to provide additional voltage amplification. This was achieved by inserting a separate preamplifier between the pickup and the main amplifier, which in those days was of course a valve amplifier. Gradually the separate preamplifier evolved into a control centre for the entire hi-fi system, providing not only the necessary preamplification for magnetic cartridges and microphones, but also tone equalization for a variety of inputs. Since the trend in hi-fi following the advent of the transistor

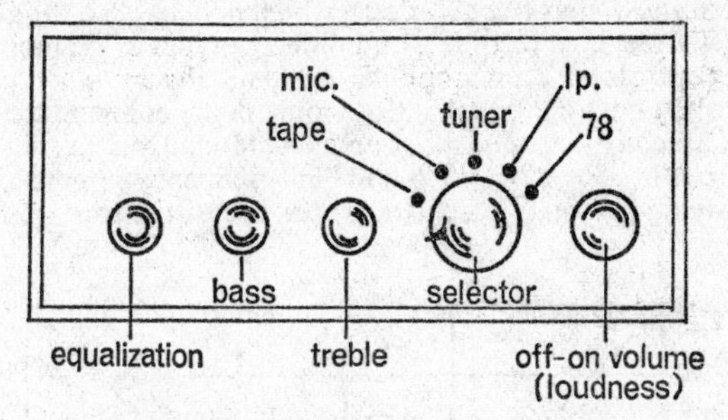

Fig. 168. Typical hi-fi preamplifier

is towards compactness, separate preamplifiers are rarely encountered today, but for clarity we shall treat the preamp stage as a separate unit. Note that with a valve-operated preamplifier that really is a separate unit, the output is almost inevitably a cathode-follower stage as described on page 115 so that the output impedance matches that of the relatively long shielded cable by which it is connected to the main (power) amplifier.

A typical preamplifier is illustrated in Fig. 168. The volume control usually combines the function of an ON-OFF power switch with a volume (audio gain) control. The simplest type of volume control sets the voltage gain of the preamplifier equally for all frequencies, leaving all tone compensation for the individual tone controls. Somewhat more sophisticated controls provide for

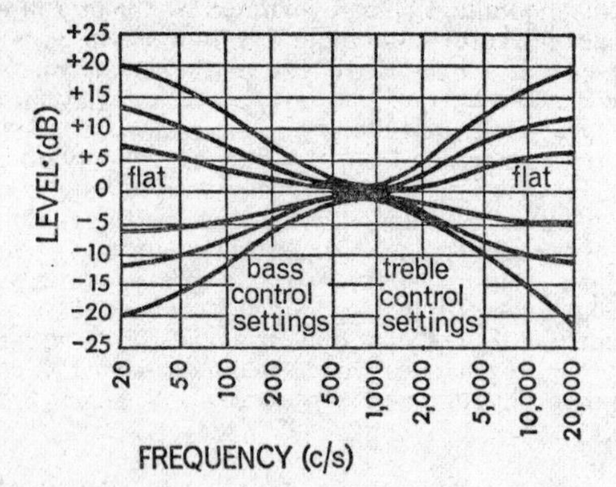

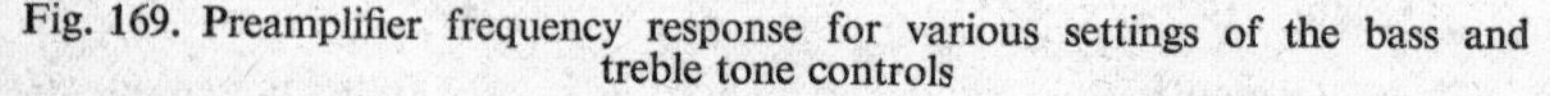
Fig. 169. Preamplifier frequency response for various settings of the bass and treble tone controls

'bass compensation', i.e. they boost the low-frequency response as the volume level goes down to compensate for the falling off in the ear's response. (See Fig. 160, threshold of hearing curve.) The cleverest kind of control is the loudness control, which compensates for the ear's deficiencies at both low and high frequencies by inverting the ear's loudness contours at a particular volume level. (See Fig. 160, upper and lower loudness contours.)

As illustrated in Fig. 161, a number of audio input devices may be plugged into the rear of a preamplifier. The selector switch on the preamplifier front panel selects the desired input device, whose output is to be amplified. The selector switch shown has five positions, which may be used for a microphone, radio tuner, TV set, tape recorder, and different types of gramophone record.

The tone controls of a preamplifier are generally of two types: bass and treble controls, and tone equalization controls to compensate for different recording characteristics. The bass and treble controls permit boosting or attenuating (cutting down) the low- and high-frequency response, respectively. Besides satisfying personal taste, the proper use of the tone controls permits

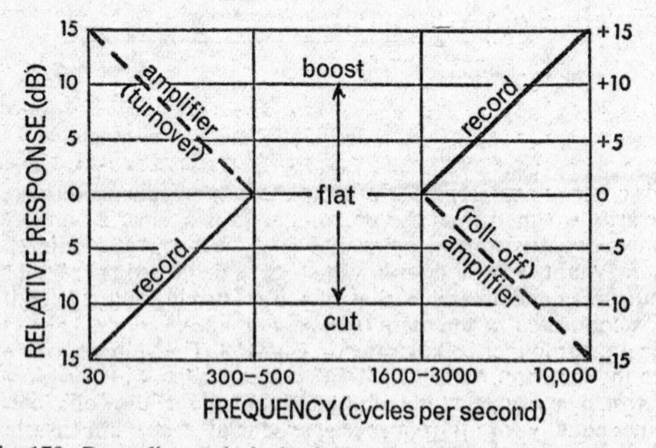

Fig. 170. Recording and playback characteristics of modern records

achieving the proper tonal balance of the sound, taking into account the deficiencies of the system, loudspeakers, room acoustics, etc. A typical set of frequency response characteristics for various settings of the bass and treble controls is illustrated in Fig. 169. The centre setting of each control usually provides 'flat' (uniform) frequency response, while the extreme settings provide maximum boost or attenuation. The particular curves shown provide up to 20 dB maximum boost or attenuation of the bass and treble response. This means that the output voltage of the preamplifier is up to ten times as great at the extreme bass or treble frequency compared with the mid-frequency (1,000 c/s) response for a boost, or only one-tenth of the mid-frequency response for maximum attenuation. While this range may appear somewhat excessive, a control range of 12 to 15 dB minimum is considered desirable.

When a gramophone record is being made, the bass frequencies are progressively attenuated ('cut') and the treble frequencies are progressively boosted, as shown by the full line in Fig. 170. This is a necessary part of the recording process because the motion of the cutting stylus increases in amplitude as the frequency goes down, and at very low frequencies the stylus might cut into an adjacent groove. Furthermore, the high-frequency response is deliberately boosted in order to give a higher signal-to-noise ratio on playback and thereby reduce surface noise: this is called high-frequency pre-emphasis. To obtain a uniformly flat frequency response during playback, therefore, the preamplifier must 'de-attenuate' the bass notes and 'de-boost' the treble; in other words it has to boost the bass and cut the treble, as shown by the broken line in Fig. 170. Note that the idealized amplifier characteristic in Fig. 170 is the exact inverse of the idealized recording characteristic. In the early 1950s when long-playing (microgroove) records were introduced, each recording company employed its own response curve, so it was highly desirable for a playback preamplifier to include a variable *equalization control* to cater for the score or more of different recording characteristics that might be met with. Nowadays practically every major company ensures that its recordings comply with a standardized curve known as the RIAA (Recording Industry Association of America) response. Since all records have the same response there is no longer the need for variable equalization; instead, fixed equalization to RIAA response is achieved by a resistor–capacitor network at the input socket. Further equalization can often be provided by switching in a *scratch filter* to reduce surface noise, a *rumble filter* to reduce very low frequency interference, and a *loudness control* to boost low-level sounds near the ends of the audible range.

MAIN AMPLIFIER

When a separate preamplifier is available, the function of the main or basic amplifier is to strengthen faithfully the output of the preamplifier to the level required for driving the loudspeaker. (If, as is usually the case nowadays, the preamplifier and main amplifier are combined, the single unit must also perform the functions of preamplification, tone equalization, and input selection.) Since all controlling is done from the preamplifier, the main amplifier is usually a simple box-like unit that has only audio input and output connexions, a main inlet, and possibly one or two adjustment controls for 'balancing' the output valves. The basic amplifier must provide, usually, for several stages of voltage amplification as well as sufficient power amplification to drive the speaker. The question is, what is sufficient power amplification? Speaker systems will work with audio output powers from a fraction of a watt to 30 W or more. Commercial amplifiers are available with power output ratings from 3 to 100 W, with the cost rising steeply as the power goes up.

Ironically, the level of speech and music performed in the home is usually no more than about one half-watt, and often only a few milliwatts. Wildly conflicting claims are issued in audio circles about how much power is required for proper reproduction, with the high-power, expensive amplifiers being favoured.

First of all, a power rating without specifying the total distortion at maximum power, is meaningless. Even cheap amplifiers can be driven to substantially greater outputs than their ratings, if you are willing to accept about 10% distortion in the bargain. The maximum distortion for which the amplifier power output is specified should never be more than 1–2%. (This includes harmonic as well as intermodulation distortion.) Secondly, an amplifier must have considerable reserve power to accommodate the tremendous dynamic range between the loudest crescendos of a symphony orchestra and the softest whisper of a human voice. (See Fig. 160.) Although this range is about 120 dB, you could never reproduce such loudness variations within the confines of your living room and, hence, there is no point in providing it. Moreover, the power output of an amplifier is to some extent a question of individual preference and depends on the average level or reproduction, as well as its peaks.

The biggest source of distortion in a well-designed amplifier is usually the output transformer. As we have seen (page 99), a step-down transformer is

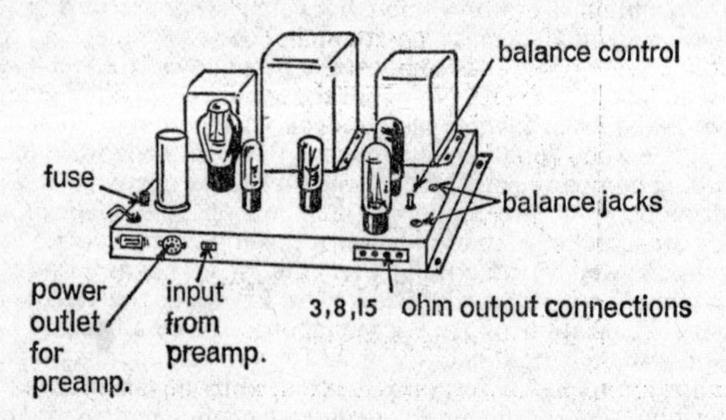

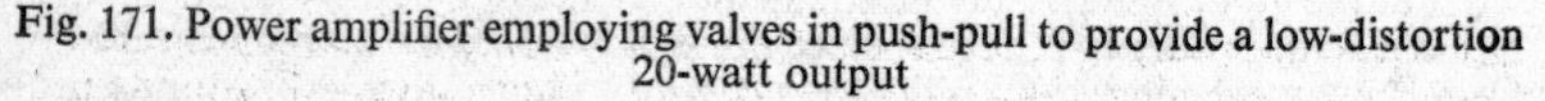
Fig. 171. Power amplifier employing valves in push-pull to provide a low-distortion 20-watt output

always used with conventional valve amplifiers to match the anode resistance of the output valve to the impedance of the loudspeaker. A good quality output transformer is, unfortunately, not only one of the most expensive items in an amplifier, but also one of the bulkiest. This is another reason for the decline of the valve amplifier, since the transistorized equivalent can easily be designed without the need for an output transformer.

A stereophonic amplifier is in principle made up of two quite separate conventional amplifiers, and hence presents no problems from the electronics point of view. Both amplifiers (one for each channel) are fed from the same well-smoothed power supply; the only other shared component is the *balance control*, a potentiometer which governs the ratio of voltage amplification (gain) between the two channels.

LOUDSPEAKERS AND ENCLOSURES

A loudspeaker is an electromechanical device for converting a varying audio voltage into corresponding sound waves. Being electromechanical, loudspeakers are the weak links of any hi-fi system. Despite all the care taken in preserving the fidelity of an audio signal from pickup to the output of the amplifier, loudspeakers invariably add some form of distortion to the signal. You can buy loudspeakers from simple units costing a few shillings to multiple-speaker systems costing hundreds of pounds, but chances are that even the most expensive system will not have an audio frequency response as 'flat' as that of an amplifier. Loudspeaker enclosures, or baffles, as they are often called, assist the speaker in doing its job properly and, thus, are almost equally important.

Basic moving-coil loudspeaker. Practically all loudspeakers at the present time are of the moving-coil or dynamic type, in which a speech coil carrying the audio signal moves in and out of a strong magnetic field produced by a permanent magnet. Fig. 172 illustrates the construction of the moving-coil speaker.

As is apparent from the figure, a large permanent magnet with a central pole piece is used to provide the magnetic field. In the narrow airgap between

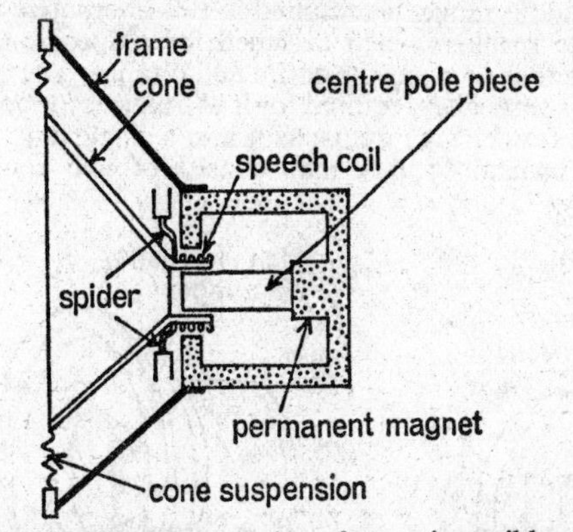

Fig. 172. Simplified construction of a moving-coil loudspeaker

the central pole piece and the outer poles, where the field is most concentrated, a small speech coil is suspended, consisting of a few turns of wire wound around a paper, plastic, or aluminium cylinder. This coil carries the audio currents from the output of the audio amplifier. At one end of the speech coil is mounted a fairly large paper or cloth cone, which radiates the sound. The cone is held at its edges by a flexible suspension ring that is fastened to the metal frame of the speaker. Another suspension at the other end, mounted at the junction of the cone and speech coil, keeps the coil centred with respect to the pole piece. This suspension, known as the spider, is made of flexible material and permits forward or backward (longitudinal) motion of the voice coil, but no sideways (lateral) motion. The entire cone and voice-coil assembly, thus, can move freely in or out of the central pole piece, but not sideways.

When audio currents flow through the speech coil, a magnetic field is produced around the voice coil that is at right angles to the field of the permanent magnet and proportional to the strength of the audio currents. The two fields attract or repel each other, depending on the instantaneous polarity of the audio currents. Since the position of the permanent magnet is fixed, the voice coil and cone assembly moves inward or outward from its central position in proportion to the attraction or repulsion of the two magnetic fields—that is, in proportion to the momentary strength and polarity of the speech-coil currents. The resulting cone vibrations produce air-pressure variations (sound) in correspondence with the audio signal.

The frequency response of such a basic speaker is generally quite irregular, with a number of resonant peaks and valleys, and rarely extends beyond a range of about 60–8,000 c/s. By using a fairly large (12–15 inch diameter), heavy cone, the low-frequency response of a single speaker can be extended downward to 45 or even 30 c/s, but then the high-frequency response suffers. In contrast, by using a small (3–5 inch diameter), light speaker cone, the high-frequency response of a single speaker can be pushed to the limits of audibility, but then the bass response will cut off around 100–150 c/s because of the insufficient mass of the speaker cone. It is almost impossible to make a single speaker perform well over the entire audio range. Nevertheless, various tricks and ingenuous designs have made available in recent years single-unit wide-range speakers that will perform well over an audio-frequency range from about 40 to 15,000 c/s.

Coaxial and triaxial speakers. The difficulty in designing a single speaker to cover the whole audio range has resulted in two alternative approaches. One is to use separate speakers, each designed for a specific audio-range, and connect them together to cover the entire audio range. The other approach is to combine the large speaker required for low-frequency reproduction and the small one needed for the high frequencies into a single unit, mounted in line, or coaxially. A coaxial speaker may consist of two completely separate

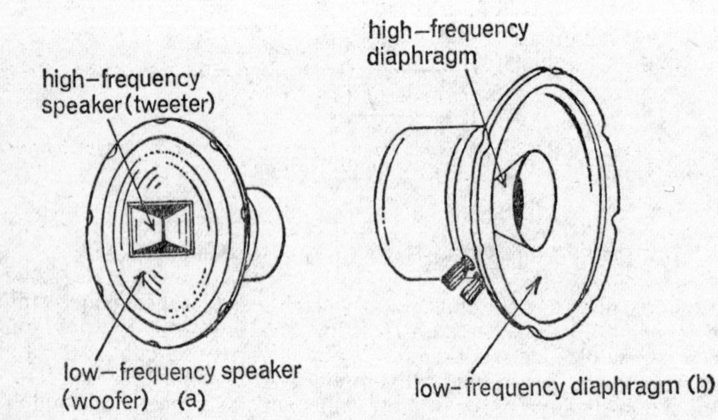

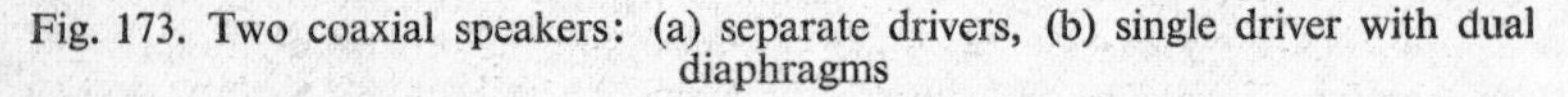

Fig. 173. Two coaxial speakers: (a) separate drivers, (b) single driver with dual diaphragms

speakers mounted on the same axis, as shown in Fig. 173a. Alternatively, a single electromechanical driver may be employed to operate two differently sized cones or diaphragms, one for low and the other for high frequencies, as illustrated in Fig. 173b. A mechanical or electrical crossover network is associated with the speaker to divide the audio input signal into the proper low- and high-frequency bands, which are fed separately to each speaker unit.

The coaxial speaker principle may be extended to three speakers by further subdividing the audio range. Either three separate speakers or a single driver with three diaphragms may be mounted along the same axis in the triaxial speaker. Fig. 174 illustrates a triaxial unit that has a single driver for a low- and medium-frequency diaphragm and, in addition, a separate high-frequency 'tweeter'. Each unit performs at maximum efficiency in a particular frequency range.

Multiple-speaker systems. The best way to overcome the limited frequency response of a single speaker is a multiple speaker system made up of completely separate speakers, each designed for maximum efficiency in a specific frequency band. The simplest type of multiple speaker system is a two-way system, consisting of a large, low-frequency speaker, or woofer, a small (horn-type) high-frequency speaker, or tweeter, and a frequency-dividing crossover network. The speakers must be matched so that the tweeter supplements the

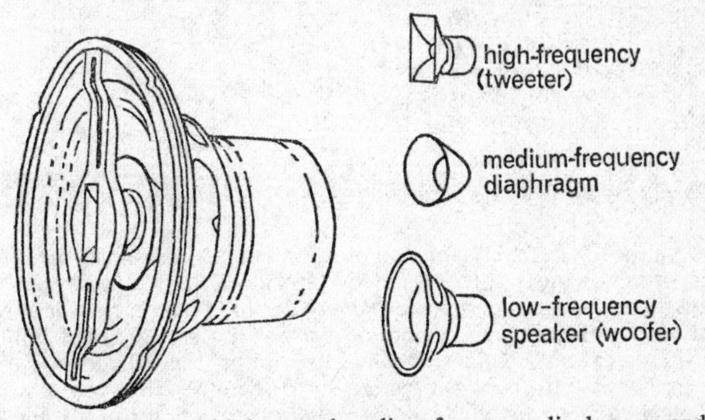

Fig. 174. Triaxial speaker with low- and medium-frequency diaphragms and separate high-frequency 'tweeter'

frequency range of the woofer and the whole audio range is covered smoothly. The crossover network is designed to split the audio range smoothly at the crossover point, so that each speaker works only with frequencies for which it is designed and is not overloaded. Depending on the system, the woofer may be designed to reproduce frequencies from about 30 c/s to a crossover point anywhere between 400 and 3,500 c/s. The crossover network, which may be a simple high-pass filter for a two-way system, will then feed the balance of the audio frequencies up to 15,000 c/s or higher to the tweeter. A typical two-way set-up is illustrated in Fig. 175.

A somewhat more ambitious approach to hi-fi reproduction is the three-way system, illustrated in Fig. 176. This particular system has been derived from the two-way set-up of Fig. 175 by adding a dual-horn high-frequency tweeter and an *L–C* crossover network, which limits the woofer to audio frequencies below 600 c/s or less. The tweeter from the previous set-up (Fig. 175) is now used as mid-range speaker for frequencies of about 600 c/s to approximately 4,000 c/s. The previously used high-pass filter is inserted between the crossover network and the dual-horn tweeter, but is adjusted to pass only frequencies above 4,000 c/s (approx.) to the tweeter. The tweeter then reproduces the range from 4,000 c/s up to 20,000 c/s. A dual-horn or other speaker with a wide flare is preferred for the tweeter so as to spread the high-frequency 'beams' over as wide an angle as possible. Still more elaborate and expensive multiple-speaker set-ups are used sometimes to achieve uniform response.

Loudspeaker enclosures. The sound you hear from a speaker is produced by the combination of the loudspeaker and its enclosure or baffle. A loudspeaker cannot perform its job properly without a suitable baffle to direct the sound energy. You can discover this for yourself by listening to the tinny sound emerging from even the best hi-fi speaker, when it is simply held in the hand. The bass response, which gives 'body' to the sound, is almost completely absent in a hand-held speaker.

The reason for this phenomenon is simple. Any loudspeaker radiates sound from both the front and the rear of the speaker cone. When the speaker cone

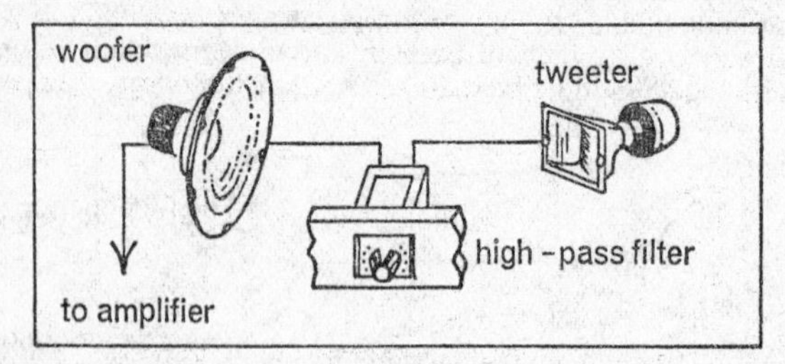

Fig. 175. Two-way speaker system consisting of woofer, tweeter, and high-pass filter

is pushing the air in front of it, thus compressing it, the air in the rear of the cone is simultaneously thinned out or rarefied, and vice versa. Since sound compressions and rarefactions are 180° out of phase with each other, the sound from the front of the speaker is out of phase with that radiated from the rear. When these two out-of-phase sound waves meet, they effectively produce a short circuit, with the rear rarefaction sucking in the front compression, and they cancel each other out. This cancellation affects primarily the low-frequency sound (long wavelengths), since the high-frequency sound is radiated from the front and rear in the form of beams, which do not meet.

The obvious answer to this situation is to put some sort of obstacle around the speaker to prevent the front and rear sound waves from reaching each other. An obstacle sufficiently large to do this completely is rarely practical

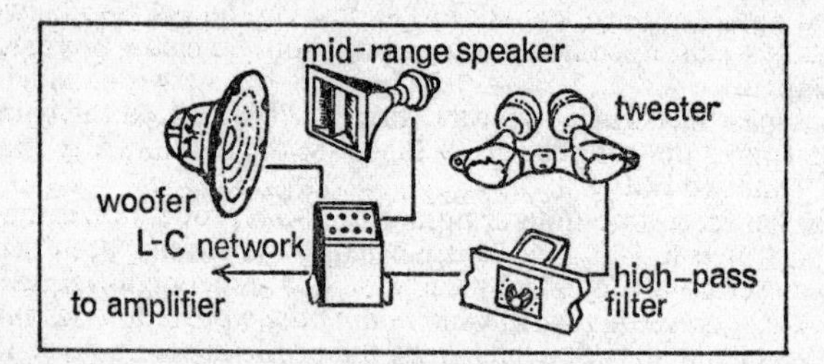

Fig. 176. Three-way speaker system consisting of woofer, tweeter, mid-range speaker, and crossover networks

nor required. It is only necessary to elongate the sound path from front to the rear of the speaker sufficiently to build up a sound wave in front of the speaker, before it has a chance to be pulled back to the rear. The longer the path length from front to rear compared to the wavelength of the sound to be radiated, the less cancellation takes place. When the wavelength of the sound becomes comparable to the front-to-rear path length of the baffle, cancellation occurs, and wavelengths longer than this (lower frequencies) will not be radiated to any extent.

This is the sort of reasoning that led to mounting the speaker on a simple flat wooden board, or 'baffle', shown in Fig. 177a. The term 'baffle' originated from this open structure; complete enclosures were not used in the early days of hi-fi. The flat, open baffle works well, but has the drawback of requiring a very large size to prevent cancellation at low frequencies. To radiate frequencies in the 30–50 c/s range, for example, the baffle board would have to be from 20 to 40 feet in length and width. That is obviously not practicable.

The attempt to reduce the physical size of the baffle while retaining the required path length from front to rear has led to the design of a variety of structures. The simplest is the open-back enclosure (Fig. 177b), which is

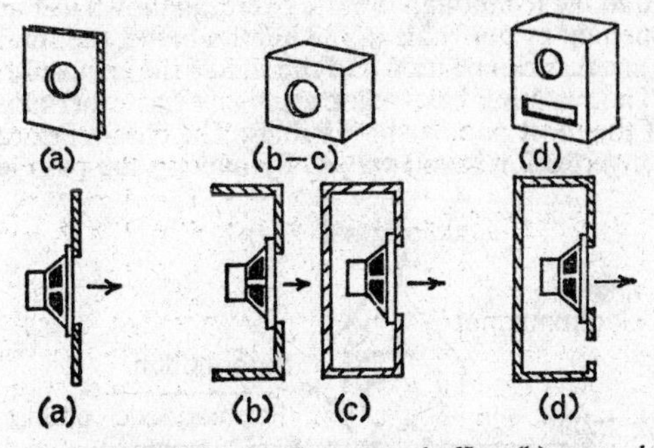

Fig. 177. Basic types of baffles: (a) flat open baffle, (b) open-back enclosure, (c) infinite baffle, (d) bass-reflex enclosure

essentially a flat baffle with its edges folded back. This open box or cabinet used to house practically all commercial radiograms has the serious defect of cabinet resonance, which makes it unsuitable for hi-fi reproduction. The entire box resonates for low-frequency sounds reproduced by the speaker, generally in the range from 100 to 200 c/s. The resonant peaking of these frequencies produces a characteristic boominess, which masks the balance of the bass response.

By closing the rear opening of the box, the rear sound wave is entirely suppressed, and an 'infinite' front-to-rear path length results. This infinite baffle (c) suppresses cabinet resonance and would at first appear to be ideal, except for some other deficiencies. Apart from wasting the entire sound output from the rear of the speaker cone, the infinite baffle has some undesirable damping effects on the speaker. The compressed air in the baffle box tends to raise the fundamental resonant frequency of the speaker assembly in the low-frequency range. Since a speaker reproduces very little below its resonant frequency (generally between 30 and 50 c/s in a hi-fi woofer), the raising of this frequency by the infinite baffle results in the elimination of a portion of the bass response. The response to this lower limit is very smooth, however. A very large box avoids this defect, but the size required again makes it impracticable. The best approach to an infinite baffle, consequently, is to mount the speaker either in a large unused cupboard or in the wall between two rooms.

More recently, the inherent defects of a small infinite baffle have been made a virtue of by a revolutionary principle of speaker and baffle design known as acoustic suspension. The elastic suspensions of the conventional loudspeaker cannot reproduce the large excursions demanded by low bass frequencies without considerable distortion. Acoustic suspension gets around this difficulty by vastly increasing the 'compliance' (flexibility) of the mechanical spring suspensions and then making up for the lost spring action through a pneumatic spring consisting of the compressed air in the sealed box. The speaker is acoustically suspended inside a small infinite baffle, which is calculated to provide the proper amount of pneumatic spring action through its sealed-in air. Not only do the dimensions of the box turn out much smaller than for an infinite or any other baffle, but the pneumatic spring action is linear, and hence distortion-free. Through this method it has become possible to reproduce the lowest organ notes (about 30 c/s) without distortion in a small 'bookcase' enclosure.

An alternative, earlier approach consisted of providing an additional opening, or air vent, in the front of an infinite baffle, as shown in Fig. 177d. By providing some opening or port hole in the infinite baffle, the sound energy from the rear of the speaker can be used and the size of the enclosure can be drastically reduced. The resulting bass-reflex enclosure has other advantages which make it one of the most popular hi-fi baffles. The bass-reflex cabinet is essentially a phase inverter for sound waves. By making the path length from the

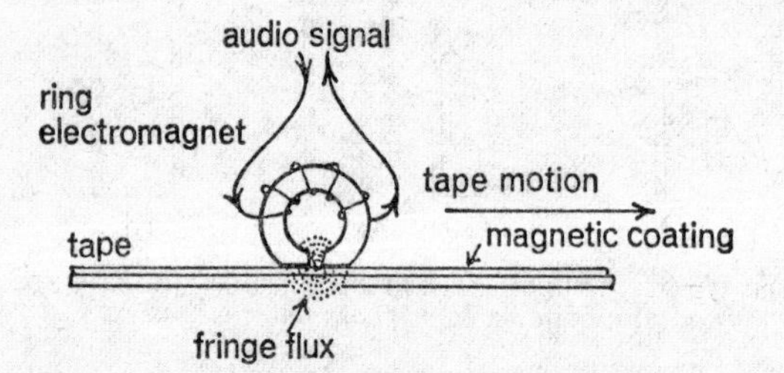

Fig. 178. Longitudinal recording on magnetic tape

rear of the speaker to the port just right, the rear wave can be delayed sufficiently (by one half-wavelength) so that it emerges from the port in phase with the front wave and thus reinforces it. As a result of this reinforcing action, the low-frequency output is twice that of an infinite baffle. Moreover, the bass-reflex enclosure makes use of its own resonant frequency to extend and smooth out the low-frequency response of the speaker. By designing the box so that its own cabinet resonance frequency is somewhat below the fundamental resonance of the speaker that it houses, the resonant peak of the speaker is damped out and the bass response is broadened.

You may have gathered from this description that the bass-reflex cabinet and its speaker must be carefully matched to secure the proper response. A definite relationship must be maintained between the size of the enclosure, the diameter of the port, and the resonant frequency of the speaker. This is the reason why many speaker manufacturers specify a particular type of bass-reflex or other enclosure when stating the speaker response. There is a way of matching a bass-reflex enclosure to a particular speaker by 'tuning' the port for the proper resonant response, but this goes beyond the scope of our discussion. You can find out about tuning a port by consulting one of the many excellent books on high-fidelity, which will also acquaint you with a variety of additional loudspeaker enclosures.

TAPE RECORDERS

Nowadays everybody is familiar with the recording of sound on magnetic tape, either for technical purposes or merely for domestic entertainment.

Fig. 178 illustrates the essentials upon which modern tape recording is based. The audio signal that is to be recorded is fed into the winding of a ring-shaped electromagnet, called the recording head. Between the two ends, or poles, of the magnetic recording head is a small airgap, from which a fringing flux is emitted with a magnitude and polarity proportional to the electrical signal. The tape, which is drawn past the poles of the ring magnet, has a thin magnetic coating of iron oxide consisting of myriads of small magnetic particles, each acting as a tiny magnet. In the unmagnetized tape these magnetic particles have a random distribution and, hence, their combined magnetic field cancels out. When the tape is moved past the poles of the recording head, however, the individual particles are subjected to a magnetizing force along the direction of tape travel and they are turned or aligned in that direction.

The constantly varying audio signal in the recording head thus produces corresponding degrees of magnetization along the moving tape. This is known as longitudinal magnetization.

The magnetized tape itself emits a small magnetic field proportional in polarity and magnitude to the original signal. Consequently, when the tape is

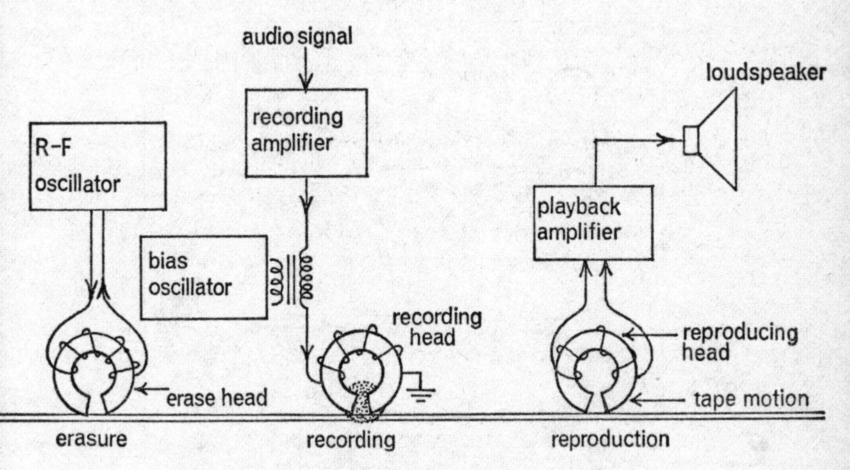

Fig. 179. Block diagram of tape recorder functions

drawn past the airgap of a similar reproducing head, the magnetic field variations induce changing voltages in the winding of the ring magnet, corresponding to the originally recorded signal. The audio signal may then be amplified and reproduced by a loudspeaker.

A practical tape recorder must perform three basic functions, as illustrated in Fig. 179. These are: recording, reproduction, and erasure. Erasure is required before a new recording can be made on a previously magnetized tape.

In tape recording, a low-level audio signal from a microphone, radio tuner, or other source, is first strengthened by the recording amplifier to the proper recording level and then is fed to the recording head. Also applied to the recording head is a bias signal from a high-frequency oscillator. The bias places the recording signal into the linear portion of the tape's magnetization curve, which brings about improved linearity, lower distortion, and a reduction of tape noise. It has been found that an a.c. signal does this job better than steady d.c. bias. The frequency of the bias signal must be above audibility (usually between 30 and 150 kc/s) to avoid interference with the audio signal.

The reproduction process is the inverse of recording. The weak signal induced in the reproducing head by the travelling magnetic tape must first be amplified by the playback amplifier before it can be reproduced by a loudspeaker. Since the requirements for optimum performance differ for recording and playback, completely separate amplifiers and magnetic heads are preferred. However, in the cheaper machines the two amplifiers are always combined in a single unit, with proper equalization being switched in for recording and playback. Furthermore, the recording and playback functions are very often performed by the same head, a compromise which is perfectly acceptable for most purposes.

Finally, the tape recorder must have provision for erasing previous recordings. For this, a strong magnetic field from a third magnetic head, the 'erase'

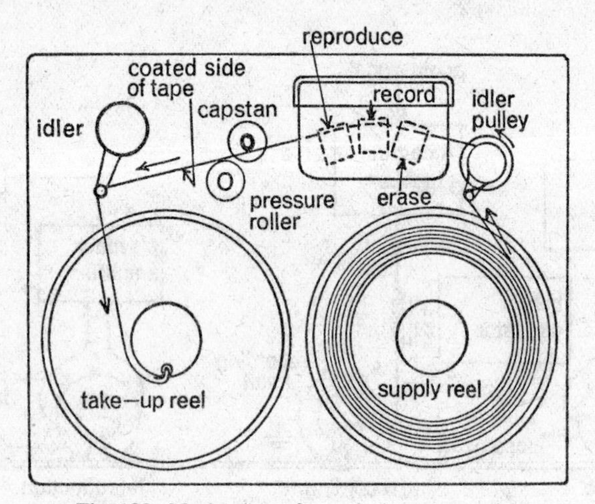

Fig. 180. Mechanism of a typical tape recorder

head, is required. Again, an r.f. field is preferred to d.c. for uniform, low-noise erasure. The field may be produced by a separate r.f. oscillator with added amplification, or the bias oscillator may be used with an extra stage of amplification switched in. In either case, the frequency for erasure must be ultrasonic—i.e. above audibility.

Tape mechanism. Some sort of mechanical drive must be provided to move the tape at an even speed past the heads. The speed with which this is done affects the high-frequency response of the recorder, higher speeds permitting an extended treble response. Hi-fi machines have tape speeds of $7\frac{1}{2}$ inches per second (in/s), or even 15 in/s, while those used primarily for voice (dictation) employ speeds as low as $1\frac{7}{8}$ in/s. Tape mechanisms are apt to be complicated affairs. The bare essentials of a typical machine are shown in Fig. 180.

The tape to be recorded is drawn off a supply reel, threaded around an idler pulley and then is passed over the magnetic heads. The tape must be pulled at constant speed, which is achieved by the combination of capstan and pressure roller. The capstan is a rubberized rotor, which is driven by an electric motor with sufficient torque to draw the tape at constant speed. The pressure roller is adjusted to press the tape evenly against the capstan surface to prevent tape slippage. The tape then travels past a second idler, which keeps it under constant tension, and is taken up by the take-up reel. The functions of the reels are reversed, of course, during playback. While this explanation applies specifically to the reel-to-reel tape machine, the mechanism of the cassette recorder is very similar in principle.

STEREOPHONIC REPRODUCTION

In recent years the hi-fi industry has achieved a major advance in adding realism to high fidelity—stereophonic reproduction of sound (frequently called 'stereo' for short). Stereo sound reproduction requires recording (or broadcasting) and reproduction over two completely separate channels, to simulate the characteristics of hearing with two ears. (It is, for this reason, sometimes called binaural reproduction.) Our ability to hear direction and depth of sound is based on the simple fact that we have two ears. You can convince yourself of that by plugging up one ear for a few hours. Sound will soon become shallow and confused and appear to lack realism. Stereo satisfies our instinctive need for two-eared, three-dimensional hearing. The older, conventional method of

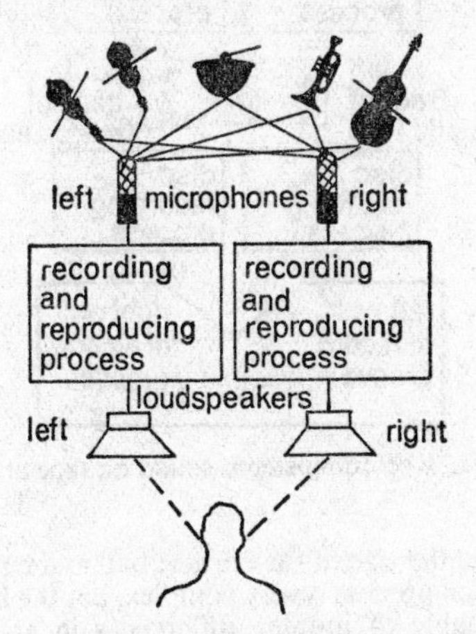

Fig. 181. The essentials of stereophonic reproduction

monophonic or monaural reproduction over a single channel cannot preserve the spatial relationships or depth of the sound, which we normally observe through sound waves striking our two ears at slightly different times. Mono reproduction makes no distinction between left and right, up and down, or front and back of a sound source. These spatial characteristics of direction and depth can be restored only by using two or more separate sound channels for recording or broadcasting and reproduction. This is what stereo is designed to do. Stereo, thus, is a completely different technique of sound reproduction. It is not necessarily high fidelity, but if it is, stereo hi-fi is the best method of sound reproduction that has yet been devised.

To understand the essentials of the stereo technique, consider what happens when listening to a 'live' concert. If you were sitting in the centre of the hall in front of the orchestra, the sound reaching your two ears would—in spite of the central position—not be identical. Instruments playing to the left of centre are heard by the left ear a fraction of a second sooner than by the right ear. Instruments to the right of centre, in contrast, are heard sooner by the right ear. Only the sound of instruments playing in dead centre reaches both ears at the same instant. These differences in time of arrival of various sounds enable us to distinguish the direction of each source. In addition, we hear sounds reflected from the walls of the hall, giving rise to reverberation or an echo effect. Again these reverberated sounds are sorted out by the two ears, blending with the earlier-received direct sounds, and these directional echo

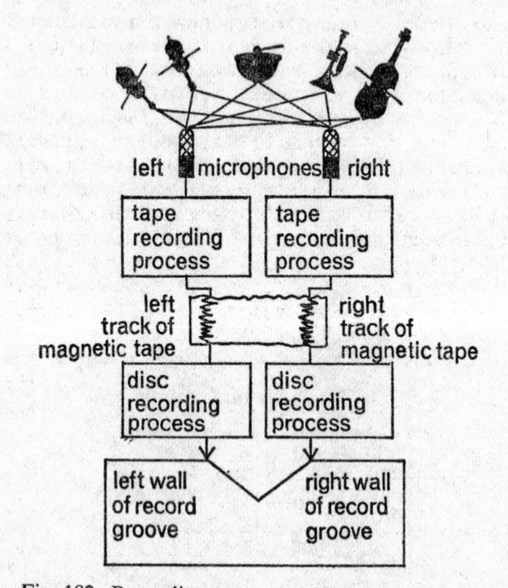

Fig. 182. Recording stereo sound on tape and disc

effects help us to judge the size of the concert hall and contribute to the depth of the sound. The actual process is very complex, but the important idea is that the two ears are capable of judging differences in arrival time of sounds emanating from various sources, either directly from the instruments or indirectly through reflection and echo effects.

By placing two microphones in front of the orchestra, one taking the place of the left ear, the other that of the right ear, the effects of binaural hearing are simulated, as shown by the differences in length of the sound-path lines in Fig. 181. (In practice, several microphones may be used in place of each ear.) The sound pattern picked up by the left microphone is fed to its own recording and reproducing channel, which may be a tape recorder or a radio transmitter and receiver. This pattern is reproduced by the left loudspeaker. Similarly, the sound pattern picked up by the right microphone is fed to a separate recording and reproducing channel and is heard on the right loudspeaker. The listener sitting in the centre between the two stereo loudspeakers then hears the differences between the two sound patterns, just as if he were listening in the concert hall.

How stereo is recorded. Until 1957, stereo sound could be recorded only on tape. Two or more microphones were set up in front of the sound source in the best acoustical positions. The sound picked up by each microphone was then recorded on two separate magnetic tracks of a single tape. (See Fig. 182.) Since most tape recorders had provisions to record two separate, parallel tracks (one forward, one reverse), it was not very hard to add a separate electronic channel for simultaneous stereo recording on the two magnetic sound tracks. Both tracks start and end together on the same tape reel, of course, but the tracks are separate sound recordings nevertheless. The parallel sound tracks are then played back simultaneously through two separate electronic amplifiers and separate loudspeakers, which reproduce the stereo illusion. In practice, usually four separate sound tracks are used on a single tape, two stereo channels in one direction and two in the other. Thus,

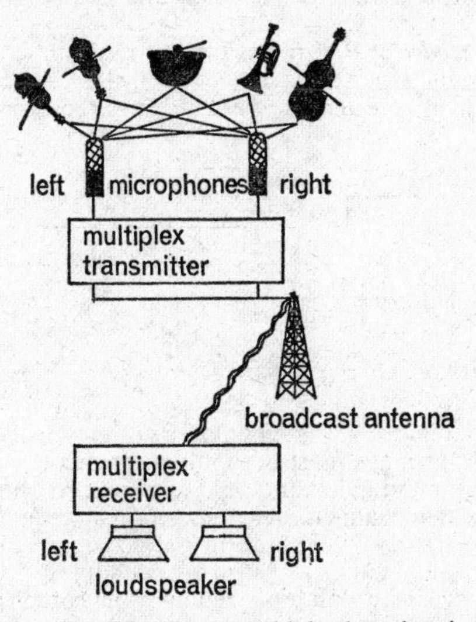

Fig. 183. F.M. stereo multiplex broadcasting

four-track stereo tape is reversible, just like a two-track monophonic tape. This is still the method used today to make 'master' recordings at a live recording session. However, it is relatively difficult to make quantity tape recordings from masters for commercial distribution because of speed differences between commercial and home machines. But, for those who can afford them, commercial stereo tapes are considered the ultimate obtainable in hi-fi reproduction.

Stereo really became popular when it was finally adapted, in 1957, to the gramophone record by the Westrex system of disc recording. The Westrex system permits recording the two separate stereo channels on the two side walls of a single V-shaped record groove (see Fig. 182). In practice, the two tracks of a stereo master tape are re-recorded on the left and right walls of the V-grooved master disc, which is then pressed in quantity for commercial distribution.

The pickup cartridges that have been developed to reproduce the complex information recorded in a stereo record groove are veritable marvels of audio engineering. Conventional monophonic cartridges were required only to respond to the lateral (sideways) vibrations recorded on a mono disc. Stereo cartridges must respond to both lateral and vertical motions (shifted by 45° to fit the walls) recorded in a single groove of a stereo disc. The stereo cartridge must not only translate these complex mechanical vibrations in two planes into the corresponding electrical vibrations, but it must keep the vibrations from each wall completely separate. As if this were not enough, the same pick-up cartridge must also be able to play single-channel or monophonic records with equal fidelity. Essentially, the stylus of such a cartridge performs motions which are the vector resultants of the vertical and lateral side wall excursions. The cartridge then resolves the resultant into the equivalent electrical vectors of the separate stylus excursions and feeds these separate signals into the left and right channels of the stereophonic reproduction system. As we said, the process is quite complex and is really a minor miracle that it works as well as it does.

How stereo is broadcast. Radio was the last medium to take advantage of stereophonic reproduction. This is only natural, since by definition two transmitters and two radio receivers are needed for stereo broadcasting; hence it

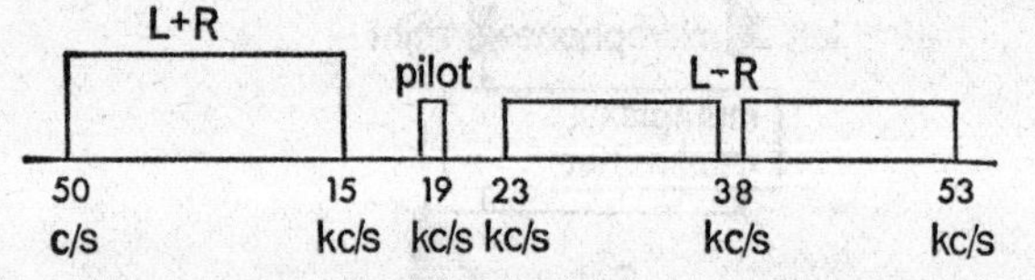

Fig. 184. The specially coded f.m. multiplex signal

would look as if stereo broadcasting might be a very costly business. But the ingenious system known as f.m. stereo multiplex actually uses only one transmitter to broadcast both the left and right channels. As shown in the illustration (Fig. 183), the two channels are combined in a predetermined manner in the multiplex transmitter before transmission and are then separated and restored to their original patterns by the f.m. multiplex receiver. This is done automatically by recent f.m. multiplex equipment; earlier models required a multiplex adaptor in addition to the f.m. receiver.

Fig. 184 indicates how f.m. multiplex works. During a stereo broadcast the multiplex station transmits an $L+R$ component (the left channel added to the right), a specially coded $L-R$ component (the left channel minus the right), and a 19-kc/s pilot signal to synchronize the receiver. The f.m. multiplex receiver picks up and detects all three components. (A conventional f.m. receiver detects only the $L+R$ component, which is monophonic.) Special filters then separate the $L+R$ and $L-R$ components. To obtain the left channel, the two components are added together, since $(L+R)+(L-R)=2L$. The right channel is recovered by subtracting the $(L-R)$ component from the $(L+R)$, since $(L+R)-(L-R)=2R$. The left and right components are then separately amplified and reproduced.

SUMMARY

The response of the human ear to sound is logarithmic: to obtain equal increases in loudness, the intensity of the sound must go up by a factor of 10 each time. Sound intensity and loudness are measured in decibels, a logarithmic unit.

Loudness contours show that the response of the ear falls off at both low and high frequencies for sound reproduction at levels lower than the original. The lows and highs must be boosted, therefore, when reproducing sound at lower levels than the original.

A radio tuner is a radio receiver without audio amplifier. A tuner should pass the complete audio bandwidth, should have high sensitivity and adjacent-channel selectivity, incorporate automatic frequency control (a.f.c.), and have cathode-follower output.

A hi-fi transcription record player usually consists of a drive mechanism and heavy non-magnetic turntable for rotating records at uniform speed (without wow or flutter), a long, non-resonant tone arm with minimum tracking error and adjustable stylus pressure, and a magnetic or high-quality ceramic cartridge with a diamond stylus.

The preamplifier is the hi-fi control centre that ties up the various components. It must provide tone equalization for different recording characteristics, as well as loudness, bass, and treble controls.

The main or basic amplifier provides additional voltage amplification and sufficient power amplification to drive the speaker system.

The moving-coil loudspeaker translates an audio signal into sound waves through the interaction (repulsion and attraction) of the magnetic field of a speech coil, carrying the audio currents, with a strong magnetic field produced by a permanent magnet.

Since it is very difficult to design a single speaker to reproduce the entire audio band, coaxial or triaxial speakers with separate diaphragms or separate drivers, or completely separate multiple-speaker systems are employed. These consist generally of a woofer to reproduce the low frequencies, a tweeter to reproduce the highs, and sometimes also a mid-range speaker for medium frequencies. Crossover networks must be provided to separate the frequency bands to be fed to each speaker.

Baffles and loudspeaker enclosures prevent the sound radiated from the front of the speaker from reaching the sound radiated from the back, which would result in cancellation of low frequencies.

The flat, open baffle works well, but is too large in size; the infinite baffle is an enclosed box that completely prevents the front-radiated sound from reaching the rear; it also raises the resonant frequency of the speaker.

The bass-reflex enclosure provides a port to release rear-radiated sound that has been inverted in phase. This augments the bass response of the speaker and makes it more uniform.

A tape recorder employs a ring electromagnet (recording head) to magnetize an iron-oxide coated tape longitudinally in proportion to the amplitude and polarity of the audio currents flowing through the winding of the electromagnet.

Stereophonic (binaural) reproduction requires two completely separate channels to simulate the effect of hearing with two ears.

CHAPTER SEVENTEEN

TELEVISION

Television is one of those miracles which we have come to accept as part of our everyday world without the slightest sense of mystification. It is taken for granted that electronics can achieve seeing at a distance (the literal translation of 'television'), just as we have long ago accepted hearing at a distance. In other words we have become conditioned to the expectation that electronics can do anything, and we have stopped wondering. If the following explanations can bring into focus the magnitude of the marvel accomplished by television, it may help to restore your sense of wonderment.

PHYSICAL BASIS

Long before the advent of television, the cinema had taken advantage of the persistence of vision of the human eye to deceive us into seeing motion, when there was none. As every schoolboy knows, the movies display a series of still pictures in rapid sequence, each picture or frame showing a slightly more advanced phase of the continuous action. When this is done more often than 16 times per second (it is done 24 times per second in professional movies), the eye is no longer capable of separating the individual pictures because of its persistence of vision, and we obtain the impression of a smoothly blended, continuously progressing motion. Television uses this same deception of conveying moving pictures by sending a rapid series of changing still pictures. Although the motion of an actual scene adds to the complications, the basic problem of television really is the transmission and reception of a still picture.

When we look at an actual scene we see a continuum of light and shade, and colours of various wavelengths. This is no longer true when we look at a (black-and-white) photograph of the same scene. The photographic print has a limited (though huge) number of fine silver grains, each being 'developed' to a brightness corresponding to that of the same spot in the scene. By distributing a tremendous number of these silver grains of varying brightness over the picture area, the correct proportions of light and shade in the actual scene are reproduced in the image. You cannot see the little grains or dots in the picture, because there are so many of them, but when the picture is greatly enlarged they become visible. Moreover, when a photographic print is 'screened' for reproduction in books or newspapers (photoengraving), the image is broken down into a much smaller number of picture elements of varying light and shade than the fine grain of the original print, and then these picture elements become clearly visible. By looking at a newspaper picture, which employs a fairly coarse, clearly visible screen, you will discover that the picture is actually composed of many black dots, the dark areas containing large, closely spaced dots, while the light areas consist of smaller, more widely separated dots. Photographic reproductions in books use a finer screen, and you may have to look at them with a magnifying glass to discover the picture dots. The dot structure of an enlarged portion of a picture that has been screened is shown in Fig. 185.

A further example illustrates that images may be composed by assembling a large number of individual picture elements, or dots. Fig. 186a shows the outline of a cross composed of relatively few black dots with white spaces between them. You are not fooled by this, since the dots are clearly evident. However, if you walk about ten feet away from this crude picture, the dots will appear to blend into a solid greyish figure. In (b) of Fig. 186 we have heightened the illusion by providing many more black dots with fewer white

Fig. 185. Dot structure of enlarged portion of a screened picture

spaces between them. The cross now appears to be a solid grey figure, even at an ordinary reading distance. You must look closely to discover the separate dots.

The basic problem of television now becomes evident: it must break down a distant scene at the transmitter into many small picture elements of varying brightness, send these out in sequence using radio waves, and then reassemble all the elements at the receiver in their proper sequence to create a replica of the original picture. There must be a sufficient number of elements and they must be transmitted so fast, that the eye can neither detect their presence nor the process of reassembly. Moreover, a sufficient number of complete images must be sent each second so that the persistence of vision of the eye will blend

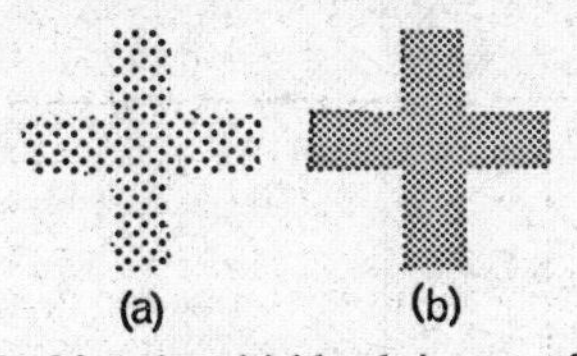

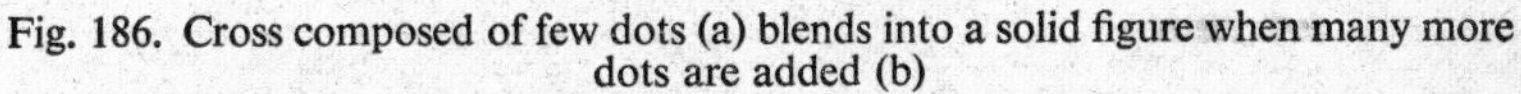

Fig. 186. Cross composed of few dots (a) blends into a solid figure when many more dots are added (b)

them into continuous motion. When you think of the tremendous number of picture dots required to make an image and the large number of images to be sent each second (25 per second in television), you will realize that the time allotted to form each picture element is in the order of millionths of a second. Only electrons can carry out a task as quickly as that.

COMPLETE TELEVISION SYSTEM

Fig. 187 is a simplified presentation of a complete television system for the transmission and reception of picture and sound signals. This figure and the following descriptions are based on the British 625-line (BBC 2) system.

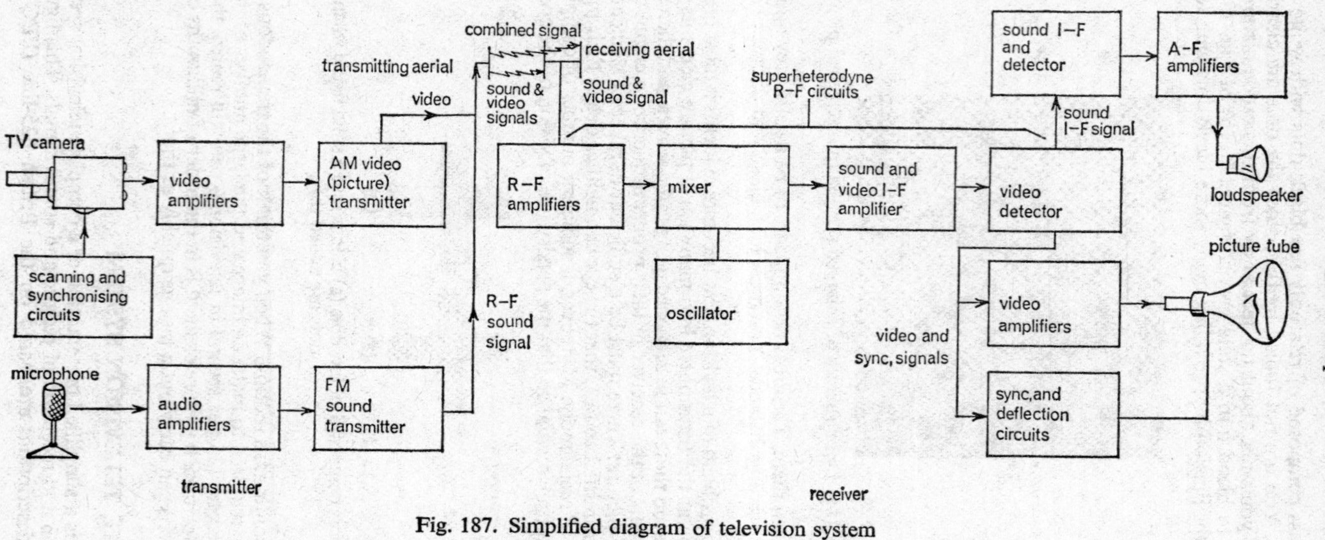

Fig. 187. Simplified diagram of television system

The British 405-line system differs in a number of important respects, but since it is planned to discontinue the 405-line standard, these differences will not be dealt with here. The 625-line standard is the one used in most of Europe and in Australia; television in the United States employs a 525-line standard, but its operation is similar in principle to that discussed here.

The television station sends out two separate r.f. carriers from a single aerial, one carrier being frequency-modulated by the sound (audio) signal, while the other is amplitude-modulated by the picture information or video signal. The two carriers are spaced 6·0 Mc/s apart.

At the television transmitter the picture and sound signals are handled separately. The television camera focuses an optical picture of the scene on to an electronic tube, which scans or breaks down the image into its picture elements and converts the varying brightness of the individual elements into a corresponding electrical, or video signal. It also adds several synchronizing signals to the video information, which are designed to keep the reassembly of the picture at the receiver in step with the scanning at the transmitter. This composite video signal is then strengthened by a number of video amplifiers (see Chapter 9) to a level sufficient to amplitude-modulate a radio-frequency transmitter. The carrier with its video modulation is sent out over the TV transmitting aerial.

The sound portion is a conventional frequency-modulation transmitter. The sound picked up by the microphone is strengthened by an audio amplifier, which frequency-modulates an r.f. transmitter with a carrier (centre) frequency 6·0 Mc/s above the video carrier. The frequency-modulated sound carrier is sent out over the same transmitting aerial used for the video carrier.

The television receiver, too, is a combination of the old and the new. The r.f. sound and video signals picked up by the receiving aerial are handled at first together by conventional superheterodyne receiving circuits. The desired television channel is selected by tuned circuits and the sound and video signals are strengthened together by a radio-frequency amplifier with sufficient bandwidth to pass both carriers and their modulation sidebands. The r.f. signal is then heterodyned in the mixer with a locally generated frequency to produce a lower intermediate frequency equal to the difference between the two signals (usually 39·5 Mc/s for the picture). The sound and video intermediate-frequency signals are amplified by several stages of i.f. amplification and then applied to a video detector.

The video detector has two functions:

1. it demodulates the composite video signal by means of a diode detector, just as is done in an a.m. broadcast receiver;
2. it separates the sound and video i.f. signals. The separation of sound and video is accomplished by beating together (heterodyning) the frequency-modulated sound i.f. signal and the amplitude-modulated video i.f. signal, which are spaced 6 Mc/s apart. Because of the detector's partially non-linear characteristic, it performs this mixing function automatically. The heterodyning produces a 6 Mc/s frequency-modulated difference frequency, which is the sound i.f. signal. Filter circuits in the output of the detector separate this 6 Mc/s sound i.f. signal from the demodulated composite video signal.

The sound i.f. signal is applied to the separate sound portion of the receiver, which is identical to the corresponding circuits in an f.m. broadcast receiver. The sound signal passes in succession through an i.f. amplifier, a limiter and discriminator (or a ratio detector), one or two stages of audio amplification and a loudspeaker.

The demodulated composite video signal from the output of the video detector is applied to the video portion of the receiver. The video signal is amplified by a video amplifier and then reassembled by the electron beam of a cathode-ray tube into a visible image on the screen. The composite video signal is also fed to a 'sync' separator, where the synchronizing signals are separated from the remainder of the video signal. The sync signals are then applied to the beam-deflection circuits to keep the electron beam that reassembles the image on the screen in step with the scanner at the transmitter.

The TV receiver discussed here and indicated in Fig. 187 is known as the intercarrier type because of the way the sound i.f. signal is obtained by heterodyning the video and sound carriers. In the older types of 405-line receiver, called split-sound receivers, the sound signal is split off at the mixer and then handled completely separately.

It is evident from Fig. 187 that we are already familiar with the f.m. sound transmitting and receiving circuits (see Chapter 15), video amplifiers (Chapter 9), and the superheterodyne receiving circuits for sound and video (Chapter 15). Let us concentrate, therefore, for the remainder of the chapter on the new and unfamiliar portions, such as the scanning, synchronizing, and deflection circuits, the TV camera and the picture tube.

TELEVISION CAMERAS

In the TV camera the video signal begins its long journey to the picture tube in the receiver. The camera must 'see' the actual scene to be televised and convert the optical image of the scene into an equivalent electrical image. The

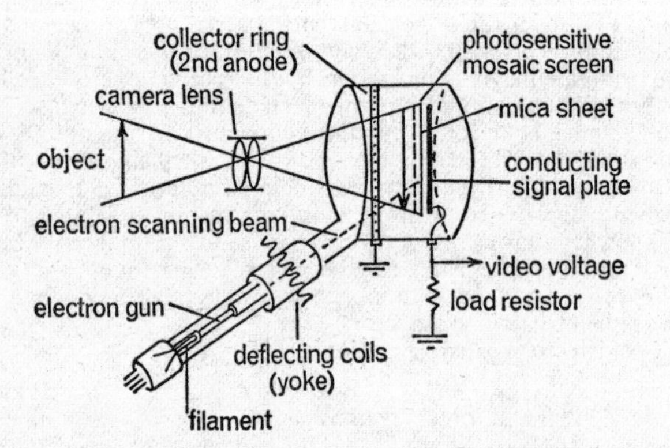

Fig. 188. Elements of an iconoscope camera tube

picture elements of this electrical image must then be 'scanned' to provide a video signal whose instantaneous magnitude corresponds to the brightness of the individual elements. One form of camera tube, called the iconoscope, is shown in schematic form in Fig. 188.

In brief, the action of an iconoscope is as follows. Light from the illuminated scene (an arrow, in this case) is focused by means of optical lenses on to a photo-sensitive screen, called the mosaic. The mosaic is a coating of millions of light-sensitive caesium globules deposited on one side of a thin sheet of mica. Each photo-sensitive globule is about one-thousandth of an inch in size and is insulated from all neighbouring globules by the mica. The other side of the mica sheet, the signal plate, is coated with a conducting film of graphite. The globules insulated by the mica from the metallic coating form myriads of tiny electric capacitors, all having the mica dielectric and the metallic signal plate in common. Each light-sensitive globule, therefore, emits electrons and charges up its individual capacitor in accordance with the intensity of the light striking it. (Since electrons are emitted or lost, each capacitor is charged positively.) The entire mosaic plate, thus, has a charge distribution corresponding to the variations in light and shade of the original picture. The upshot is that the mosaic plate stores in its charged globules an electrical image of the optical picture focused upon it.

Obtaining a video signal. The electrical image stored on the mosaic screen cannot be transmitted as a whole, but the individual picture elements must be scanned one at a time by discharging the globule-capacitors in an orderly sequence. This is accomplished by an electron scanning beam formed by the electron gun in the narrow elbow of the tube. The action of this electron gun is identical to that of the conventional cathode-ray tube discussed in Chapter 6. The gun contains an electronic lens system of charged electrodes, which produce a sharply focused electron beam. This beam is aimed at the mosaic through the attraction of the highly positive (about 1,000 V) second anode, which consists of a metallic coating on the inside of the glass tube, known as collector ring. Horizontal and vertical deflecting coils, mounted at right angles in a yoke around the neck of the tube, provide magnetic deflection of the electron beam to scan the electrical image on the mosaic. As we shall see later, this is done in an orderly fashion from left to right and top to bottom of the mosaic, one line at a time.

When the scanning electron beam strikes each globule, the electrons fill in the 'holes' left by the previous photoelectric emission of electrons. The beam thus neutralizes the previous positive charge due to photoemission and, in effect, discharges the globule-capacitor. At the instant of discharge a rush of current flows through the load resistor, which is equal to the positive charge stored on the globule and, hence, is proportional to the light illumination of the picture element represented by the globule. This discharge current flowing through the load resistor builds up the video voltage, which is fed to the succeeding video amplifier. As the entire mosaic is scanned, the electrical image stored on it is converted successively into a video voltage of varying instantaneous magnitude, which corresponds to the illumination on the individual globules.

Image orthicon. The video output of the iconoscope is rather low and it requires a brightly illuminated picture to be useful. The iconoscope has been largely replaced, therefore, by another camera tube, the image orthicon,

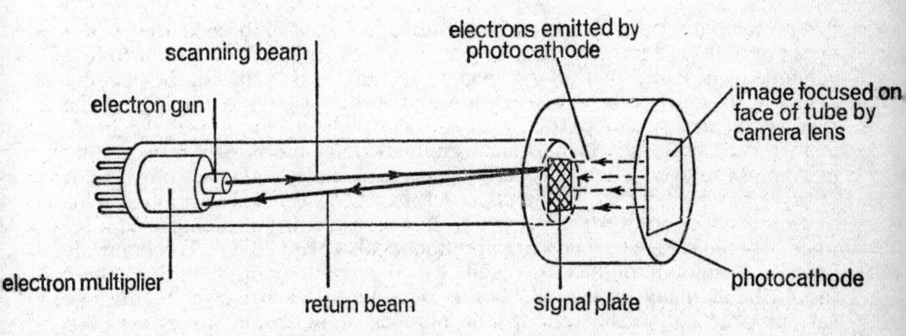

Fig. 189. Elements of an image orthicon camera tube

which is far more sensitive and can televise anything that is visible to the naked eye. The image orthicon owes its exceptional sensitivity to the electron multiplier action of a series of secondary-emission electrodes, or dynodes (Fig. 189). We have already discussed the electron multiplier in Chapter 6, and other features of the orthicon are similar to those of the iconoscope. There is a third camera tube, known as vidicon, which is simpler than the other cameras, but it provides less fineness of detail (resolution) and, hence, is used primarily for televising from film.

SCANNING AND SYNCHRONIZING

We appreciate by now the need for scanning and understand in a general way how it is done, but we have not yet explored the manner of deflecting an electron beam to obtain the desired scanning pattern, and how 'sync' signals are used to keep the transmitter and receiver scanning exactly in step. Evidently, if an image is to be assembled on the screen of the TV receiver at the same time as the picture elements of the actual scene are being televised, the scanning at the transmitter and receiver must be done in exactly the same manner and in perfect synchronism.

Progressive scanning. Scanning may be carried out in the same way as reading a page in a book. You start at the top, read all the words in the first line from left to right, then return rapidly to the left to read the next line, and so on, until you arrive at the bottom of the page. Progressive horizontal scanning, illustrated in Fig. 190, is done in this fashion. As shown in (a), the electron beam at the transmitter and receiver is made to sweep across a horizontal line, covering all the picture elements, and is then quickly returned to the left to scan the next line. Both camera and picture tubes are blanked out during

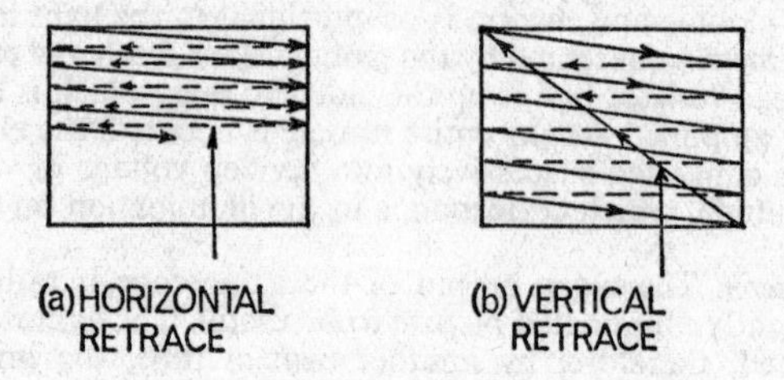

Fig. 190. Progressive horizontal scanning

this horizontal retrace period (shown dashed in Fig. 190) to make the retrace lines invisible. The retraces must be very rapid, of course, in order not to lose valuable picture information. As each horizontal line is scanned, the position of the beam must be progressively lowered so that the same line will not be repeated. This is accomplished by a vertical scanning motion of the beam from top to bottom, which is superimposed upon the horizontal scanning motion. Moreover, after the scanning beam completes the bottom line, it must be rapidly returned to the upper left-hand corner, as shown in (b). This motion is called the vertical retrace to distinguish it from the horizontal retrace between each line. The vertical retrace is also blanked out. To obtain the maximum amount of picture detail, called resolution or definition, there should be as many horizontal lines as possible for each image. Because of various practical considerations, the number of horizontal lines has been standardized at a total of 625 per image or frame.

Interlaced scanning. To produce the illusion of motion the individual still pictures or frames must be displayed so rapidly that the persistence of vision of the eye will blend them smoothly. In television, the frame repetition rate has been standardized at 25 per second. Despite this high repetition rate, a certain amount of annoying flicker is still present. To eliminate this residual flicker, each frame is shown twice on the face of the picture tube. To accomplish this, a scanning pattern slightly different from progressive scanning must be used. Fig. 191 illustrates the interlaced horizontal scanning pattern, which divides the total number of lines into two groups of lines, called fields.

As you can see, during the presentation of the first field only the odd-numbered lines are scanned, while during the second field all even-numbered lines are scanned. Halfway along the bottom line of the first field (line 7 in the figure), the vertical retrace returns the scanning beam to the top of the image

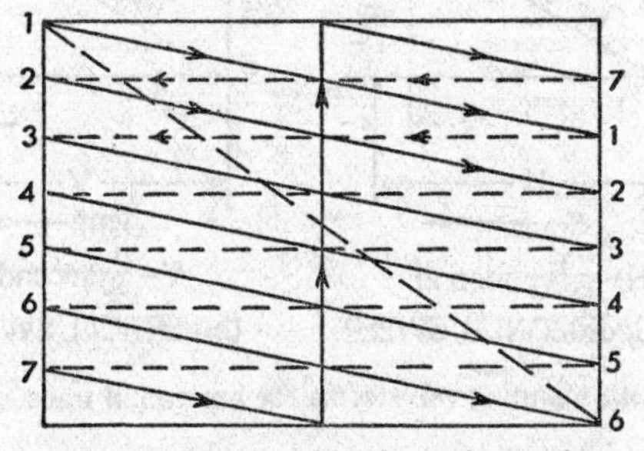

Fig. 191. Interlaced horizontal scanning

and completes the unfinished lines. The remainder of the even-numbered lines are then scanned during the second field. For clarity only seven lines are shown in Fig. 191. Actually, of course, each field consists of exactly one-half of the 625-line total per frame, or 312·5 lines. Since two fields are shown for each frame, the repetition rate of the fields is 50 per second, or twice that of the frame repetition rate. Being equal to the a.c. mains frequency, the 50-cycle scanning frequency simplifies the design of the receiver and transmitter power-supply filters.

Note that the scanning patterns illustrated in Figs. 190 and 191 always have the same ratio of picture width to picture height. This ratio of width to height is called the aspect ratio and it has been standardized at 4:3; that is, the width is 1·33 times height of the picture frame. This ratio remains the same, whether the actual picture is 2 ft × $1\frac{1}{2}$ ft or possibly 20 ft × 15 ft.

How scanning is carried out. With 625 horizontal lines included in each frame, and 25 frames being scanned per second, the total number of horizontal lines scanned per second is 625 × 25, or 15,625 lines per second. The horizontal repetition rate, or line scanning frequency, is thus 15,625 c/s. Similarly, the frame scanning frequency is 25 c/s, and since each frame is scanned in two fields, the field scanning frequency is 25 × 2 or 50 c/s.

Scanning is accomplished by deflecting electron beams in synchronism at the transmitter and receiver. To do this, a deflection voltage or current must pull the electron beam horizontally across the tube and vertically up and down at the line and field scanning frequencies, respectively. (A deflection voltage is used for electrostatic deflection while a current must be used for magnetic deflection.) The amplitude of this voltage or current must rise linearly to provide equal increments in amplitude for equal intervals of time. At the end of each scanning line or field, the deflection voltage (or current) must snap back to zero in minimum time; that is, the retrace must be as short as possible. A waveform that provides this relatively slow linear rise in amplitude and rapid retrace to zero has the appearance of the teeth of a saw and, hence, is called a saw-tooth waveform. The saw-tooth scanning waveforms for horizontal and vertical scanning (called sweep) are shown in Fig. 192.

Note that the appearance of the saw-tooth scanning waveforms is the same for horizontal and vertical sweep and that only the time relationships have

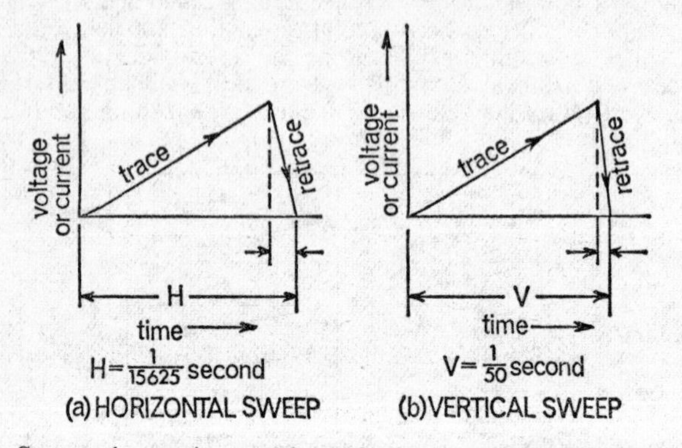

Fig. 192. Saw-tooth scanning waveforms for horizontal trace and vertical sweep

been changed. The time allowable for sweeping a horizontal line (Fig. 192a) is 1/15,625 second. This includes the retrace, which takes about 15% of the total time. In contrast, the time permissible for sweeping a vertical field (Fig. 192b) is considerably longer, being equal to 1/50 second. The retrace for vertical sweep is only about 7% of the trace time.

The required linear saw-tooth scanning waveforms may be obtained either through the approximately linear rise in the current through an inductance, or the linear rise in voltage across a capacitor that is being charged. Actually, neither of these waveforms is quite linear, but is curved in an 'exponential' manner. One can get around this by using only a small portion of the exponential waveform, which is approximately linear. Thus, by only partially charging a capacitor through a very high resistance, the voltage across the capacitor will build up approximately in the manner of a linear saw-tooth waveform. Then, without waiting for the capacitor to become fully charged and acquire its exponential charging curvature, the capacitor is suddenly discharged through a relatively low resistance and its voltage drops quickly to zero, as required by the retrace portion of the saw-tooth waveform. In practice, the rapid switching from charge to discharge is accomplished by a valve oscillator with the aid of the synchronizing signals, as we shall see presently.

SYNCHRONIZATION

To maintain the correct timing of the vertical and horizontal sweep motion and keep the receiver and transmitter locked in step, synchronizing signals must be sent out along with the picture information. These 'sync' signals, which have the form of rectangular pulses, are sent out periodically at the line- and field-scanning frequencies. To keep the horizontal line scanning synchronized, a horizontal sync pulse is transmitted for each horizontal line;

hence, the horizontal sync frequency is 15,625 c/s. Similarly, a vertical sync pulse is sent out for each field to keep the vertical scanning in step; the vertical sync frequency, thus, is 50 c/s. To prevent interference with the picture on the TV screen, the sync pulses are transmitted during the scanning retrace time, when the screen is blanked out and no picture information is transmitted. Another pulse, the blanking pulse or pedestal, must be transmitted to accomplish this blanking out. Fig. 193 illustrates the appearance of a composite

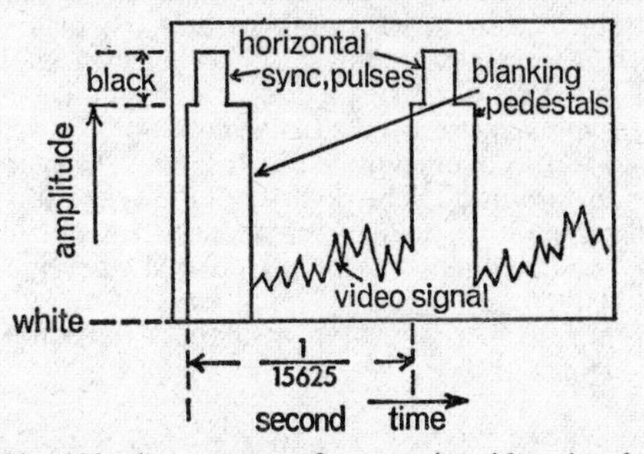

Fig. 193. Appearance of composite video signal

video signal (after detection); it contains the video signal as well as horizontal sync and blanking pulses. The vertical pulses are similar in appearance, but their action is somewhat more complicated, as we shall see later on.

Note, first of all, that the lowest values of the signal amplitude correspond to the brightest parts of the picture (white), while the highest amplitudes correspond to the darkest parts, or black. This is known as inverse or negative modulation, since the instantaneous video signal amplitude is inversely proportional to the light intensity. The chief reason why this system has been adopted is its relative freedom from noise interference. With negative modulation, a strong noise pulse will simply blank out the screen momentarily (and probably unnoticeably), while with positive modulation a similar pulse can cause annoying light streaks across the receiver screen.

The video signal for two successive horizontal lines is shown in Fig. 193. A blanking pulse, or pedestal, is interposed during the retrace interval between

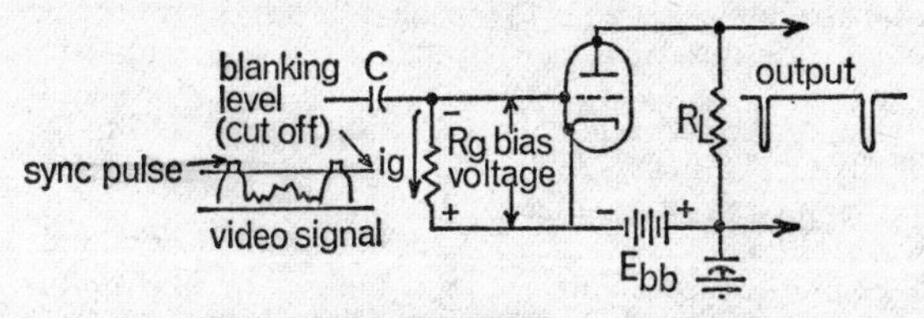

Fig. 194. Circuit of triode base clipper

each line, with an amplitude sufficient to drive the electron scanning beam to cut-off (i.e. into the black). On top of each blanking pedestal is perched a horizontal (or vertical) synchronizing pulse, which starts the horizontal (or vertical) retrace of the beam. The sync pulses do not affect the picture, since their amplitude is beyond the black level, where the picture is blanked out.

Sync separation. To utilize the sync pulses for controlling the timing of the beam deflection (sweep) oscillator, they must first be separated from the composite video signal. This is done by a sync separator circuit, which shaves or clips off all signals below the blanking level, so that only the sync signals are permitted to pass. A typical triode 'base clipper' for accomplishing this, is shown in Fig. 194.

The triode derives its bias from a grid-leak resistor, R_g, placed in its grid circuit. When the composite video signal is applied through capacitor C to the input of the valve, the grid is driven positive and a grid current flows through R_g, which makes the grid-connected end negative with respect to the cathode. The value of the grid resistor is chosen so that the bias voltage developed across it drives the triode to cut-off at the blanking level, as shown (inset). As a result, the valve is cut off at all times, except during the positive sync pulses, whose amplitudes are larger than the blanking level. The sync pulses permit a momentary flow of anode current, indicated by the negative output pulses across R_L in the anode circuit. The polarity of the sync output pulses does not matter, since pulses of either polarity may be used to control the timing of the sweep oscillator. If necessary, the polarity can be reversed by means of a phase-inverter circuit.

Synchronizing pulses. The appearance of the sync pulses after separation

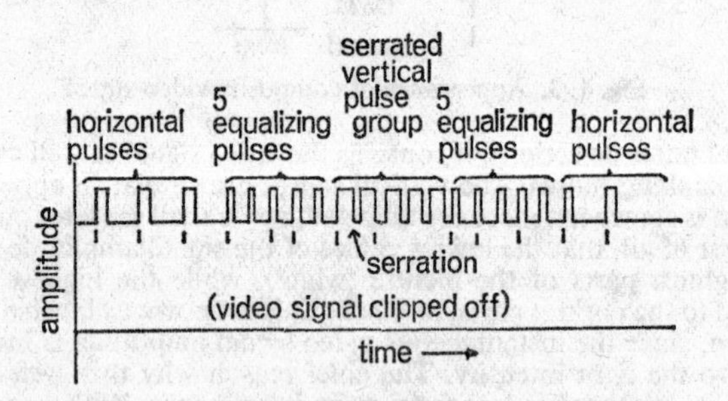

Fig. 195. Vertical and horizontal synchronizing pulses

form the composite video signal is illustrated in Fig. 195. The pulses have been made positive in polarity by means of a phase inverter.

The illustration shows from left to right:

1. three horizontal sync pulses for the last three horizontal lines of field;
2. five equalizing pulses, somewhat smaller in width than the horizontal pulses;
3. a prolonged vertical pulse group, consisting of five individual pulses separated by five serrations;
4. another group of five equalizing pulses;
5. three more horizontal pulses for the first three horizontal lines of the next field (second field of a frame). To complete the illustration you would have to add another 622 horizontal pulses, after which the sequence of equalizing and vertical pulses would be repeated. Note that all pulses in Fig. 195 have the same amplitude, but that they differ in width, or the length of time each lasts. Moreover, each vertical pulse extends over a period of three horizontal lines, the serrations being inserted at half-line intervals.

Though not immediately apparent, there are good reasons for this peculiar arrangement of horizontal, vertical, and equalizing pulses. The vertical pulse is made long so that it can easily be separated from the multitude of horizontal pulses. The serrations are inserted into the vertical pulse, so that horizontal

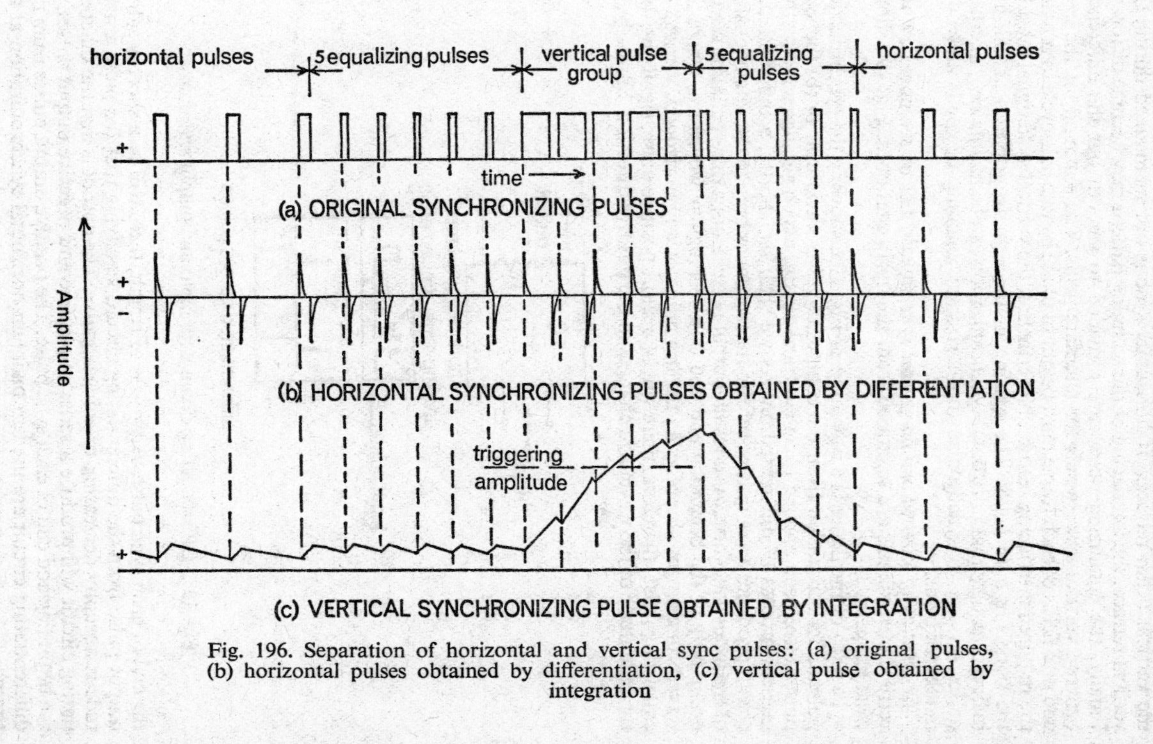

Fig. 196. Separation of horizontal and vertical sync pulses: (a) original pulses, (b) horizontal pulses obtained by differentiation, (c) vertical pulse obtained by integration

synchronization can be maintained over the three-line interval, during which the vertical beam retraces. If horizontal sync is not maintained during the vertical retrace, the interlacing of the scanning pattern may become distorted. Finally, the equalizing pulses are provided to smooth out the differences between vertical sync signals for alternate fields. The scanning for the even and odd fields of each frame must be out of phase by one half-cycle to provide the required interlacing feature. Nevertheless, the vertical sync signal must be the same for both fields. The spacing of the equalizing pulses at half-line intervals assures that horizontal synchronization can be maintained, alternate pulses serving to synchronize the horizontal scanning of even- and odd-numbered fields.

Separation of vertical and horizontal sync signals. Let us see now how the vertical sync signals are separated from the horizontal ones. Fig. 196 shows what happens.

Part (a) of the figure is a repetition of Fig. 195, showing the synchronizing pulses after separation from the composite video waveform. In (b) the sync pulses have been applied to an electrical circuit, which simulates the mathematical process of differentiation. As we shall see shortly, a differentiating circuit produces an output waveform that responds only to the rate of change of the input waveform. As a result, output pulses from a differentiating circuit occur only at the beginning and end of a rectangular waveform, where the waveform changes abruptly. At the beginning of any horizontal, vertical, or equalizing pulse, therefore, there will be a short positive pulse, signifying that the amplitude of the waveform changes in the positive direction. Similarly, at

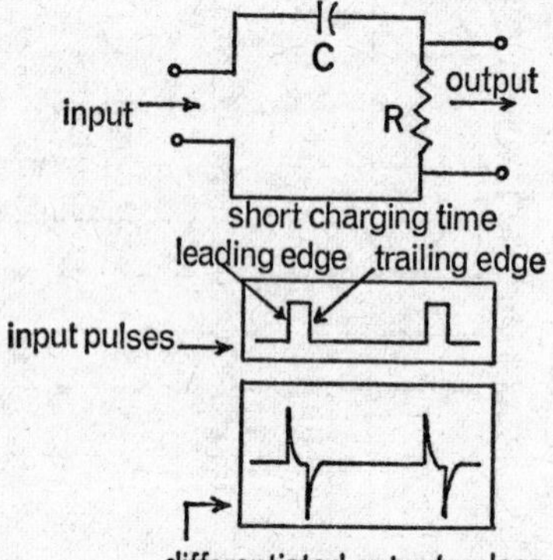

Fig. 197. Differentiating circuit with input and output waveforms

the end of each sync pulse, regardless of length, there will be a short negative output pulse from the differentiating circuit, signifying that the pulse amplitude is abruptly decreasing or going negative. The upshot is that the differentiating circuit will produce a series of positive and negative output pulses, at the beginning and end of each sync pulse. The positive output pulses from the differentiating circuit are used to maintain horizontal synchronization at all times.

In (c) of Fig. 196 the sync pulses have been applied to an integrating circuit, which performs the inverse of differentiation. During integration the total area (height times width) occupied by the pulses is added up, regardless of their individual shape or rate of change. However, as we shall see, the integrating circuit can perform this addition only for fairly wide, closely spaced pulses, such as the vertical pulse group. For brief widely spaced pulses, such as the horizontal and equalizing pulses, no addition takes place. As a result, only the vertical sync pulses produce an integrated sync pulse of sufficiently large amplitude for 'triggering' the vertical sweep oscillator to initiate vertical scanning.

Differentiating circuit. Fig. 197 illustrates a differentiating circuit in its simplest form. It is made up of a capacitor–resistor combination, with the input applied across both in series and the output being taken from the resistor. Any input pulse applied across the *R–C* combination, obviously, charges up the capacitor through the resistor, and, if the charging time of the capacitor is made short compared with the length of the applied pulses, the circuit will respond only to changes in the pulse waveform, thus fulfilling the requirement for differentiation. Thus, if a horizontal sync pulse comes along, as shown in Fig. 197, the capacitor will charge rapidly to the peak voltage of the leading edge of the pulse, resulting in a brief, positive charging pulse through output resistor *R*. The capacitor will hold its charge during the remainder of the input pulse, and no current flows until the trailing edge comes along. With the input voltage suddenly removed, the capacitor discharges rapidly through *R*, in the opposite direction. Thus, a negative discharge pulse flows through *R* during the trailing edge of the input pulse.

The result of the differentiating action is a series of brief positive and negative output pulses, respectively, for the leading and trailing edges of the input sync pulses, regardless of their length. The negative output pulses are rejected or clipped off, and the positive pulses are used to synchronize the horizontal

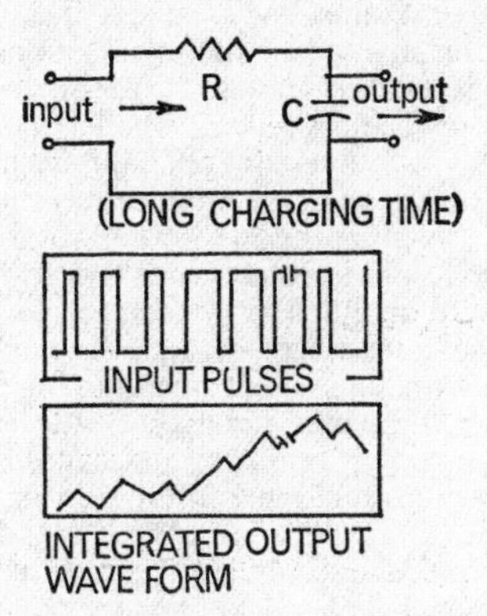

Fig. 198. Integrating circuit with input and output waveforms

sweep oscillator. It thus becomes evident that even the small breaks (serrations) in the vertical pulse group are sufficient to produce output pulses, which will permit the horizontal oscillator to stay in step. The only condition which must be fulfilled is to keep the charging time of the capacitor short compared with the pulse length. This can be attained by choosing the values of the resistor and capacitor so that their product, known as time constant, is small compared with the pulse length in microseconds. (The time constant, obtained by multiplying the resistance in ohms by the capacitance in microfarads, expresses the approximate charging time in microseconds.)

Integrating circuit. By taking the output from the capacitor rather than the resistor of the previous circuit, the simple *R–C* combination becomes an integrating circuit. (Fig. 198.) Since it is desired to accumulate or integrate the effect of a number of pulses, the charging time (time constant) of the capacitor must now be made long compared with the duration of the input pulses, so that no single pulse can charge up the capacitor completely.

Fig. 198 illustrates the action of an integrating circuit on a series of pulses of different length. The first three pulses, which are similar to equalizing pulses, are widely spaced and of short duration compared with the time constant of the circuit. As a consequence, each pulse partially charges the capacitor to a relatively low voltage and even this small charge leaks off through the resistance in the relatively long intervals between pulses. No total charge, and hence no voltage, is accumulated on the capacitor during these brief, widely spaced pulses. The picture changes radically, however, when the spacing between the pulses is less than their duration (width); this is illustrated by the next three pulses, which are similar to the vertical sync group. Because of its longer duration each input pulse now charges the capacitor to a considerably higher voltage; moreover, very little of the accumulated charge can leak off through the resistor during the brief time intervals between pulses. The result is a steady accumulation of charge, or voltage, until the next series of brief, widely spaced pulses arrives; these initiate a progressive discharge of the capacitor. Before this happens, however, the output voltage across *C* has reached a sufficient amplitude to trigger the vertical sweep oscillator into operation. (See Fig. 196c.)

We are now able to understand the effect of the equalizing pulses in smoothing out the differences in the vertical sync signals between alternate fields. As mentioned earlier, the horizontal sync pulses of alternate (even- and odd-numbered) fields must be spaced one half-cycle out of phase with each other to obtain the required interlaced scanning pattern. (See Fig. 191.) Because of this phase difference, the spacing between the last horizontal sync pulse and the first equalizing pulse also differs by one half-cycle. Thus, the capacitor of the integrating circuit starts to charge up a little earlier during the equalizing pulses of one field than for the other. If it were not for the equalizing pulses, the triggering voltage that initiates the vertical sweep would be reached a little sooner for one field than for the other, which would result in a poorly interlaced pattern. The equalizing pulses act as a 'buffer' between the last horizontal and first vertical sync pulse to compensate for this difference in timing. Though the capacitor accumulates a charge a little earlier during one field, this does not matter, since the equalizing pulses lose more charge than they contribute, so that any excess charge initially present is dissipated before the first vertical pulse comes along. The integrating action for the vertical pulses therefore starts with a clean slate and the triggering amplitude is reached at the same time during alternate fields.

SWEEP OSCILLATOR

Let us now look briefly at a popular oscillator circuit that is used to sweep the electron scanning beam of a picture tube either horizontally or vertically. The blocking oscillator illustrated in Fig. 199 is used for vertical deflection, but, by changing the component values, it becomes equally suitable for horizontal sweep.

The blocking oscillator is essentially a simple valve oscillator of the type discussed in Chapter 11, but it has a few refinements added to make it suitable as a linear-deflection (sweep) oscillator. As you can see, the anode circuit of the triode is coupled back to the grid circuit through the transformer, whose windings are common to both circuits. Because of the 180° phase reversal

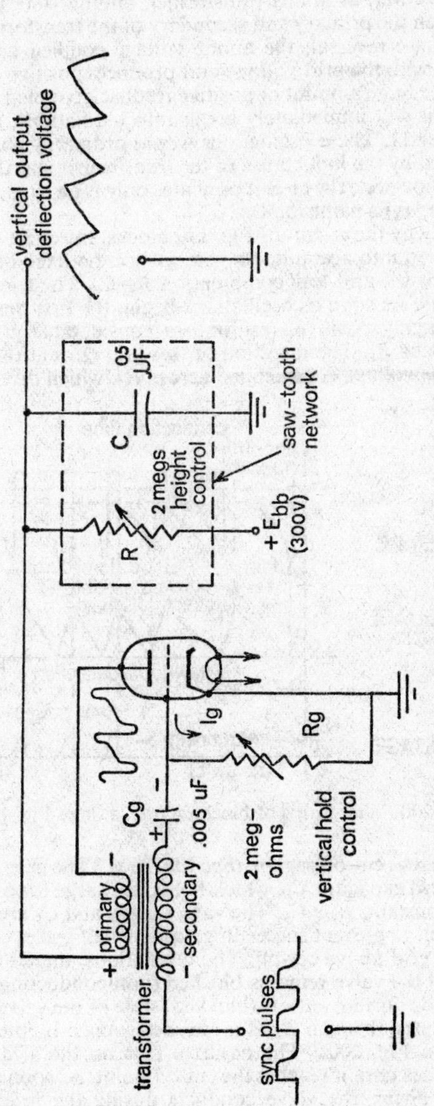

Fig. 199. Typical blocking oscillator for vertical deflection

occurring within the triode, the anode voltage is out of phase with the grid input voltage. However, as in any transformer, another 180° phase inversion takes place between the primary and secondary of the transformer. As a result of this double phase reversal, the anode voltage coupled back to the grid circuit is in phase with the grid voltage and produces positive or regenerative feedback. With the large amount of positive feedback coupled in by the transformer, the circuit will immediately break into oscillations, as we have explained in Chapter 11. These oscillations would ordinarily continue at a frequency determined by the inductance of the transformer and the stray capacitances. Actually, however, the circuit oscillates only for a single cycle and then 'blocks' for a time, as its name implies.

To understand why the circuit of Fig. 199 blocks, the effect of the grid-leak bias has to be taken into account. As you can see, no fixed bias is provided: its place is taken by the grid-leak combination R_g–C_g. The triode is, thus, initially without bias. As soon as oscillations begin, the first positive half-cycle drives the grid highly positive, resulting in considerable grid-current flow through grid resistor R_g. The direction of this grid current (i_g) is such that a large negative bias voltage is developed across R_g, which drives the valve far

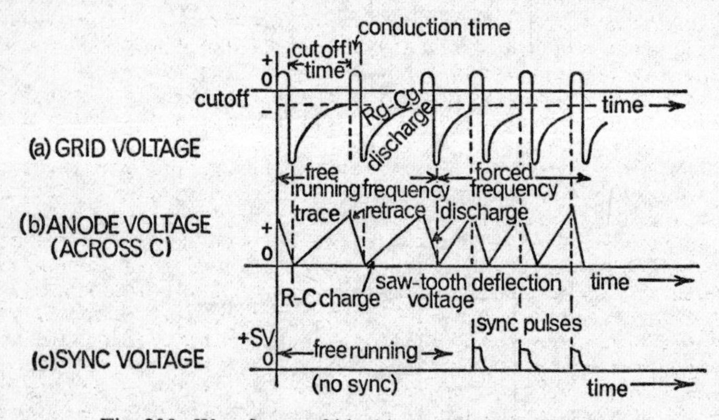

Fig. 200. Waveforms of blocking oscillator of Fig. 199

into the anode-current cut-off region. (See Fig. 200a.) The negative bias voltage also charges up grid capacitor C_g, which holds its charge for a period depending on the time constant, $R_g \times C_g$. The values of R_g and C_g are chosen so that the bias is sufficient to prevent succeeding positive half-cycles of the oscillation from driving the grid above cut-off. The oscillations, therefore, die out after the first cycle and the valve remains blocked (non-conducting) for a time.

During the anode-current cut-off (blocked) state of the valve, grid capacitor C_g starts to discharge through R_g at a rate determined by the time constant, as is illustrated in Fig. 200a. The negative bias on the grid, consequently, gradually diminishes until it reaches the cut-off point. As soon as this happens, oscillations start again, the valve conducts during the brief positive grid-voltage pulse, and is again blocked by the action of the grid-leak bias. The entire cycle described above then repeats. The upshot is that the valve alternately conducts a short anode-current pulse and then cuts off or blocks, as shown by the waveform of Fig. 200a. The value of the product of R_g and C_g (i.e. the time constant) in Fig. 199 has been chosen to obtain a pulse repetition frequency in the 50 c/s range, as required for vertical scanning. Grid resistor

R_g is variable to permit adjusting the repetition frequency (vertical 'hold'). For a horizontal deflection oscillator, the values of R_g and C_g must be selected to produce a pulse repetition frequency of 15,625 c/s.

Producing the saw-tooth sweep voltage. To adapt the basic blocking oscillator (Fig. 199) for deflecting an electron beam, an R–C network must be added in the anode circuit to produce the required saw-tooth voltage. Note that the R–C network in the anode circuit of Fig. 199, consisting of a 2-megohm variable resistor and 0·05-microfarad capacitor, has a time constant that is ten times that of the grid-circuit network R_g–C_g. The large time constant (corresponding to about 0·1 s) is chosen so that anode capacitor C will charge only to a fraction of its full charge during each grid-voltage cycle. As a result, the capacitor operates only over a small, approximately linear portion of its exponential charging curve and, thus, produces a linear trace.

The action of the R–C saw-tooth network is simple. Whenever the blocking oscillator is cut off, the capacitor is charged through 'height' control R by the positive anode-supply voltage, E_{bb}. This forms the linear-deflection voltage or trace, illustrated in Fig. 200b. The amplitude of the deflection voltage, and hence the vertical picture size, is adjustable by the variable 'height' resistor, which controls the amount of charging current into C. (A similar control for adjusting the picture width is present in the horizontal deflection oscillator.) As soon as the valve conducts during the positive grid-voltage pulse, capacitor C discharges rapidly through the relatively low anode resistance of the triode, and the retrace (discharge) portion of the saw-tooth wave is formed. (Fig. 200b.) The cycle continues to repeat, with capacitor C alternately charging through R during the blocked intervals of the valve, and discharging through the valve during the conducting periods. The charge and discharge intervals are equal to the repetition frequency of the blocking oscillator, so that the saw-tooth deflection voltage has the correct field (or line) scanning frequency.

The circuit of Fig. 199 is not as efficient as is desirable, since it combines the functions of a blocking and saw-tooth oscillator, which results in some interaction between the grid- and anode-circuit controls. The blocking oscillator, therefore, is frequently separated from the saw-tooth network through a separate discharge valve. The discharge valve is another triode, whose grid and cathode is connected in parallel with the blocking oscillator valve, and the saw-tooth network is inserted into the anode circuit of the discharge valve. Since the grid of the discharge valve is connected to that of the blocking oscillator, it has the same waveform (Fig. 200a), and hence is switched on and off at the frequency of the blocking oscillator. The anode-circuit action of the discharge valve is identical with that described for the single-valve oscillator. The blocking oscillator and discharge valve are usually combined in the single envelope of a double triode.

Obtaining synchronization. Although the free-running frequency of the blocking oscillator is controlled by its grid-circuit network (R_g–C_g), this frequency is neither exact nor stable, and hence some means must be provided to lock the oscillator exactly in step with the scanning frequency of the TV transmitter. This is done by the sync signals, as previously mentioned. In practice, the positive sync pulses are applied in series with the grid winding of the transformer (Fig. 199), so that their amplitude subtracts from the negative bias voltage produced by the oscillator. The triggering action is illustrated in Fig. 200c. Here the first vertical sync pulse comes along at a time when the negative bias voltage has declined considerably, but has not as yet reached the cut-off value. The slight additional positive voltage (about 5 V) of the sync pulse is sufficient to raise the valve above cut-off and trigger the next cycle of the oscillator. The oscillator then goes through its complete charge and discharge cycle, until the next sync pulse again triggers it into action, just before

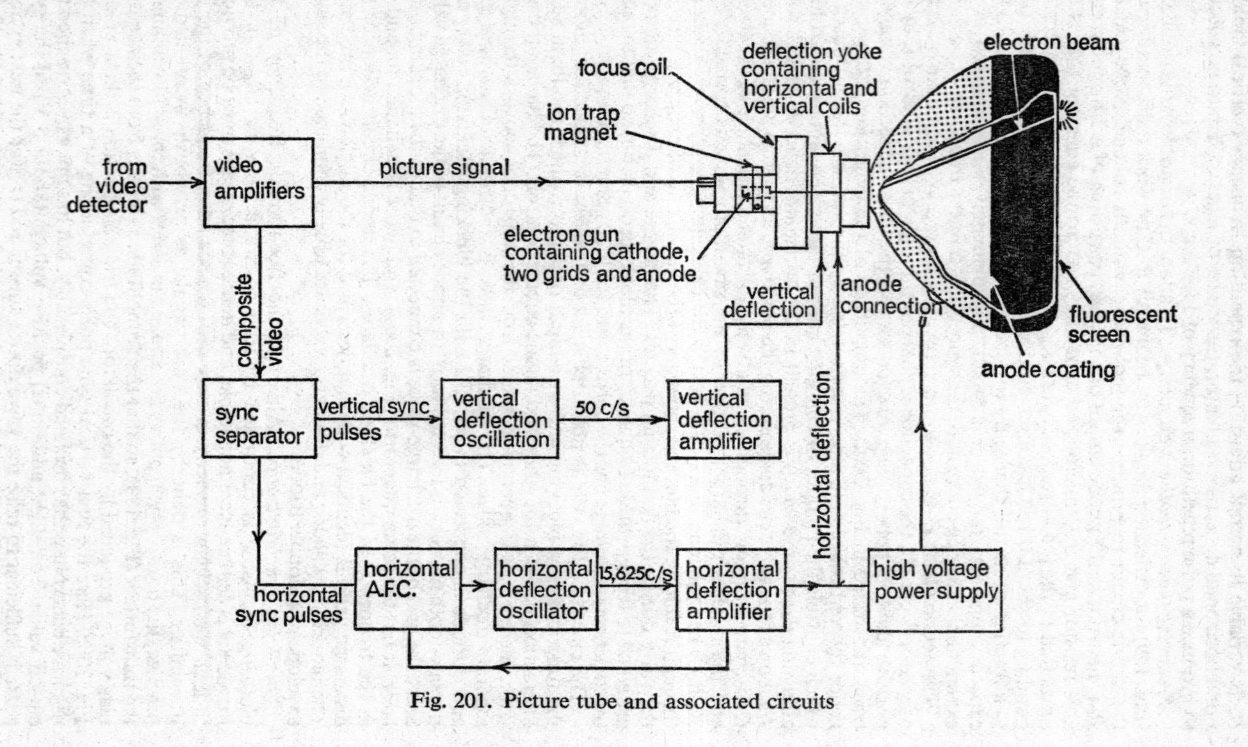

Fig. 201. Picture tube and associated circuits

reaching the cut-off-bias value. Note that the sync triggers always occur slightly ahead of the time when the oscillator would start up by itself and, hence, the forced frequency due to sync action is somewhat higher than the free-running frequency of the oscillator. To attain the proper timing of the sync pulses, therefore, the free-running frequency of the oscillator must be set at a value lower than the sync frequency. The 'vertical hold' control R_g in the grid circuit permits adjusting the free-running oscillator frequency so that it will lock in with the arriving sync pulses.

Note also, in Fig. 200, that only positive sync pulses can raise the grid of the valve above cut-off. The negative horizontal sync pulses produced by differentiation (see Fig. 196b) are, therefore, automatically rejected by the horizontal blocking oscillator. As a matter of fact, since the horizontal line scanning frequency of 15,625 c/s must be held extremely accurately, the horizontal sync circuits are somewhat more complicated than those for the vertical deflection oscillator. We shall go into this matter shortly.

PICTURE TUBE AND ASSOCIATED CIRCUITS

The crux of any television system is the picture tube, which displays the video information sent out by the TV transmitter. The picture tube in practically all modern television receivers is a magnetic-deflection cathode-ray tube, which is discussed at length in Chapter 6 (Fig. 35). We need not go into details, therefore, of how the horizontal and vertical deflecting coils in the yoke around the neck of the tube produce the proper scanning of the beam produced by the electron gun. Instead let us pick up some loose ends connected with the picture tube and its associated circuits. Fig. 201 shows in block-diagram form the video and deflection signals applied to the tube and the circuits generating them.

Scanning raster. Before any picture information can be displayed on the face of the tube, a picture frame or scanning raster must be provided, which corresponds to the interlaced scanning pattern illustrated in Fig. 191. This is accomplished by the synchronizing and deflection circuits we have described in detail. The demodulated composite video signal from the video detector is amplified by one or more video amplifiers and then applied to the sync separator. Here a base-clipper circuit separates the sync pulses from the video information, an integrator obtains vertical pulses, and a differentiator circuit provides the horizontal sync pulses. The vertical sync pulses are then applied to the vertical-deflection oscillator to keep its pulse repetition (sweep) frequency in step with the 50 c/s field-scanning frequency at the transmitter. The synchronized vertical-deflection (saw-tooth) voltage from the output of the oscillator is strengthened by a power amplifier stage, the vertical-deflection amplifier, and then fed to the vertical-deflecting coils in the yoke around the picture tube. The vertical coils produce the up-and-down deflection of the electron scanning beam, the maximum height of the raster depending on the amplitude of the saw-tooth current in the coils.

AUTOMATIC FREQUENCY CONTROL

The horizontal sync and deflection circuits are similar to the vertical circuits, except that automatic frequency control (a.f.c.) is generally added. The horizontal-deflection oscillator is especially vulnerable to noise voltages, which it might mistake for sync pulses, resulting in faulty synchronization. Hence, rather than using any single pulse, the horizontal oscillator frequency is controlled by the average of a number of horizontal-sync pulses. This is done by an automatic frequency control circuit, similar to the one discussed in

Chapter 15 in connexion with f.m. transmitters (Fig. 135). The a.f.c. circuit contains a phase detector which compares the difference in frequency or phase between the horizontal-deflection voltage and the horizontal-sync pulses. The output of the phase detector is a d.c. control voltage that is proportional to this frequency or phase difference. The d.c. control voltage is applied to the grid of the horizontal blocking oscillator and thus changes its bias by the amount necessary to bring the oscillator back into synchronism with the average frequency of the horizontal sync pulses. The 15,625 c/s horizontal sweep frequency is then amplified by the horizontal-deflection amplifier and finally applied to the horizontal-deflection coils in the yoke. This produces the width dimension of the scanning raster in accordance with the amplitude of the horizontal saw-tooth deflection current.

Picture signal. Now that we have the proper scanning raster on the face of the picture tube, we are ready to display the picture. This is accomplished by modifying the brightness of the electron scanning beam in accordance with the instantaneous amplitude of the picture signal. As illustrated in Fig. 201, the output from the video amplifiers is applied to the grid-cathode circuit in the electron gun of the picture tube. The grid will thus control the instantaneous beam current impinging on the fluorescent screen in exact relation to the amplitude of the picture signal. As a result, a replica of the individual picture elements scanned at the transmitter is produced by the changes in brightness of the scanning spot on the fluorescent screen of the picture tube.

Picture-tube accessories. The magnetic-deflection picture tube illustrated in Fig. 201 contains a few accessories that are worth noting. A magnet, known as ion trap, is placed around the neck of the tube in the vicinity of the electron gun. This magnet literally traps the negative gas ions, which are inevitably present in any large, evacuated tube. Because of their relatively large mass, these negative ions are not easily deflected, and if allowed to reach the screen, they will eventually burn a brown spot in the centre of the screen. The ion-trap magnet produces a sufficiently strong magnetic field to divert the ions to the side of the tube and thus prevent them from reaching the screen.

Although the electron gun produces an electron beam that comes to a focus at the crossover point (see Fig. 31), the beam diverges rapidly again after this point. To focus the beam at a point on the screen, an additional focus coil or permanent magnet is placed around the neck of the tube. The focusing action is explained in Chapter 6.

High-voltage power supply. Every television set contains a conventional rectifier power supply to provide the proper voltages to the electrodes of the valves, including the electron gun of the picture tube. This is called the low-voltage power supply to distinguish it from the high-voltage power supply that supplies from 5,000 to 10,000 V to the anode of the picture tube. A high voltage of this order is required to produce an intense electron beam that strikes the screen with sufficient force to attain a bright picture over the entire area.

Note that this final accelerating anode takes the form of a conductive coating (a form of graphite called Aquadag) on the inside wall of the tube near the flared portion, almost up to the screen. Since the anode coating is on the side wall of the tube, the electron beam—though accelerated by it—does not strike it. What actually happens is that the electrons strike the screen (in accordance with the deflection voltages), where they produce considerable secondary electron emission due to the force of the impact. These secondary electrons are immediately attracted to the highly positive anode coating from where they are returned to the cathode via the high-voltage power supply. Thus, the electron beam has a complete circuit for the round trip from cathode to screen, to the anode coating, and back to the cathode through the power supply.

In practice, a trick is used to obtain the required high voltage from the horizontal output stage. With the horizontal deflection voltage applied to the highly inductive deflection coils, a sharp 'kickback' (back e.m.f.) of several thousand volts is produced during the retrace or flyback portion of the saw-tooth current. The reason for this kick-back energy is that the combined inductance of the yoke coils and horizontal output transformer resist the sudden change in current during the retrace. By inserting an additional winding in the horizontal output transformer, the kick-back voltage may be further stepped up to the value required for the anode voltage. The output from the high-voltage secondary winding is then rectified and filtered, and finally applied to the external anode connexion on the picture tube. Because of its origin, the circuit is frequently called fly-back or kick-back power supply.

TELEVISION BANDWIDTH AND CHANNELS

We have seen that a television system must transmit and reproduce a great deal of information. Each second 15,625 horizontal lines (625 each 1/25th second) must be scanned, and each line contains a great many individual picture elements. Because of the huge amount of information to be transmitted in an extremely short period of time, very high video signal frequencies, up to 5·5 Mc/s, are produced. In addition to the video information, a complete f.m. sound channel must be added to the overall TV bandwidth. The result is that a channel 8 Mc/s wide must be provided to transmit the complete TV (picture and sound) signal. Obviously, this tremendous frequency range cannot be

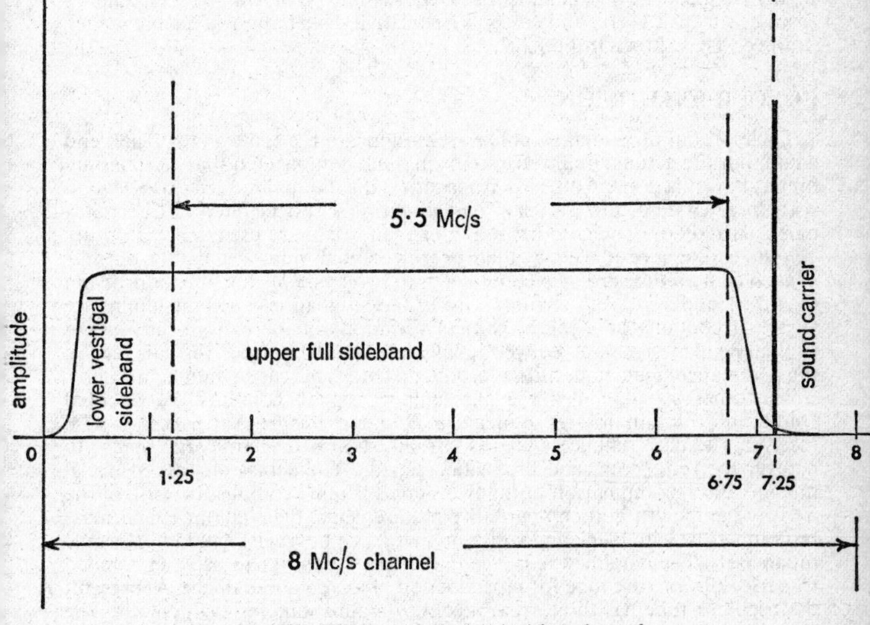

Fig. 202. Standard 8 Mc/s television channel

saddled upon a carrier in the medium-wave broadcast frequency range from 535 to 1,600 kc/s. Television is in fact broadcast at very high frequencies (v.h.f.) and ultra-high frequencies (u.h.f.). Each station is allocated a 'channel' of frequencies by international agreement, and there are 44 such channels between 470 Mc/s and 854 Mc/s in the u.h.f. bands.

Vestigial-sideband transmission. If television produces video frequencies up to 5·5 Mc/s, the two sidebands of the amplitude-modulated video carrier alone would embrace 11 Mc/s, or more than the entire video-plus-sound channel. To conserve space, therefore, the video carrier is transmitted on a single side-band, a process which we have touched upon in Chapter 15. To avoid distortion and simplify the receiver circuits, however, a small vestigial portion of the lower video sideband is also transmitted. This is known as vestigial-sideband transmission, and it is the process used in the British 625-line and the U.S. 525-line television.

Fig. 202 illustrates the distribution of the standard 8 Mc/s television channel. If the limit of the lower sideband is arbitrarily labelled zero frequency, the vestigial sideband extends up to 1·25 Mc/s, where the video carrier is located. The upper video sideband extends at full amplitude from 1·25 to 6·75 Mc/s, a range of 5·5 Mc/s, and at reduced amplitude up to 7·25 Mc/s. The f.m. sound carrier is positioned 6 Mc/s above the video carrier at a centre frequency of 7·25 Mc/s, and has two sidebands extending over a maximum of 200 kc/s. To attain this small bandwidth, narrow-band frequency modulation is used, with a frequency deviation of ± 50 kc/s maximum. (You will recall that the side-bands exceed the deviation from the centre frequency.) The small frequency band to the right of the sound channel is left open to avoid interference with an adjacent TV channel.

In practice, the 8 Mc/s TV channel range is placed, of course, at a much higher frequency. Thus, channel 21 extends from 470 to 478 Mc/s, channel 39 from 614 to 622 Mc/s, and so on. The relative distribution of each channel, however, is as shown in Fig. 202.

COLOUR TELEVISION

The essential principles of colour television are the same as for black and white, but the actual circuits are so much more complicated that we can only hint at how it is done. At the TV transmitter, the illuminated scene is televised with three camera tubes, each provided with an optical filter to transmit a particular colour. The colours, red, green, and blue are used, since all other colours can be reproduced by their proper combination. The colour outputs of each camera tube are then combined into two basic signals for transmission over the standard 8 Mc/s channel. One of these signals is called the luminance signal, it contains only the brightness variations of the picture, just as the ordinary video signal. A conventional black-and-white receiver can receive this luminance signal and, thus, reproduce a colour telecast in monochrome.

The other signal, called the chrominance signal, contains the essential colour information for reproducing a coloured image. At the colour TV receiver, the chrominance signal is combined with the luminance signal to recover the red, green, and blue video signals. The actual process of trans-mission and recombination of the two signals is quite complex because of the narrow bandwidth and compatibility requirements. (The colour signal must be compatible with black and white, so that it can be received in monochrome on any set.) The amplified red, green, and blue video signals are then applied to a tricolour picture tube for reproduction of the coloured image. At present, the tricolour tube has three separate electron guns, one for each colour. The three guns are so oriented that their beams pass at slightly different angles through the holes of a shadow mask and each strikes a phosphor dot that glows in the colour represented by the beam. A single-gun tube has recently been introduced and will probably replace this arrangement.

SUMMARY

Television is achieved by the consecutive transmission of still pictures, each broken up into thousands of tiny picture elements. The persistence of vision accounts for the illusion of motion.

A scanning electron beam in the TV camera tube breaks up the electrical image of the scene stored on the mosaic into the individual picture elements and generates a video signal whose instantaneous amplitude corresponds to the brightness of the elements.

The scanning pattern (or raster at the picture tube) consists of 625 interlaced horizontal lines for each picture or frame, which is broken up into two alternately even- and odd-numbered fields. 25 frames, or 50 fields, are transmitted each second. The horizontal line scanning frequency, thus, is 625×25 or 15,625 per second. The vertical field scanning frequency is 50 cycles per second.

To keep the transmitter and receiver scanning in step, horizontal and vertical sync pulses are transmitted for each line and field, respectively, at 15,625 and 50 c/s. The sync pulses together with the picture information make up the composite video signal.

The TV picture signal is transmitted over an amplitude-modulated video carrier, using a wide upper sideband and a vestigial lower sideband. The sound signal is transmitted over a frequency-modulated carrier, using narrow-band deviation.

A superheterodyne receiver is used to pick up the TV sound and video signals. In the intercarrier-type receiver, the signals are separated by heterodyning them in the video detector. Separate video and sound circuits are used from this point on.

CHAPTER EIGHTEEN

RADAR AND NAVIGATIONAL AIDS

Radar (*ra*dio *d*etection *a*nd *r*anging) detects objects or 'targets' and determines their distance (range) and direction (azimuth). It accomplishes this by sending out short bursts of radio energy that are bounced back to the sender by a distant target. The time it takes the radio wave to complete its two-way journey indicates the distance between sender and target.

Radar's basic principles are as old as radio itself. Both Hertz and Marconi experimented with reflected ultra short waves, the 'radio echoes' used in radar. Hertz demonstrated that electric waves could be reflected from plane or curved metal surfaces in accordance with the same laws that apply to light waves. By measuring the wavelengths and the frequency of the impulses, Hertz calculated their velocity and found it to be the same as that of light and all other electromagnetic radiations.

Nothing much was done after those early experiments, and it remained for Nikola Tesla to recognize and point out the highly practical application of the radio echo in the magazine *Century*, in June 1890. Thus Tesla foresaw radar in 1890 and he explained it by using the simple analogy of the sound echo. According to this analogy anyone who has ever shouted at the top of his lungs from a cliff and heard his echo bounce back, has engaged in an activity akin to radar: the sound projected forth in a certain direction has bounced off some cliff or mountain top and been returned to the sender.

If we happen to know that sound travels about 1,100 feet each second we can determine the approximate distance to the reflecting object from the length of time it took for the echo to come back to us. For example, if it takes two seconds for an echo to return after the initiation of the shouting, we will know immediately that the sound waves must have travelled a distance of $2 \times 1,100$ or 2,200 feet and, since it takes just as long for the sound to reach the reflecting object as for the echo to come back, the target distance must be 1,100 feet.

We also know, of course, the approximate direction in which the target (the cliff or mountain top) is located from the direction in which we were shouting. We could even make a crude map of the nearest hills or mountains by shouting in various directions and timing each particular echo coming back to us. This would be a rather rough way of scanning the surrounding area. If we were to shout continuously, however, while making the map, we would probably never hear the weak echoes returning to us and thus lose most of the information. A more efficient way would be to send a short burst of sound in a particular direction, wait for the echo to come back and time it, and then repeat the same process in various other directions, until we had completed the map.

SIMPLE RADAR SYSTEM

The process of sending out wave energy in short bursts is called pulsing and it is used in radar. The basic operation of radar is almost exactly the same as we have just outlined for the example of making a crude map by means of

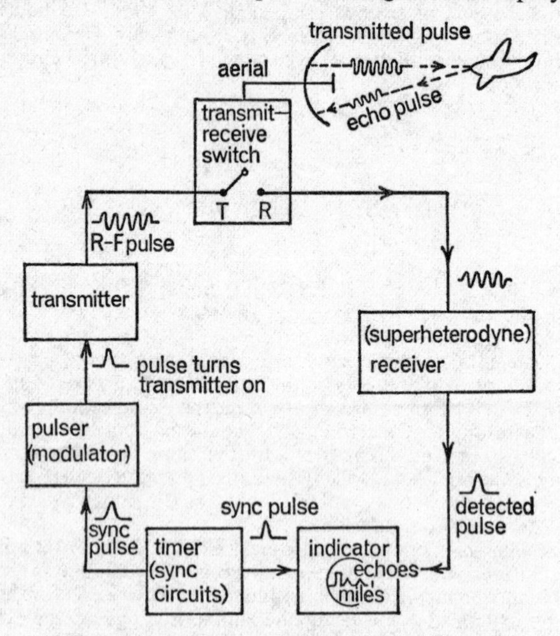

Fig. 203. Block diagram of a simple radar system

sound waves and their echoes. The essentials of a simple radar system are shown in Fig. 203. The only part that is different from the sound analogy is the indicator unit, which consists of a cathode-ray tube similar to that used in a television set. This tube displays the original transmitted pulse as well as the returned echo pulse along a horizontal base line that is marked in terms of distance. Other indicators exist that are capable of drawing a map of the area searched by the radar aerial, but the simple system shown here suffices for giving information of the range and direction of the target.

Operation. The block diagram (Fig. 203) shows the operation of a radar set at a glance. The timer assures that a single pulse from the pulser turns on the transmitter at exactly the same time as the indicator begins to draw the pulses along its scale. The transmitter, a high-power magnetron, then sends out short bursts of v.h.f. radio energy through a fast-acting transmit-receive (t.r.) switch (shown in the 'transmit' position) to the rotating aerial, which, in turn, sends it out into space. At the same time the transmitted pulse is recorded visually on the left-hand edge of the indicator scale. If the transmitted pulse happens to hit a target somewhere in space, a weak echo will be returned to the aerial. This echo signal becomes a maximum when the target is exactly in the centre of the transmitted beam—i.e. when the aerial axis is lined up with the target. In practice, this is achieved by turning the aerial until a maximum echo pulse is obtained.

In the meantime the fast-acting t.r. switch has automatically turned to the 'receive' position and guides the weak echo pulse to the receiver, where it is amplified and demodulated from its r.f. carrier by a superheterodyne circuit. The output of the receiver is applied to the cathode-ray tube, where it is displayed as a small 'pip' to the right of the large originally transmitted pulse, called the 'main bang'. The distance between the original pulse and the echo pulse at the right can be read off directly in miles or yards, thus giving the range of the target. The direction the aerial is pointing at gives the 'azimuth' of the target.

AZIMUTH AND RANGE MEASUREMENT

Since radar relies upon the principle of wave reflection, it must use the quasi-optical microwaves in the range from a 1,000 to over 10,000 Mc/s. These can be shaped by dish-like (parabolic) reflectors into sharply focused beams like those of a searchlight. This beam is made to scan the sky in a systematic fashion by precisely controlled rotation of the radar aerial and reflector. When the beam is reflected from a target, the direction in which the aerial points at that moment indicates the horizontal bearing or azimuth of the target with respect to the aerial. The horizontal azimuth in degrees is usually shown on the indicator.

To determine the distance between the radar and the target, or range, the elapsed time between the outgoing (transmitted) radar pulse and the returning echo pulse must be precisely measured. Radar waves travel through space at the velocity of light, which is about 186,284 miles per second. Thus, during each microsecond (millionth of a second) the waves travel about 984 feet, or 328 yards. Since the radar pulse must complete a round trip from transmitter to target and back, the range of the target is one-half the total distance travelled by the pulse, or about 164 yards for each microsecond of elapsed time. This elapsed time is conveniently measured by the linear (saw-tooth) sweep of an electron beam across the screen of a cathode-ray tube. By deflecting the beam horizontally at a known rate of motion (say, 0·01 inch per microsecond), the resulting trace forms a time scale or base, which may be calibrated in inches, or directly either in miles or yards of range.

As an example, assume that a radar aerial sends out a 1-microsecond long pulse at the same time as the cathode-ray indicator starts sweeping a beam horizontally across the face of the tube at a rate of 1 inch per 100 microseconds. Say the radar pulse strikes and is reflected by an aircraft (target) at a range of 32,800 yards from the radar aerial. The situation is depicted in Fig. 204.

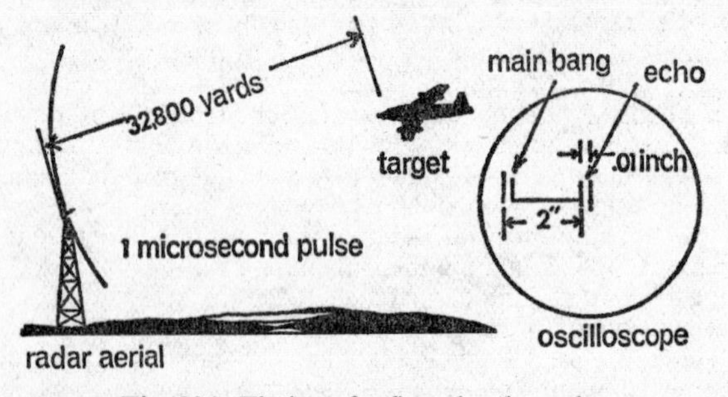

Fig. 204. Timing of reflected radar pulse

The 1-μsec. pulse is seen leaving the radar aerial simultaneously with the indication of a 0·01-inch long 'main bang' at the left edge of the radar indicator. The pulse travels the distance of 32,800 yards to the target in 100 microseconds, and, after reflection, requires another 100 microseconds to return to the aerial and receiver. Accordingly, a weak echo pulse, or pip, is recorded on the screen at a distance of 2 inches, or 200 microseconds, from the main bang.

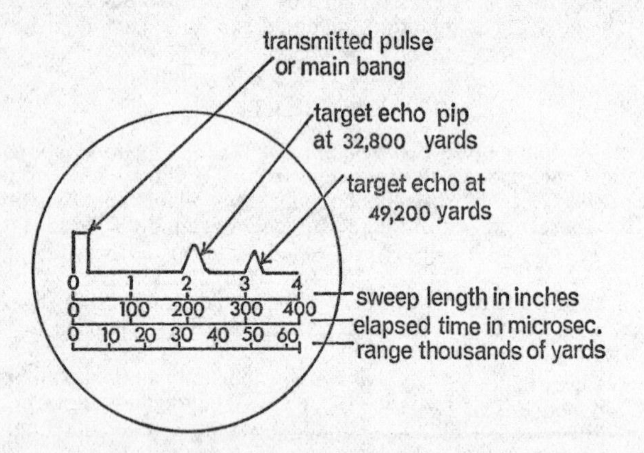

Fig. 205. Range indication of several targets on radar screen

This distance must be measured between the leading edges of the main bang and echo pulse, since the pulse itself lasts 1 μsec. and takes up 0·01 inch of the time base. Of course, the single indication of the pip after 200 microseconds is far too brief to be visible. The indication must be repeated many times each second to be perceived, by sweeping the beam across the scope at the scanning rate of the aerial (usually about 15 to 20 rev/min).

Fig. 205 illustrates how the ranges of several targets may be indicated on the face of the same cathode-ray tube. The three time bases, in inches, microseconds, and thousands of yards, are equivalent to each other, the sweep velocity being 1 inch per 100 microseconds. Usually, only the 'yard' scale is shown.

Plan position indicator (p.p.i.). The radar screen of Fig. 205 does not indicate the direction (azimuth) of the target at all. There are several methods for displaying the azimuth on the same indicator, the most useful being the plan-position indicator (p.p.i.), which draws a rough map of the area scanned by the radar beam.

As shown in Fig. 206, the electron beam on the face of the p.p.i. is deflected in successive radial lines through an entire circular area. The centre of this

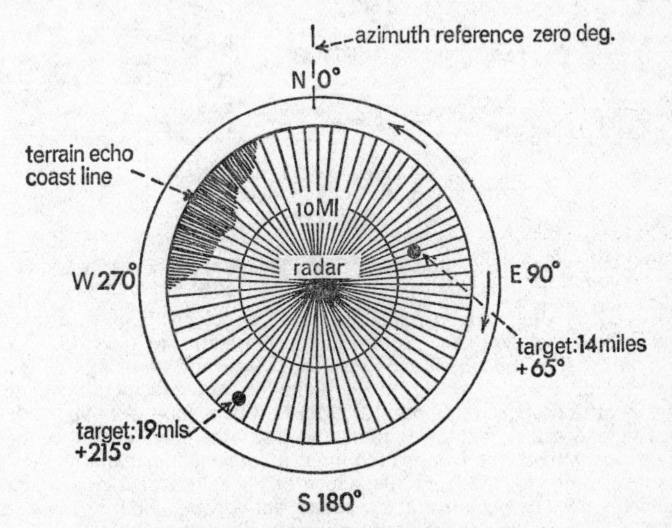

Fig. 206. Typical plan position indication (p.p.i.)

circle indicates the position of the radar. The distance from the centre along each radial line is the time base of the sweep and, hence, represents the range of a possible target. The direction of each radial line with respect to North or zero degree reference represents the azimuth or horizontal bearing of the target from the radar. To make the azimuth presentation accurate, the rotation of the scanning radar aerial must be synchronized, of course, with the direction of the radial beam deflection, so that each radial line points—at the moment it is swept—in the same direction as the radar aerial at this same instant. The motor that rotates the aerial must therefore be electrically coupled to the p.p.i. screen.

TYPES OF RADAR

Radar has been used in war and peace for a variety of applications. Its original purpose in the Second World War was to give early warning of approaching enemy planes. Early warning radars are large ground-based installations employing huge rotating parabolic reflectors or arrays of stationary dipole aerials, which look somewhat like bedsprings. (See Fig. 207.)

The war-time applications of radar were soon expanded to include ground control of interceptor aircraft and the direction of anti-aircraft artillery and searchlights. These so-called fire-control radars not only detect the enemy planes, but also automatically direct the anti-aircraft guns that fire on them. This job is done by servomechanisms. Warning and fire-control radars eventually became airborne, and are now included in most military aircraft.

Radar altimeters. Since radar can measure the distance to targets many

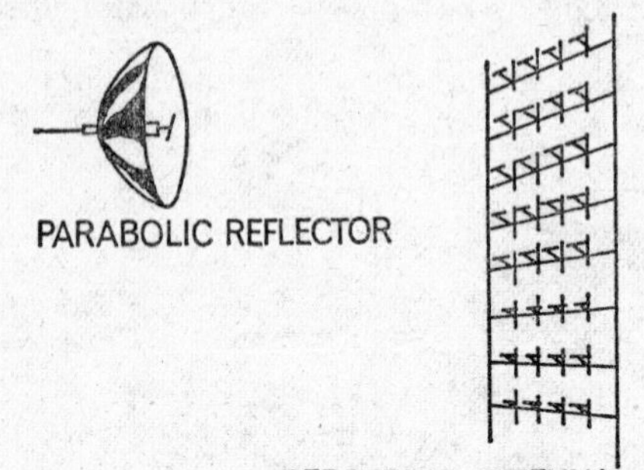

Fig. 207. Parabolic reflector and 'bedspring' antennas used for early warning radars

miles away, it obviously can do the much simpler job of measuring the distance from a plane to the ground directly beneath it. Conventional pulse-type radars may be used to measure the absolute altitude of an aeroplane above the surrounding terrain, provided the plane flies at a considerable height. At very low altitudes (below about 1,000 ft) however, the pulse-type radar becomes unsatisfactory, since for the short distances involved the echo pulse tends to merge with the transmitted main pulse (remember that a 1 μs pulse takes up a range of 984 ft on the screen of the indicator). Most radar altimeters, for this reason, do not use pulses but continuous, frequency-modulated waves. A typical set-up is shown in Fig. 208.

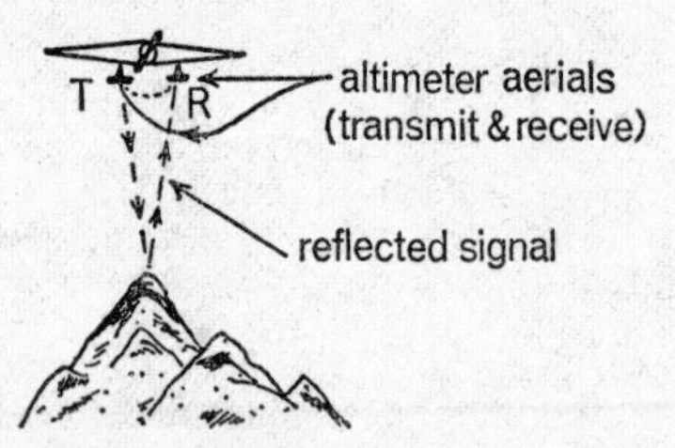

Fig. 208. Principle of radar altimeter

The transmitting aerial of the radar altimeter sends out a vertical radio beam to the ground with a continuously changing frequency. At the instant the signal leaves the aerial, its frequency will have a certain value. The signal is then reflected from the ground and returns to the receiving aerial of the altimeter. The receiver contains a phase discriminator which compares the frequency (or phase) of the returning signal with that of the signal now being transmitted. Since some time has elapsed for the echo of the original signal to return, the frequency of the signal now being transmitted has changed by some amount, of course. With the frequency deviation in cycles per second being known, the round trip of the original signal can be timed and, hence, the distance to the ground can be computed. All these computations may be made in advance, so that the indicator of the radar altimeter can be calibrated directly in thousands of feet of absolute altitude.

Radar beacons. Besides providing altitude indications, radar is also useful as an aid to the navigation of aeroplanes and ships. Radar beacons, located at known points on the map, provide positive identification markers of a city, airfield, mountain, or other specific point.

As illustrated in Fig. 209, a radar beacon consists of a ground-based radar

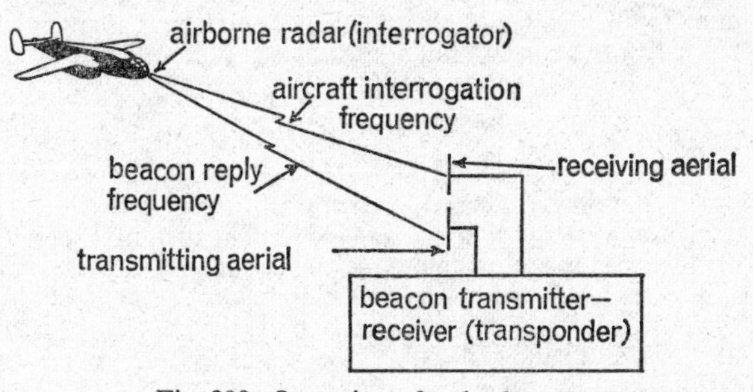

Fig. 209. Operation of radar beacon

transmitter–receiver, called a transponder, with separate receiving and transmitting aerials. Instead of sending out pulses and receiving echoes, a beacon receives a pulse from an airborne or shipborne radar set and sends back a reply. But it is not desirable that the beacon should respond to every spurious radar signal. The interrogating radars and the beacon transponder are therefore provided with coding circuits, so that only a received radar signal of the correct frequency, pulse length and spacing will 'trigger' the beacon to send out a coded reply, which identifies the station. The reply is received on the interrogating radar set and indicates the bearing and distance from the known beacon station. Since the beacon reply consists of an active retransmission rather than a passive reflection of a radar signal, the range of the airborne radar set is considerably extended.

A navigator aboard an aircraft (or ship) may obtain his position from the known locations of two or more radar beacons by measuring the distance (range) to each station. After obtaining the ranges, he need only draw a circle with the indicated range around each of the beacon stations marked on his map; the intersection point of the circles is his position. He must know, however, whether he is in front or behind the beacon stations, since the circles will usually intersect at two points.

Identification friend or foe (*i.f.f.*). By placing radar beacon transponders inside aircraft, positive identification of friendly or enemy craft can be made. I.F.F. (identification friend or foe) equipment is based on the principles illustrated in Fig. 209 for the conventional radar beacon. In this case, a ground-based radar set sends out an interrogation pulse at a certain frequency, whose pulse length and spacing has been coded in a specific manner. The beacon transponder of a friendly aircraft will be triggered by this interrogation pulse and a coded reply automatically will be sent back to the radar station, identifying the aircraft. An enemy aircraft, presumably, does not know the code and, hence, its beacon transponder (if any) will not be triggered into sending out the properly coded reply.

Ground-controlled approach (*g.c.a.*). Radar, finally, comes to the rescue of the pilot who must make a blind landing in bad weather, but does not have the special equipment and instruments to do it with. A skilled crew of ground operators can 'talk him down' by means of ground-controlled approach radar equipment. Basically, the ground equipment at the airport consists of two microwave radar sets, which are usually installed in a single trailer placed adjacent to the runway. One of the radars, known as the search system, locates all aircraft within 30 miles or so of the airport and thus provides a radar map of the vicinity on the p.p.i. indicator. The other radar, called the precision system, provides continuous information regarding the position of the incoming aircraft with respect to the runway. The aircraft may thus be safely talked down along the sloping glide path along which it must approach the runway.

SUMMARY

Radar determines the range and bearing of targets by sending out short pulses of microwave energy that are reflected by the distant target. The time it takes the radio pulses to complete their two-way journey indicates the distance between transmitter and target. The direction in which the radar aerial points indicates the horizontal bearing (azimuth) to the target.

Radar consists essentially of a timer, a pulser (modulator), a magnetron transmitter, a transmit-receive switch, a scanning aerial, superheterodyne receiver, and cathode-ray-tube indicator. The timer assures synchronism between the transmitted pulse and the indicator time base. The pulser turns on the transmitter to send out a radar pulse over the aerial, with the transmit-receive switch in 'transmit' position. The reflected echo pulse from the target is amplified and detected by the superhet receiver and is displayed on the screen of the indicator. The distance between transmitted and echo pulse along the time base indicates the range from the target. In a p.p.i. indicator, a map-like presentation is given.

A radar altimeter uses frequency-modulated continuous waves to indicate an aircraft's absolute altitude by comparing the frequency of the transmitted waves with that of the reflected (echo) waves.

A radar beacon is a transmitter–receiver (transponder) that replies to an interrogating aircraft signal and identifies itself. This permits determination of the range and bearing to the known beacon.

In ground-controlled approach (g.c.a.) radar, an aircraft is talked down to a blind landing by means of ground-based search and precision radars.

CHAPTER NINETEEN

RECENT ADVANCES IN ELECTRONICS

Electronics is now well and truly into the 'solid-state' era. Since the junction transistor (described in Chapter 7) became commercially available in the early 1950s, the thermionic valve has been steadily superseded to the extent that the only example which most people meet today is the TV picture tube. The reasons for the rapid and radical changes that have altered the whole field of electronics are not hard to find if we remember that semiconductor devices (of which the transistor is only one example) offer two enormous advantages over the valve: reliability and compactness.

It is significant that the revolution in electronics coincided with the growth of military guided-missile programmes and space exploration, particularly in the United States, for the twin factors of reliability and compactness are obviously essential requirements of aerospace equipment.

RELIABILITY

There is no theoretical limit to the life of a transistor; unless it is abused by overloading or overheating, it should last indefinitely. (For convenience the term 'transistor' as used here may be assumed to include all active semiconductor devices.) By contrast, a thermionic valve must fail sooner or later—either because the heater burns out or because the vacuum degenerates. Not only is the transistor inherently more reliable than the valve, it is also less liable to suffer mechanical damage. Moreover, although it is very sensitive to temperature changes, the transistor itself generates almost no heat. This is a tremendous advantage over any thermionic device which, by definition, dissipates heat in considerable quantities, for overheating is a major cause of failure in electronic equipment.

Overloading a transistor will usually cause irreversible damage in a fraction of a second. However, the risk of such damage is avoided by including in the circuit surge-limiting resistors (varistors) and/or an 'electronic fuse'. The latter is a simple transistorized arrangement which ceases to conduct as soon as the current exceeds some predetermined level. It is a practical illustration of the very rapid on-off switching action of transistors—the property which makes them ideally suited to computer applications.

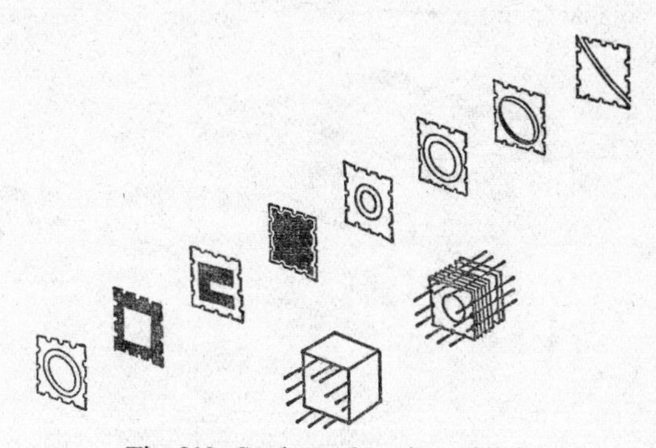

Fig. 210. Sections of a micromodule

The prime cause of equipment failure is break-down of electromechanical components such as switches, relays, push-buttons and the sliding contacts of variable resistors. In their pursuit of ever greater reliability, electronics engineers have developed a variety of methods whereby solid-state devices take the place of electromechanical components. One example of this development is the electronic fuse mentioned above, which is a marked improvement on thermal fuses and electromagnetic cut-outs. A better illustration may be seen in some of the latest laboratory measuring equipment.

Since d'Arsonval perfected his moving-coil galvanometer in 1881, this instrument above all others has been used for measuring electric currents and indirectly for measuring quantities such as voltage, resistance, temperature and pressure that can be converted into currents. Now, after almost a century, the moving-coil meter is giving way to direct-readout instruments. These have no moving parts: the reading is displayed on solid-state light-emitting diodes (l.e.d.s). Not only does the elimination of moving parts reduce the risk of failure due to wear and tear or mechanical damage, the fact that the reading appears as a numerical display greatly reduces the risk of human error in counting scale divisions.

MINIATURIZATION

Side by side with the quest for improved reliability (or, to use the engineers' term, greater mean time between failures) there has been in recent years a dramatic reduction in the size and weight of electronic equipment. These two trends are by no means unrelated, for smaller components have smaller power requirements and hence are less liable to develop electrical faults. In fact the prospect of improved reliability was one of the main reasons why miniaturization was started in the first place.

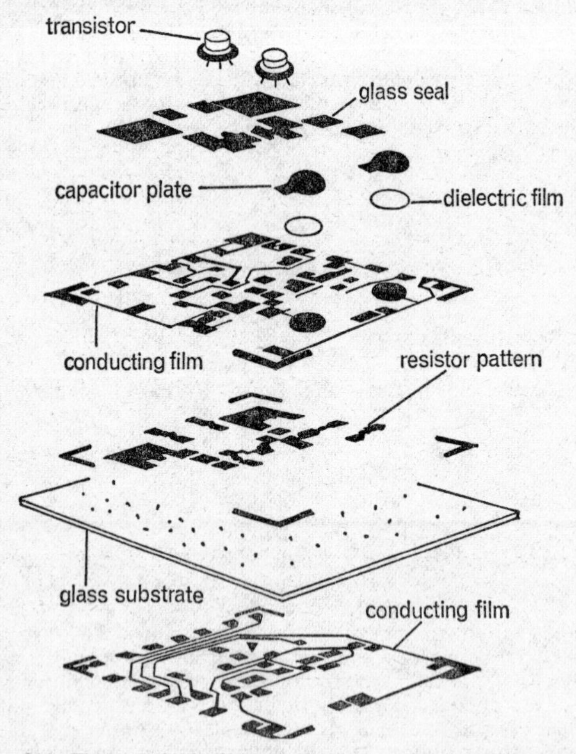

Fig. 211. The use of thin films to construct a two-dimensional circuit

The initial step on the road to miniaturization was obviously to make the various discrete components of a circuit as small as possible. Examples of this 'component-oriented' approach are the development of 'miniature' metal-film resistors and tantalum-bead capacitors. At the same time efforts were made to achieve tighter packing by the extensive use of printed-circuit boards.

The technique of modular construction, i.e. dividing a circuit into relatively small sub-sections and mounting each on individual standard-size printed-circuit boards, was devised originally for use in computers—where the com-

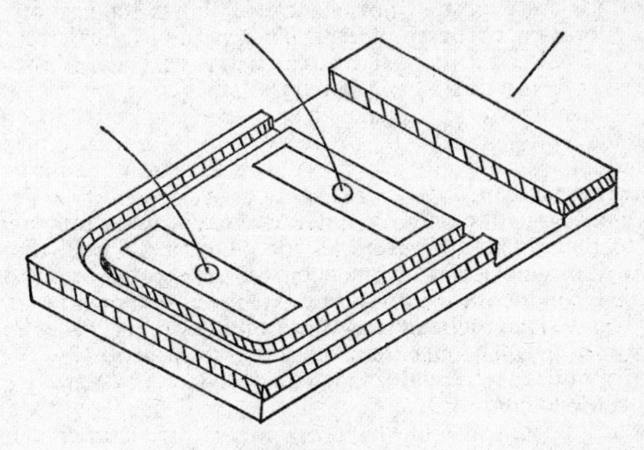

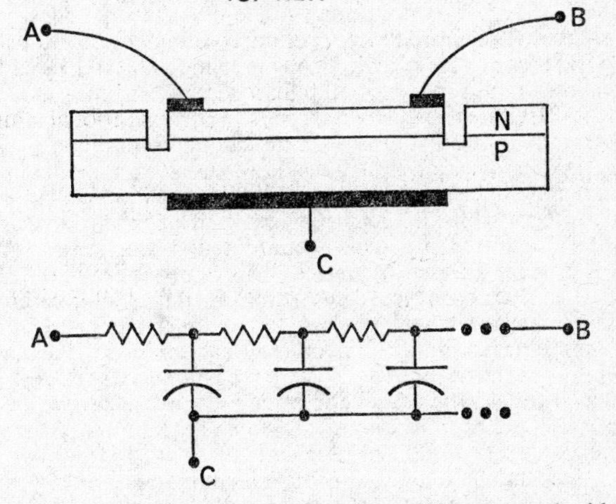

Fig. 212. A diffused-silicon block containing a phase-shifting network with distributed resistance and capacitance

ponents and interconnexions are frequently numbered somewhere in the hundreds of thousands. If a fault should occur, it is a relatively simple matter to locate and replace the defective module—certainly much simpler than attempting to trace one faulty circuit-element out of nearly a million.

Higher packing densities are achieved by designing conventional components to standardized forms and shapes (such as wafers) and stacking them tightly as shown in the exploded view in Fig. 210. An important feature of this construction is that individual stacks can be completely encapsulated in a plastic resin to seal them against atmospheric contamination.

INTEGRATED CIRCUITS

An entirely different technique, which may be described as the 'circuit-oriented' approach, is to deposit a complete electronic circuit on a flat 'substrate' of glass or ceramic. The components (resistors, capacitors, etc.) and interconnexions are deposited in the form of thin metallic films, either by vacuum evaporation or by sputtering. The circuit is thus essentially two-dimensional. Several of these flat circuits can be stacked vertically and encapsulated to form a three-dimensional equipment module. Fig. 211 shows a typical two-dimensional circuit made up of a variety of thin films. It should be noted that the 'active' components (the two transistors) are not part of the thin-film system but are added separately; this device is therefore not a completely integrated circuit.

The true integrated circuit is the outcome of the 'function-oriented' approach to miniaturization. Here all the components (and there may be hundreds of thousands) of a complete circuit are created *in situ* on a single crystal of semiconductor. The resulting structure is known as a functional block or integrated microcircuit, and the technology of its fabrication is called molecular electronics. The functional block is usually formed by the controlled diffusion of impurities into small 'chips' of silicon under a layer of metal oxide.

Fig. 212 shows a comparatively simple integrated microcircuit, actually an *R-C* phase-shifting network, in which resistive and capacitative elements have been formed in a silicon chip by vapour growth, planting, alloying and etching. It will be seen that interconnexions (in the conventional sense) have been totally eliminated. The entire device is of microscopic proportions, the correct positioning of the different circuit elements being achieved by photographic techniques rather than physical manipulation.

Among the first integrated microcircuits were operational amplifiers for analogue computers and logic circuits for digital computers. More recently the techniques of molecular electronics have been extended to audio equipment, and a complete high-fidelity stereophonic amplifier can be produced on a large-scale-integrated (l.s.i.) chip no bigger than a conventional capacitor. A virtue of these integrated circuits is that any faults which arise usually occur during manufacture and can be detected by rigorous testing. Sub-standard circuits are discarded: the rest are encapsulated in plastic or ceramic casings.

The mass-produced pocket calculator offers a pertinent illustration of the progress of miniaturization in electronics. How small the calculator can be made is dictated by the need for a display large enough to read and a keyboard large enough to touch: the actual computing element is an integrated circuit perhaps 30 × 10 × 2 mm overall, including its ceramic sheath. Using miniature discrete components, the same calculator would be as big as a filing cabinet. But if we had to go back to using the components of 25 years ago, our 'pocket' calculator would need an average-size living room to house it.

N	0	1	2	3	4	5	6	7	8	9	1	2	3	4	5	6	7	8	9
10	·0000	0043	0086	0128	0170	0212	0253	0294	0334	0374	4	8	12	17	21	25	29	33	37
11	·0414	0453	0492	0531	0569	0607	0645	0682	0719	0755	4	8	11	15	19	23	26	30	34
12	·0792	0828	0864	0899	0934	0969	1004	1038	1072	1106	3	7	10	14	17	21	24	28	31
13	·1139	1173	1206	1239	1271	1303	1335	1367	1399	1430	3	6	10	13	16	19	23	26	29
14	·1461	1492	1523	1553	1584	1614	1644	1673	1703	1732	3	6	9	12	15	18	21	24	27
15	·1761	1790	1818	1847	1875	1903	1931	1959	1987	2014	3	6	8	11	14	17	20	22	25
16	·2041	2068	2095	2122	2148	2175	2201	2227	2253	2279	3	5	8	11	13	16	18	21	24
17	·2304	2330	2355	2380	2405	2430	2455	2480	2504	2529	2	5	7	10	12	15	17	20	22
18	·2553	2577	2601	2625	2648	2672	2695	2718	2742	2765	2	5	7	9	12	14	16	19	21
19	·2788	2810	2833	2856	2878	2900	2923	2945	2967	2989	2	4	7	9	11	13	16	18	20
20	·3010	3032	3054	3075	3096	3118	3139	3160	3181	3201	2	4	6	8	11	13	15	17	19
21	·3222	3243	3263	3284	3304	3324	3345	3365	3385	3404	2	4	6	8	10	12	14	16	18
22	·3424	3444	3464	3483	3502	3522	3541	3560	3579	3598	2	4	6	8	10	12	14	15	17
23	·3617	3636	3655	3674	3692	3711	3729	3747	3766	3784	2	4	6	7	9	11	13	15	17
24	·3802	3820	3838	3856	3874	3892	3909	3927	3945	3962	2	4	5	7	9	11	12	14	16
25	·3979	3997	4014	4031	4048	4065	4082	4099	4116	4133	2	3	5	7	9	10	12	14	15
26	·4150	4166	4183	4200	4216	4232	4249	4265	4281	4298	2	3	5	7	8	10	11	13	15
27	·4314	4330	4346	4362	4378	4393	4409	4425	4440	4456	2	3	5	6	8	9	11	13	14
28	·4472	4487	4502	4518	4533	4548	4564	4579	4594	4609	2	3	5	6	8	9	11	12	14
29	·4624	4639	4654	4669	4683	4698	4713	4728	4742	4757	1	3	4	6	7	9	10	12	13
30	·4771	4786	4800	4814	4829	4843	4857	4871	4886	4900	1	3	4	6	7	9	10	11	13
31	·4914	4928	4942	4955	4969	4983	4997	5011	5024	5038	1	3	4	6	7	8	10	11	12
32	·5051	5065	5079	5092	5105	5119	5132	5145	5159	5172	1	3	4	5	7	8	9	11	12
33	·5185	5198	5211	5224	5237	5250	5263	5276	5289	5302	1	3	4	5	6	8	9	10	12
34	·5315	5328	5340	5353	5366	5378	5391	5403	5416	5428	1	3	4	5	6	8	9	10	11
35	·5441	5453	5465	5478	5490	5502	5514	5527	5539	5551	1	2	4	5	6	7	9	10	11
36	·5563	5575	5587	5599	5611	5623	5635	5647	5658	5670	1	2	4	5	6	7	8	10	11
37	·5682	5694	5705	5717	5729	5740	5752	5763	5775	5786	1	2	3	5	6	7	8	9	10
38	·5798	5809	5821	5832	5843	5855	5866	5877	5888	5899	1	2	3	5	6	7	8	9	10
39	·5911	5922	5933	5944	5955	5966	5977	5988	5999	6010	1	2	3	4	5	7	8	9	10
40	·6021	6031	6042	6053	6064	6075	6085	6096	6107	6117	1	2	3	4	5	6	8	9	10
41	·6128	6138	6149	6160	6170	6180	6191	6201	6212	6222	1	2	3	4	5	6	7	8	9
42	·6232	6243	6253	6263	6274	6284	6294	6304	6314	6325	1	2	3	4	5	6	7	8	9
43	·6335	6345	6355	6365	6375	6385	6395	6405	6415	6425	1	2	3	4	5	6	7	8	9
44	·6435	6444	6454	6464	6474	6484	6493	6503	6513	6522	1	2	3	4	5	6	7	8	9
45	·6532	6542	6551	6561	6571	6580	6590	6599	6609	6618	1	2	3	4	5	6	7	8	9
46	·6628	6637	6646	6656	6665	6675	6684	6693	6702	6712	1	2	3	4	5	6	7	7	8
47	·6721	6730	6739	6749	6758	6767	6776	6785	6794	6803	1	2	3	4	5	5	6	7	8
48	·6812	6821	6830	6839	6848	6857	6866	6875	6884	6893	1	2	3	4	4	5	6	7	8
49	·6902	6911	6920	6928	6937	6946	6955	6964	6972	6981	1	2	3	4	4	5	6	7	8
50	·6990	6998	7007	7016	7024	7033	7042	7050	7059	7067	1	2	3	3	4	5	6	7	8
51	·7076	7084	7093	7101	7110	7118	7126	7135	7143	7152	1	2	3	3	4	5	6	7	8
52	·7160	7168	7177	7185	7193	7202	7210	7218	7226	7235	1	2	2	3	4	5	6	7	7
53	·7243	7251	7259	7267	7275	7284	7292	7300	7308	7316	1	2	2	3	4	5	6	6	7
54	·7324	7332	7340	7348	7356	7364	7372	7380	7388	7396	1	2	2	3	4	5	6	6	7
	0	1	2	3	4	5	6	7	8	9	1	2	3	4	5	6	7	8	9

N	0	1	2	3	4	5	6	7	8	9	1	2	3	4	5	6	7	8	9
55	·7404	7412	7419	7427	7435	7443	7451	7459	7466	7474	1	2	2	3	4	5	5	6	7
56	·7482	7490	7497	7505	7513	7520	7528	7536	7543	7551	1	2	2	3	4	5	5	6	7
57	·7559	7566	7574	7582	7589	7597	7604	7612	7619	7627	1	2	2	3	4	5	5	6	7
58	·7634	7642	7649	7657	7664	7672	7679	7686	7694	7701	1	1	2	3	4	4	5	6	7
59	·7709	7716	7723	7731	7738	7745	7752	7760	7767	7774	1	1	2	3	4	4	5	6	7
60	·7782	7789	7796	7803	7810	7818	7825	7832	7839	7846	1	1	2	3	4	4	5	6	6
61	·7853	7860	7868	7875	7882	7889	7896	7903	7910	7917	1	1	2	3	4	4	5	6	6
62	·7924	7931	7938	7945	7952	7959	7966	7973	7980	7987	1	1	2	3	3	4	5	6	6
63	·7993	8000	8007	8014	8021	8028	8035	8041	8048	8055	1	1	2	3	3	4	5	5	6
64	·8062	8069	8075	8082	8089	8096	8102	8109	8116	8122	1	1	2	3	3	4	5	5	6
65	·8129	8136	8142	8149	8156	8162	8169	8176	8182	8189	1	1	2	3	3	4	5	5	6
66	·8195	8202	8209	8215	8222	8228	8235	8241	8248	8254	1	1	2	3	3	4	5	5	6
67	·8261	8267	8274	8280	8287	8293	8299	8306	8312	8319	1	1	2	3	3	4	5	5	6
68	·8325	8331	8338	8344	8351	8357	8363	8370	8376	8382	1	1	2	3	3	4	4	5	6
69	·8388	8395	8401	8407	8414	8420	8426	8432	8439	8445	1	1	2	2	3	4	4	5	6
70	·8451	8457	8463	8470	8476	8482	8488	8494	8500	8506	1	1	2	2	3	4	4	5	6
71	·8513	8519	8525	8531	8537	8543	8549	8555	8561	8567	1	1	2	2	3	4	4	5	5
72	·8573	8579	8585	8591	8597	8603	8609	8615	8621	8627	1	1	2	2	3	4	4	5	5
73	·8633	8639	8645	8651	8657	8663	8669	8675	8681	8686	1	1	2	2	3	4	4	5	5
74	·8692	8698	8704	8710	8716	8722	8727	8733	8739	8745	1	1	2	2	3	4	4	5	5
75	·8751	8756	8762	8768	8774	8779	8785	8791	8797	8802	1	1	2	2	3	3	4	5	5
76	·8808	8814	8820	8825	8831	8837	8842	8848	8854	8859	1	1	2	2	3	3	4	5	5
77	·8865	8871	8876	8882	8887	8893	8899	8904	8910	8915	1	1	2	2	3	3	4	4	5
78	·8921	8927	8932	8938	8943	8949	8954	8960	8965	8971	1	1	2	2	3	3	4	4	5
79	·8976	8982	8987	8993	8998	9004	9009	9015	9020	9025	1	1	2	2	3	3	4	4	5
80	·9031	9036	9042	9047	9053	9058	9063	9069	9074	9079	1	1	2	2	3	3	4	4	5
81	·9085	9090	9096	9101	9106	9112	9117	9122	9128	9133	1	1	2	2	3	3	4	4	5
82	·9138	9143	9149	9154	9159	9165	9170	9175	9180	9186	1	1	2	2	3	3	4	4	5
83	·9191	9196	9201	9206	9212	9217	9222	9227	9232	9238	1	1	2	2	3	3	4	4	5
84	·9243	9248	9253	9258	9263	9269	9274	9279	9284	9289	1	1	2	2	3	3	4	4	5
85	·9294	9299	9304	9309	9315	9320	9325	9330	9335	9340	1	1	2	2	3	3	4	4	5
86	·9345	9350	9355	9360	9365	9370	9375	9380	9385	9390	1	1	1	2	3	3	4	4	5
87	·9395	9400	9405	9410	9415	9420	9425	9430	9435	9440	0	1	1	2	2	3	3	4	4
88	·9445	9450	9455	9460	9465	9469	9474	9479	9484	9489	0	1	1	2	2	3	3	4	4
89	·9494	9499	9504	9509	9513	9518	9523	9528	9533	9538	0	1	1	2	2	3	3	4	4
90	·9542	9547	9552	9557	9562	9566	9571	9576	9581	9586	0	1	1	2	2	3	3	4	4
91	·9590	9595	9600	9605	9609	9614	9619	9624	9628	9633	0	1	1	2	2	3	3	4	4
92	·9638	9643	9647	9652	9657	9661	9666	9671	9675	9680	0	1	1	2	2	3	3	4	4
93	·9685	9689	9694	9699	9703	9708	9713	9717	9722	9727	0	1	1	2	2	3	3	4	4
94	·9731	9736	9741	9745	9750	9754	9759	9763	9768	9773	0	1	1	2	2	3	3	4	4
95	·9777	9782	9786	9791	9795	9800	9805	9809	9814	9818	0	1	1	2	2	3	3	4	4
96	·9823	9827	9832	9836	9841	9845	9850	9854	9859	9863	0	1	1	2	2	3	3	4	4
97	·9868	9872	9877	9881	9886	9890	9894	9899	9903	9908	0	1	1	2	2	3	3	4	4
98	·9912	9917	9921	9926	9930	9934	9939	9943	9948	9952	0	1	1	2	2	3	3	4	4
99	·9956	9961	9965	9969	9974	9978	9983	9987	9991	9996	0	1	1	2	2	3	3	3	4
N	0	1	2	3	4	5	6	7	8	9	1	2	3	4	5	6	7	8	9

	0′	6′	12′	18′	24′	30′	36′	42′	48′	54′	1′	2′	3′	4′	5′
0°	·0000	0017	0035	0052	0070	0087	0105	0122	0140	0157	3	6	9	12	15
1	·0175	0192	0209	0227	0244	0262	0279	0297	0314	0332	3	6	9	12	15
2	·0349	0366	0384	0401	0419	0436	0454	0471	0488	0506	3	6	9	12	15
3	·0523	0541	0558	0576	0593	0610	0628	0645	0663	0680	3	6	9	12	15
4	·0698	0715	0732	0750	0767	0785	0802	0819	0837	0854	3	6	9	12	14
5	·0872	0889	0906	0924	0941	0958	0976	0993	1011	1028	3	6	9	12	14
6	·1045	1063	1080	1097	1115	1132	1149	1167	1184	1201	3	6	9	12	14
7	·1219	1236	1253	1271	1288	1305	1323	1340	1357	1374	3	6	9	12	14
8	·1392	1409	1426	1444	1461	1478	1495	1513	1530	1547	3	6	9	12	14
9	·1564	1582	1599	1616	1633	1650	1668	1685	1702	1719	3	6	9	11	14
10	·1736	1754	1771	1788	1805	1822	1840	1857	1874	1891	3	6	9	11	14
11	·1908	1925	1942	1959	1977	1994	2011	2028	2045	2062	3	6	9	11	14
12	·2079	2096	2113	2130	2147	2164	2181	2198	2215	2233	3	6	9	11	14
13	·2250	2267	2284	2300	2317	2334	2351	2368	2385	2402	3	6	8	11	14
14	·2419	2436	2453	2470	2487	2504	2521	2538	2554	2571	3	6	8	11	14
15	·2588	2605	2622	2639	2656	2672	2689	2706	2723	2740	3	6	8	11	14
16	·2756	2773	2790	2807	2823	2840	2857	2874	2890	2907	3	6	8	11	14
17	·2924	2940	2957	2974	2990	3007	3024	3040	3057	3074	3	6	8	11	14
18	·3090	3107	3123	3140	3156	3173	3190	3206	3223	3239	3	6	8	11	14
19	·3256	3272	3289	3305	3322	3338	3355	3371	3387	3404	3	5	8	11	14
20	·3420	3437	3453	3469	3486	3502	3518	3535	3551	3567	3	5	8	11	14
21	·3584	3600	3616	3633	3649	3665	3681	3697	3714	3730	3	5	8	11	14
22	·3746	3762	3778	3795	3811	3827	3843	3859	3875	3891	3	5	8	11	13
23	·3907	3923	3939	3955	3971	3987	4003	4019	4035	4051	3	5	8	11	13
24	·4067	4083	4099	4115	4131	4147	4163	4179	4195	4210	3	5	8	11	13
25	·4226	4242	4258	4274	4289	4305	4321	4337	4352	4368	3	5	8	11	13
26	·4384	4399	4415	4431	4446	4462	4478	4493	4509	4524	3	5	8	10	13
27	·4540	4555	4571	4586	4602	4617	4633	4648	4664	4679	3	5	8	10	13
28	·4695	4710	4726	4741	4756	4772	4787	4802	4818	4833	3	5	8	10	13
29	·4848	4863	4879	4894	4909	4924	4939	4955	4970	4985	3	5	8	10	13
30	·5000	5015	5030	5045	5060	5075	5090	5105	5120	5135	3	5	8	10	13
31	·5150	5165	5180	5195	5210	5225	5240	5255	5270	5284	2	5	7	10	12
32	·5299	5314	5329	5344	5358	5373	5388	5402	5417	5432	2	5	7	10	12
33	·5446	5461	5476	5490	5505	5519	5534	5548	5563	5577	2	5	7	10	12
34	·5592	5606	5621	5635	5650	5664	5678	5693	5707	5721	2	5	7	10	12
35	·5736	5750	5764	5779	5793	5807	5821	5835	5850	5864	2	5	7	9	12
36	·5878	5892	5906	5920	5934	5948	5962	5976	5990	6004	2	5	7	9	12
37	·6018	6032	6046	6060	6074	6088	6101	6115	6129	6143	2	5	7	9	12
38	·6157	6170	6184	6198	6211	6225	6239	6252	6266	6280	2	5	7	9	11
39	·6293	6307	6320	6334	6347	6361	6374	6388	6401	6414	2	4	7	9	11
40	·6428	6441	6455	6468	6481	6494	6508	6521	6534	6547	2	4	7	9	11
41	·6561	6574	6587	6600	6613	6626	6639	6652	6665	6678	2	4	7	9	11
42	·6691	6704	6717	6730	6743	6756	6769	6782	6794	6807	2	4	6	9	11
43	·6820	6833	6845	6858	6871	6884	6896	6909	6921	6934	2	4	6	8	11
44	·6947	6959	6972	6984	6997	7009	7022	7034	7046	7059	2	4	6	8	10
	0′	6′	12′	18′	24′	30′	36′	42′	48′	54′	1′	2′	3′	4′	5′

	0′	6′	12′	18′	24′	30′	36′	42′	48′	54′	1′	2′	3′	4′	5′
45°	·7071	7083	7096	7108	7120	7133	7145	7157	7169	7181	2	4	6	8	10
46	·7193	7206	7218	7230	7242	7254	7266	7278	7290	7302	2	4	6	8	10
47	·7314	7325	7337	7349	7361	7373	7385	7396	7408	7420	2	4	6	8	10
48	·7431	7443	7455	7466	7478	7490	7501	7513	7524	7536	2	4	6	8	10
49	·7547	7559	7570	7581	7593	7604	7615	7627	7638	7649	2	4	6	8	9
50	·7660	7672	7683	7694	7705	7716	7727	7738	7749	7760	2	4	6	7	9
51	·7771	7782	7793	7804	7815	7826	7837	7848	7859	7869	2	4	5	7	9
52	·7880	7891	7902	7912	7923	7934	7944	7955	7965	7976	2	4	5	7	9
53	·7986	7997	8007	8018	8028	8039	8049	8059	8070	8080	2	3	5	7	9
54	·8090	8100	8111	8121	8131	8141	8151	8161	8171	8181	2	3	5	7	8
55	·8192	8202	8211	8221	8231	8241	8251	8261	8271	8281	2	3	5	7	8
56	·8290	8300	8310	8320	8329	8339	8348	8358	8368	8377	2	3	5	6	8
57	·8387	8396	8406	8415	8425	8434	8443	8453	8462	8471	2	3	5	6	8
58	·8480	8490	8499	8508	8517	8526	8536	8545	8554	8563	2	3	5	6	8
59	·8572	8581	8590	8599	8607	8616	8625	8634	8643	8652	1	3	4	6	7
60	·8660	8669	8678	8686	8695	8704	8712	8721	8729	8738	1	3	4	6	7
61	·8746	8755	8763	8771	8780	8788	8796	8805	8813	8821	1	3	4	6	7
62	·8829	8838	8846	8854	8862	8870	8878	8886	8894	8902	1	3	4	5	7
63	·8910	8918	8926	8934	8942	8949	8957	8965	8973	8980	1	3	4	5	6
64	·8988	8996	9003	9011	9018	9026	9033	9041	9048	9056	1	3	4	5	6
65	·9063	9070	9078	9085	9092	9100	9107	9114	9121	9128	1	2	4	5	6
66	·9135	9143	9150	9157	9164	9171	9178	9184	9191	9198	1	2	3	5	6
67	·9205	9212	9219	9225	9232	9239	9245	9252	9259	9265	1	2	3	4	6
68	·9272	9278	9285	9291	9298	9304	9311	9317	9323	9330	1	2	3	4	5
69	·9336	9342	9348	9354	9361	9367	9373	9379	9385	9391	1	2	3	4	5
70	·9397	9403	9409	9415	9421	9426	9432	9438	9444	9449	1	2	3	4	5
71	·9455	9461	9466	9472	9478	9483	9489	9494	9500	9505	1	2	3	4	5
72	·9511	9516	9521	9527	9532	9537	9542	9548	9553	9558	1	2	3	4	4
73	·9563	9568	9573	9578	9583	9588	9593	9598	9603	9608	1	2	2	3	4
74	·9613	9617	9622	9627	9632	9636	9641	9646	9650	9655	1	2	2	3	4
75	·9659	9664	9668	9673	9677	9681	9686	9690	9694	9699	1	1	2	3	4
76	·9703	9707	9711	9715	9720	9724	9728	9732	9736	9740	1	1	2	3	3
77	·9744	9748	9751	9755	9759	9763	9767	9770	9774	9778	1	1	2	3	3
78	·9781	9785	9789	9792	9796	9799	9803	9806	9810	9813	1	1	2	2	3
79	·9816	9820	9823	9826	9829	9833	9836	9839	9842	9845	1	1	2	2	3
80	·9848	9851	9854	9857	9860	9863	9866	9869	9871	9874	0	1	1	2	2
81	·9877	9880	9882	9885	9888	9890	9893	9895	9898	9900	0	1	1	2	2
82	·9903	9905	9907	9910	9912	9914	9917	9919	9921	9923	0	1	1	2	2
83	·9925	9928	9930	9932	9934	9936	9938	9940	9942	9943	0	1	1	1	2
84	·9945	9947	9949	9951	9952	9954	9956	9957	9959	9960	0	1	1	1	1
85	·9962	9963	9965	9966	9968	9969	9971	9972	9973	9974	0	0	1	1	1
86	·9976	9977	9978	9979	9980	9981	9982	9983	9984	9985	0	0	1	1	1
87	·9986	9987	9988	9989	9990	9990	9991	9992	9993	9993	0	0	0	1	1
88	·9994	9995	9995	9996	9996	9997	9997	9997	9998	9998					
89	·9998	9999	9999	9999	9999	1·000	1·000	1·000	1·000	1·000					
	0′	6′	12′	18′	24′	30′	36′	42′	48′	54′	1′	2′	3′	4′	5′

SUBTRACT

	0′	6′	12′	18′	24′	30′	36′	42′	48′	54′	1′	2′	3′	4′	5′
0°	1·0000	1·000	1·000	1·000	1·000	1·000	9999	9999	9999	9999					
1	·9998	9998	9998	9997	9997	9997	9996	9996	9995	9995					
2	·9994	9993	9993	9992	9991	9990	9990	9989	9988	9987					
3	·9986	9985	9984	9983	9982	9981	9980	9979	9978	9977					
4	·9976	9974	9973	9972	9971	9969	9968	9966	9965	9963					
5	·9962	9960	9959	9957	9956	9954	9952	9951	9949	9947					
6	·9945	9943	9942	9940	9938	9936	9934	9932	9930	9928	0	1	1	1	2
7	·9925	9923	9921	9919	9917	9914	9912	9910	9907	9905	0	1	1	2	2
8	·9903	9900	9898	9895	9893	9890	9888	9885	9882	9880	0	1	1	2	2
9	·9877	9874	9871	9869	9866	9863	9860	9857	9854	9851	0	1	1	2	2
10	·9848	9845	9842	9839	9836	9833	9829	9826	9823	9820	1	1	2	2	3
11	·9816	9813	9810	9806	9803	9799	9796	9792	9789	9785	1	1	2	2	3
12	·9781	9778	9774	9770	9767	9763	9759	9755	9751	9748	1	1	2	3	3
13	·9744	9740	9736	9732	9728	9724	9720	9715	9711	9707	1	1	2	3	3
14	·9703	9699	9694	9690	9686	9681	9677	9673	9668	9664	1	1	2	3	4
15	·9659	9655	9650	9646	9641	9636	9632	9627	9622	9617	1	2	2	3	4
16	·9613	9608	9603	9598	9593	9588	9583	9578	9573	9568	1	2	2	3	4
17	·9563	9558	9553	9548	9542	9537	9532	9527	9521	9516	1	2	3	3	4
18	·9511	9505	9500	9494	9489	9483	9478	9472	9466	9461	1	2	3	4	5
19	·9455	9449	9444	9438	9432	9426	9421	9415	9409	9403	1	2	3	4	5
20	·9397	9391	9385	9379	9373	9367	9361	9354	9348	9342	1	2	3	4	5
21	·9336	9330	9323	9317	9311	9304	9298	9291	9285	9278	1	2	3	4	5
22	·9272	9265	9259	9252	9245	9239	9232	9225	9219	9212	1	2	3	4	6
23	·9205	9198	9191	9184	9178	9171	9164	9157	9150	9143	1	2	3	5	6
24	·9135	9128	9121	9114	9107	9100	9092	9085	9078	9070	1	2	4	5	6
25	·9063	9056	9048	9041	9033	9026	9018	9011	9003	8996	1	3	4	5	6
26	·8988	8980	8973	8965	8957	8949	8942	8934	8926	8918	1	3	4	5	6
27	·8910	8902	8894	8886	8878	8870	8862	8854	8846	8838	1	3	4	5	7
28	·8829	8821	8813	8805	8796	8788	8780	8771	8763	8755	1	3	4	6	7
29	·8746	8738	8729	8721	8712	8704	8695	8686	8678	8669	1	3	4	6	7
30	·8660	8652	8643	8634	8625	8616	8607	8599	8590	8581	1	3	4	6	7
31	·8572	8563	8554	8545	8536	8526	8517	8508	8499	8490	2	3	5	6	8
32	·8480	8471	8462	8453	8443	8434	8425	8415	8406	8396	2	3	5	6	8
33	·8387	8377	8368	8358	8348	8339	8329	8320	8310	8300	2	3	5	6	8
34	·8290	8281	8271	8261	8251	8241	8231	8221	8211	8202	2	3	5	7	8
35	·8192	8181	8171	8161	8151	8141	8131	8121	8111	8100	2	3	5	7	8
36	·8090	8080	8070	8059	8049	8039	8028	8018	8007	7997	2	3	5	7	9
37	·7986	7976	7965	7955	7944	7934	7923	7912	7902	7891	2	4	5	7	9
38	·7880	7869	7859	7848	7837	7826	7815	7804	7793	7782	2	4	5	7	9
39	·7771	7760	7749	7738	7727	7716	7705	7694	7683	7672	2	4	6	7	9
40	·7660	7649	7638	7627	7615	7604	7593	7581	7570	7559	2	4	6	8	9
41	·7547	7536	7524	7513	7501	7490	7478	7466	7455	7443	2	4	6	8	10
42	·7431	7420	7408	7396	7385	7373	7361	7349	7337	7325	2	4	6	8	10
43	·7314	7302	7290	7278	7266	7254	7242	7230	7218	7206	2	4	6	8	10
44	·7193	7181	7169	7157	7145	7133	7120	7108	7096	7083	2	4	6	8	10

SUBTRACT

	0′	6′	12′	18′	24′	30′	36′	42′	48′	54′	SUBTRACT 1′	2′	3′	4′	5′
45°	·7071	7059	7046	7034	7022	7009	6997	6984	6972	6959	2	4	6	8	10
46	·6947	6934	6921	6909	6896	6884	6871	6858	6845	6833	2	4	6	8	11
47	·6820	6807	6794	6782	6769	6756	6743	6730	6717	6704	2	4	6	9	11
48	·6691	6678	6665	6652	6639	6626	6613	6600	6587	6574	2	4	7	9	11
49	·6561	6547	6534	6521	6508	6494	6481	6468	6455	6441	2	4	7	9	11
50	·6428	6414	6401	6388	6374	6361	6347	6334	6320	6307	2	4	7	9	11
51	·6293	6280	6266	6252	6239	6225	6211	6198	6184	6170	2	5	7	9	11
52	·6157	6143	6129	6115	6101	6088	6074	6060	6046	6032	2	5	7	9	12
53	·6018	6004	5990	5976	5962	5948	5934	5920	5906	5892	2	5	7	9	12
54	·5878	5864	5850	5835	5821	5807	5793	5779	5764	5750	2	5	7	9	12
55	·5736	5721	5707	5693	5678	5664	5650	5635	5621	5606	2	5	7	10	12
56	·5592	5577	5563	5548	5534	5519	5505	5490	5476	5461	2	5	7	10	12
57	·5446	5432	5417	5402	5388	5373	5358	5344	5329	5314	2	5	7	10	12
58	·5299	5284	5270	5255	5240	5225	5210	5195	5180	5165	2	5	7	10	12
59	·5150	5135	5120	5105	5090	5075	5060	5045	5030	5015	3	5	8	10	13
60	·5000	4985	4970	4955	4939	4924	4909	4894	4879	4863	3	5	8	10	13
61	·4848	4833	4818	4802	4787	4772	4756	4741	4726	4710	3	5	8	10	13
62	·4695	4679	4664	4648	4633	4617	4602	4586	4571	4555	3	5	8	10	13
63	·4540	4524	4509	4493	4478	4462	4446	4431	4415	4399	3	5	8	10	13
64	·4384	4368	4352	4337	4321	4305	4289	4274	4258	4242	3	5	8	11	13
65	·4226	4210	4195	4179	4163	4147	4131	4115	4099	4083	3	5	8	11	13
66	·4067	4051	4035	4019	4003	3987	3971	3955	3939	3923	3	5	8	11	13
67	·3907	3891	3875	3859	3843	3827	3811	3795	3778	3762	3	5	8	11	13
68	·3746	3730	3714	3697	3681	3665	3649	3633	3616	3600	3	5	8	11	14
69	·3584	3567	3551	3535	3518	3502	3486	3469	3453	3437	3	5	8	11	14
70	·3420	3404	3387	3371	3355	3338	3322	3305	3289	3272	3	5	8	11	14
71	·3256	3239	3223	3206	3190	3173	3156	3140	3123	3107	3	6	8	11	14
72	·3090	3074	3057	3040	3024	3007	2990	2974	2957	2940	3	6	8	11	14
73	·2924	2907	2890	2874	2857	2840	2823	2807	2790	2773	3	6	8	11	14
74	·2756	2740	2723	2706	2689	2672	2656	2639	2622	2605	3	6	8	11	14
75	·2588	2571	2554	2538	2521	2504	2487	2470	2453	2436	3	6	8	11	14
76	·2419	2402	2385	2368	2351	2334	2317	2300	2284	2267	3	6	8	11	14
77	·2250	2233	2215	2198	2181	2164	2147	2130	2113	2096	3	6	9	11	14
78	·2079	2062	2045	2028	2011	1994	1977	1959	1942	1925	3	6	9	11	14
79	·1908	1891	1874	1857	1840	1822	1805	1788	1771	1754	3	6	9	11	14
80	·1736	1719	1702	1685	1668	1650	1633	1616	1599	1582	3	6	9	11	14
81	·1564	1547	1530	1513	1495	1478	1461	1444	1426	1409	3	6	9	12	14
82	·1392	1374	1357	1340	1323	1305	1288	1271	1253	1236	3	6	9	12	14
83	·1219	1201	1184	1167	1149	1132	1115	1097	1080	1063	3	6	9	12	14
84	·1045	1028	1011	0993	0976	0958	0941	0924	0906	0889	3	6	9	12	14
85	·0872	0854	0837	0819	0802	0785	0767	0750	0732	0715	3	6	9	12	14
86	·0698	0680	0663	0645	0628	0610	0593	0576	0558	0541	3	6	9	12	15
87	·0523	0506	0488	0471	0454	0436	0419	0401	0384	0366	3	6	9	12	15
88	·0349	0332	0314	0297	0279	0262	0244	0227	0209	0192	3	6	9	12	15
89	·0175	0157	0140	0122	0105	0087	0070	0052	0035	0017	3	6	9	12	15

SUBTRACT

°	0′	6′	12′	18′	24′	30′	36′	42′	48′	54′	1′	2′	3′	4′	5′
0	0·0000	0017	0035	0052	0070	0087	0105	0122	0140	0157	3	6	9	12	15
1	0·0175	0192	0209	0227	0244	0262	0279	0297	0314	0332	3	6	9	12	15
2	0·0349	0367	0384	0402	0419	0437	0454	0472	0489	0507	3	6	9	12	15
3	0·0524	0542	0559	0577	0594	0612	0629	0647	0664	0682	3	6	9	12	15
4	0·0699	0717	0734	0752	0769	0787	0805	0822	0840	0857	3	6	9	12	15
5	0·0875	0892	0910	0928	0945	0963	0981	0998	1016	1033	3	6	9	12	15
6	0·1051	1069	1086	1104	1122	1139	1157	1175	1192	1210	3	6	9	12	15
7	0·1228	1246	1263	1281	1299	1317	1334	1352	1370	1388	3	6	9	12	15
8	0·1405	1423	1441	1459	1477	1495	1512	1530	1548	1566	3	6	9	12	1
9	0·1584	1602	1620	1638	1655	1673	1691	1709	1727	1745	3	6	9	12	1
10	0·1763	1781	1799	1817	1835	1853	1871	1890	1908	1926	3	6	9	12	1
11	0·1944	1962	1980	1998	2016	2035	2053	2071	2089	2107	3	6	9	12	1
12	0·2126	2144	2162	2180	2199	2217	2235	2254	2272	2290	3	6	9	12	1
13	0·2309	2327	2345	2364	2382	2401	2419	2438	2456	2475	3	6	9	12	1
14	0·2493	2512	2530	2549	2568	2586	2605	2623	2642	2661	3	6	9	12	1
15	0·2679	2698	2717	2736	2754	2773	2792	2811	2830	2849	3	6	9	13	1
16	0·2867	2886	2905	2924	2943	2962	2981	3000	3019	3038	3	6	9	13	1
17	0·3057	3076	3096	3115	3134	3153	3172	3191	3211	3230	3	6	10	13	1
18	0·3249	3269	3288	3307	3327	3346	3365	3385	3404	3424	3	6	10	13	1
19	0·3443	3463	3482	3502	3522	3541	3561	3581	3600	3620	3	7	10	13	1
20	0·3640	3659	3679	3699	3719	3739	3759	3779	3799	3819	3	7	10	13	1
21	0·3839	3859	3879	3899	3919	3939	3959	3979	4000	4020	3	7	10	13	1
22	0·4040	4061	4081	4101	4122	4142	4163	4183	4204	4224	3	7	10	14	1
23	0·4245	4265	4286	4307	4327	4348	4369	4390	4411	4431	3	7	10	14	1
24	0·4452	4473	4494	4515	4536	4557	4578	4599	4621	4642	4	7	11	14	1
25	0·4663	4684	4706	4727	4748	4770	4791	4813	4834	4856	4	7	11	14	1
26	0·4877	4899	4921	4942	4964	4986	5008	5029	5051	5073	4	7	11	15	1
27	0·5095	5117	5139	5161	5184	5206	5228	5250	5272	5295	4	7	11	15	1
28	0·5317	5340	5362	5384	5407	5430	5452	5475	5498	5520	4	8	11	15	1
29	0·5543	5566	5589	5612	5635	5658	5681	5704	5727	5750	4	8	12	15	1
30	0·5774	5797	5820	5844	5867	5890	5914	5938	5961	5985	4	8	12	16	2
31	0·6009	6032	6056	6080	6104	6128	6152	6176	6200	6224	4	8	12	16	2
32	0·6249	6273	6297	6322	6346	6371	6395	6420	6445	6469	4	8	12	16	2
33	0·6494	6519	6544	6569	6594	6619	6644	6669	6694	6720	4	8	13	17	2
34	0·6745	6771	6796	6822	6847	6873	6899	6924	6950	6976	4	9	13	17	2
35	0·7002	7028	7054	7080	7107	7133	7159	7186	7212	7239	4	9	13	18	2
36	0·7265	7292	7319	7346	7373	7400	7427	7454	7481	7508	5	9	14	18	2
37	0·7536	7563	7590	7618	7646	7673	7701	7729	7757	7785	5	9	14	18	2
38	0·7813	7841	7869	7898	7926	7954	7983	8012	8040	8069	5	9	14	19	2
39	0·8098	8127	8156	8185	8214	8243	8273	8302	8332	8361	5	10	15	20	2
40	0·8391	8421	8451	8481	8511	8541	8571	8601	8632	8662	5	10	15	20	2
41	0·8693	8724	8754	8785	8816	8847	8878	8910	8941	8972	5	10	16	21	2
42	0·9004	9036	9067	9099	9131	9163	9195	9228	9260	9293	5	11	16	21	2
43	0·9325	9358	9391	9424	9457	9490	9523	9556	9590	9623	6	11	17	22	28
44	0·9657	9691	9725	9759	9793	9827	9861	9896	9930	9965	6	11	17	23	29
	0′	6′	12′	18′	24′	30′	36′	42′	48′	54′	1′	2′	3′	4′	5′

0′	6′	12′	18′	24′	30′	36′	42′	48′	54′	1′	2′	3′	4′	5′
1·0000	0035	0070	0105	0141	0176	0212	0247	0283	0319	6	12	18	24	30
1·0355	0392	0428	0464	0501	0538	0575	0612	0649	0686	6	12	18	25	31
1·0724	0761	0799	0837	0875	0913	0951	0990	1028	1067	6	13	19	25	32
1·1106	1145	1184	1224	1263	1303	1343	1383	1423	1463	7	13	20	26	33
1·1504	1544	1585	1626	1667	1708	1750	1792	1833	1875	7	14	21	28	34
1·1918	1960	2002	2045	2088	2131	2174	2218	2261	2305	7	14	22	29	36
1·2349	2393	2437	2482	2527	2572	2617	2662	2708	2753	8	15	23	30	38
1·2799	2846	2892	2938	2985	3032	3079	3127	3175	3222	8	16	24	31	39
1·3270	3319	3367	3416	3465	3514	3564	3613	3663	3713	8	16	25	33	41
1·3764	3814	3865	3916	3968	4019	4071	4124	4176	4229	9	17	26	34	43
1·4281	4335	4388	4442	4496	4550	4605	4659	4715	4770	9	18	27	36	45
1·4826	4882	4938	4994	5051	5108	5166	5224	5282	5340	10	19	29	38	48
1·5399	5458	5517	5577	5637	5697	5757	5818	5880	5941	10	20	30	40	50
1·6003	6066	6128	6191	6255	6319	6383	6447	6512	6577	11	21	32	43	53
1·6643	6709	6775	6842	6909	6977	7045	7113	7182	7251	11	23	34	45	56
1·7321	7391	7461	7532	7603	7675	7747	7820	7893	7966	12	24	36	48	60
1·8040	8115	8190	8265	8341	8418	8495	8572	8650	8728	13	26	38	51	64
1·8807	8887	8967	9047	9128	9210	9292	9375	9458	9542	14	27	41	55	68
1·9626	9711	9797	9883	9970	**0057**	**0145**	**0233**	**0323**	**0413**	15	29	44	58	73
2·0503	0594	0686	0778	0872	0965	1060	1155	1251	1348	16	31	47	63	78
2·1445	1543	1642	1742	1842	1943	2045	2148	2251	2355	17	34	51	68	85
2·2460	2566	2673	2781	2889	2998	3109	3220	3332	3445	18	37	55	73	91
2·3559	3673	3789	3906	4023	4142	4262	4383	4504	4627	20	40	60	79	99
2·4751	4876	5002	5129	5257	5386	5517	5649	5782	5916	22	43	65	87	108
2·6051	6187	6325	6464	6605	6746	6889	7034	7179	7326	24	47	71	95	119
2·7475	7625	7776	7929	8083	8239	8397	8556	8716	8878	26	52	78	104	130
2·9042	9208	9375	9544	9714	9887	**0061**	**0237**	**0415**	**0595**	29	58	87	116	144
3·0777	0961	1146	1334	1524	1716	1910	2106	2305	2506	32	64	97	129	161
3·2709	2914	3122	3332	3544	3759	3977	4197	4420	4646	36	72	108	144	180
3·4874	5105	5339	5576	5816	6059	6305	6554	6806	7062	41	81	122	163	203
3·7321	7583	7848	8118	8391	8667	8947	9232	9520	9812	46	93	139	186	232
4·0108	0408	0713	1022	1335	1653	1976	2303	2635	2972	53	107	160	214	267
4·3315	3662	4015	4373	4737	5107	5483	5864	6252	6646	62	124	186	248	310
4·7046	7453	7867	8288	8716	9152	9594	**0045**	**0504**	**0970**	73	146	220	293	366
5·1446	1929	2422	2924	3435	3955	4486	5026	5578	6140	87	175	263	350	438
5·671	5·730	5·789	5·850	5·912	5·976	6·041	6·107	6·174	6·243					
6·314	6·386	6·460	6·535	6·612	6·691	6·772	6·855	6·940	7·026	Differences untrustworthy here				
7·115	7·207	7·300	7·396	7·495	7·596	7·700	7·806	7·916	8·028					
8·144	8·264	8·386	8·513	8·643	8·777	8·915	9·058	9·205	9·357					
9·51	9·68	9·84	10·02	10·20	10·39	10·58	10·78	10·99	11·20					
11·43	11·66	11·91	12·16	12·43	12·71	13·00	13·30	13·62	13·95					
14·30	14·67	15·06	15·46	15·89	16·35	16·83	17·34	17·89	18·46					
19·08	19·74	20·45	21·20	22·02	22·90	23·86	24·90	26·03	27·27					
28·64	30·14	31·82	33·69	35·80	38·19	40·92	44·07	47·74	52·08					
57·29	63·66	71·62	81·85	95·49	114·6	143·2	191·0	286·5	573·0					
0′	6′	12′	18′	24′	30′	36′	42′	48′	54′	1′	2′	3′	4′	5′

The black type indicates that the integer changes.

APPENDIX:

GLOSSARY OF TERMS USED IN ELECTRONICS

Italicized terms within individual definitions are defined elsewhere (in alphabetical order) in the Glossary.

A

AMPLIFICATION FACTOR—Ratio of the small change in anode voltage to the small change in control-grid voltage such that both changes produce equal effects on the anode current. Symbol: μ.

AMPLIFIER—Device containing one or more valves or transistors, the output signal of which is a magnified replica of the input signal.

AMPLITUDE—Peak value of any voltage or current (or voltage × current = power) that varies with time in a repeating pattern.

AMPLITUDE MODULATION—Imposition of a signal on to a radio wave by varying the *amplitude* of the wave in sympathy with the amplitude of the signal.

ANODE—The most positive electrode in a valve and therefore the one to which electrons are attracted. Also called *plate*.

ANODE CURRENT—Flow of electrons from cathode to anode across a valve. But the conventional current is a flow of imaginary positive charges in the opposite direction, and this is the anode current indicated on circuit diagrams.

ANODE RESISTANCE—Ratio of a small change made in the voltage on the anode of a valve to the resulting change in anode current. It is an indication of the valve's opposition to electrons flowing through it.

AQUADAG—Trade name for the graphite (a form of carbon) used as a conducting coating over the inner surfaces of cathode-ray tubes and some other valves.

A.V.C.—Automatic volume control. Method of keeping the level of output signals from an amplifier substantially constant by using part of the output from one of the output stages to bias the grid of an earlier stage. Also known as automatic gain control (a.g.c.).

B

BANDWIDTH—Difference between the highest and lowest frequencies that any particular circuit can handle.

BASE—In a junction transistor, the region between the *emitter* and the *collector*.

BIAS—Voltage put on to the grid of a valve in order to make the grid more positive or more negative and thereby control the flow of electrons across the valve.

BYPASS CAPACITOR—Capacitor whose function is to provide an easier alternative path for alternating currents around some circuit component such as a resistor.

C

CAPACITOR—System comprising a pair or pairs of conductors (the plates) separated by an insulator (the dielectric) which has the property of storing charges. Use is also made of the fact that it presents less opposition to high-frequency currents than it does to low-frequency currents. The CAPACITANCE of a capacitor is the ratio of the charge stored to the voltage between the plates.

CATHODE—The most negative electrode in a valve and in effect the source of electrons passing across the valve.

CHARACTERISTIC CURVE—Graph obtained by plotting any two currents or voltages applied to a valve or transistor. E.g. anode current/grid voltage; collector current/collector voltage.

CLASS A AMPLIFIER—Amplifier in which the signal voltage and grid bias are such that the anode current flows throughout the entire cycle of the signal voltage.

CLASS B AMPLIFIER—Amplifier in which anode current flows for only half of each cycle of signal voltage, because the grid bias is made equal to the anode-current *cut-off* value.

CLASS AB AMPLIFIER—Amplifier in which anode current flows for more than half but less than the entire cycle of signal voltage.

CLASS C AMPLIFIER—Amplifier in which anode current flows for less than half of each cycle of signal voltage.

COLLECTOR—That part of a transistor to which the major charge carriers drift, i.e. where the 'holes' collect in a P-N-P junction transistor and where the electrons collect in an N-P-N transistor.

COLPITTS OSCILLATOR—Oscillator in which the tank circuit contains two capacitors in series, their junction being earthed.

CONTROL GRID—The grid in a multielectrode valve which is used to control the flow of electrons across the valve.

CRYSTAL-CONTROLLED OSCILLATOR—Oscillator, usually in a transmitter, whose frequency is locked to the natural frequency of vibration of a *piezoelectric* crystal.

CURRENT GAIN—Ratio of the current in the *collector* circuit of a transistor to the current in the *emitter* circuit. Symbol: α.

CUT-OFF—Minimum negative bias applied to the *control grid* of a valve that will prevent the flow of *anode current*.

C.W. TRANSMISSION—Continuous-wave transmission: broadcasting by means of waves that do not vary in amplitude or frequency. Continuous waves may be used for sending Morse code by chopping them into long pulses (dashes) and short pulses (dots).

D

DAMPED OSCILLATION—Oscillation in which the amplitude of successive cycles decreases.

DEMODULATION—Process of extracting the information carried by radio waves. Also called DETECTION.

DIODE—Device such as a valve containing two electrodes (anode and cathode) only. Allows current to flow in one direction but not in reverse.

DIRECT HEATING—Use of the heater or filament in a valve as the cathode or source of electrons. When direct heating is employed the filament must be supplied with direct current.

DISCRIMINATOR—Circuit which senses the information carried by a radio wave as *frequency* variations and converts this information to *amplitude* variations. Used in the detector circuits of f.m. radios.

DISTORTION—Unwanted changes in the 'shape' of a wave, i.e. any kind of defect in its frequency, amplitude or phase.

E

EARTH—Any part of a circuit that is connected to the Earth and therefore is at the same electrical potential as the Earth. Also called GROUND.

ELECTRON—Smallest possible quantity of electricity, for convenience regarded as a negatively charged particle, present in various numbers in every atom.

ELECTRON-COUPLED OSCILLATOR—Oscillator in which the main valve has more than one grid so that the stream of electrons passing through the grids couples the output circuit to the oscillating circuit.

ELECTRON MULTIPLIER—Light-sensitive valve having several anodes arranged in a cascade. Electrons emitted when light falls on the light-sensitive surface cause *secondary emission* in each of the anodes so that a greatly magnified flow of electrons results at the output. Also called PHOTO-MULTIPLIER.

ELECTRON GUN—Part of a cathode-ray tube that emits electrons, accelerates them and focuses them into a narrow beam.

EMITTER—That part of a transistor from which the drift of major charge carriers begins, i.e. the source of 'holes' in a P-N-P junction transistor and the source of electrons in an N-P-N transistor.

ENERGY—Capacity for doing work. KINETIC ENERGY is energy of motion, e.g. the kinetic energy of the rapidly moving electrons in cathode rays. POTENTIAL ENERGY is energy due to position, e.g. the potential energy of charges stored on the plates of a capacitor.

F

FEEDBACK—The process of feeding some of the output from an amplifier back to the input. In NEGATIVE FEEDBACK the power fed back reduces the input power and thereby stabilizes the amplifier. In POSITIVE FEEDBACK the power fed back reinforces the input power and thereby leads to instability.

FIELD—The region in which a magnetic or an electric effect can be detected. Also used to describe the group of lines in a television picture that is scanned in one sequence, usually half the total number of lines forming the picture.

FILAMENT—Fine wire which becomes hot when current passes through it. Used to heat the cathode of thermionic valves.

FILTER CIRCUIT—Circuit which allows certain frequencies of current to pass more easily than others.

FREQUENCY—The number of times an oscillation repeats itself in one second.

FREE ELECTRON—An electron that is bound only loosely to a particular atom and can therefore move from atom to atom carrying electricity. Conductors contain large numbers of free electrons; insulators contain very few.

FREQUENCY MODULATION—Imposition of a signal on to a radio wave by varying the *frequency* of the wave in sympathy with the *amplitude* of the signal.

FULL-WAVE RECTIFIER—Device that converts every half-cycle of an alternating current into direct current.

G

GAIN—Increase in signal power.

GRID—An electrode in a valve which has an open construction to allow the electron stream to pass. The charge put on to the grid drastically affects the size of the electron stream.

GROUND WAVE—Part of a radio transmission that travels close to the surface of the Earth.

H

HALF-WAVE RECTIFIER—Device that converts either the positive or the negative (but not both) half-cycles of an alternating current into direct current.

HARTLEY OSCILLATOR—Oscillator in which the *tank circuit* is connected so that its *inductor* is in effect common to both the grid circuit and the anode circuit of an oscillating valve (or common to the *base* circuit and the *collector* circuit of an oscillating transistor).

HEATER—*Filament* of an indirectly heated valve.

HETERODYNE—Method used in radio receivers whereby a 'local' signal (frequency f_1) is generated and mixed with the broadcast signal (frequency f_2). The result is a new signal of frequency f_2–f_1, that is detected etc. to extract the broadcast information.

I

IGNITRON—Heavy duty mercury-arc rectifying valve using a pool of mercury as cathode.

IMPEDANCE—Opposition presented by a circuit to the passage of alternating current. Measured from the ratio alternating voltage/alternating current.

INDIRECT HEATING—Use of a *filament* to heat the cathode of a valve without there being any electrical contact between the two.

INDUCTOR—Coil of wire, which may have an empty (air) core or a solid core made of some magnetic material. Any coil possesses what is known as INDUCTANCE, the property whereby a voltage is induced in it by a changing current in the same coil or one near to it.

INVERTER—Device for converting direct current to alternating current.

ION—Atom or molecule that has acquired electrical charge by gaining or losing electrons.

K

KLYSTRON—Special valve for the amplification or generation of microwaves.

L

LIMITER—Part of an f.m. receiver in which any variations in the *amplitude* of the frequency-modulated wave are removed.

LINES OF FORCE—Imaginary lines showing the direction of a *field*. A line of magnetic force is the path an isolated north pole would take if it were free to travel: a line of electric force is the path an isolated electron would take if it were free to travel.

M

MAGIC EYE—Device similar in action to a small cathode-ray tube in which the size of a 'shadow' on a fluorescent screen is governed by the voltage applied to it. Used as tuning indicators and, in tape recorders, as volume or level indicators.

MAGNETRON—Special valve used in radar for generating high-power high-frequency waves. In use it is situated between the poles of a powerful magnet.

MODULATION—Process of imposing a signal on to a radio wave: the information contained in the signal may be transferred to the radio wave either by changing (modulating) the amplitude (a.m.) or the frequency (f.m.) of the latter.

N

N-TYPE SEMICONDUCTOR—Semiconductor that has been 'doped' with an impurity to give it an excess of electrons. The N stands for negative.

O

OHM'S LAW—Fundamental law of electricity, which states that the current in a circuit is directly proportional to the voltage applied to the circuit. Mathematically, the current is equal to the voltage divided by resistance of the circuit.

OSCILLATOR—Circuit for producing alternating voltages. Usually contains one or more valves or transistors and a *tank circuit*.

P

P-TYPE SEMICONDUCTOR—Semiconductor that has been 'doped' with an impurity to give it an excess of 'holes', i.e. a deficiency of electrons. The P stands for positive.

P-N JUNCTION—Boundary between a piece of p-type semiconductor and a piece of n-type semiconductor. Such a junction allows current to flow easily in one direction only; important in the *transistor*.

PENTODE—Five-electrode valve consisting of anode, cathode and three grids.

PHOSPHOR—Substance used for coating the screen of a cathode-ray tube. When electrons fall on the phosphor, it emits light, and the glow persists for a short time after the electron bombardment has ceased.

PHOTOELECTRIC EMISSION—Process whereby certain substances emit *electrons* when light falls on them.

PHOTOMULTIPLIER—See ELECTRON MULTIPLIER.

PIEZOELECTRIC EFFECT—Process whereby certain crystals develop electrical charges on their surfaces when they are mechanically distorted. The same crystals become mechanically distorted when electrical charges are put on to them.

PLATE—American name for the *anode* in a valve.

POTENTIAL—An electrical measure of the potential *energy* of any point. The difference of potential (p.d.) between two points is the *voltage* between them.

PUSH-PULL AMPLIFIER—Amplifier using two identical valves connected in parallel: the cathodes are connected together and to earth. The signal to be amplified is split and fed to the two grids 'out of phase', i.e. when one grid receives a positive half-cycle of signal, the other receives a negative half-cycle, and vice versa.

PULSE SIGNAL—Form of signal (current or voltage) that rises to a definite amplitude, stays at that amplitude for a definite length of time, and then falls again.

Q

Q FACTOR—Quality factor. Measure of the 'goodness' of a component. The Q factor of a coil, for instance, is the ratio of its *reactance* at any frequency to its total resistance at that frequency.

R

RATIO DETECTOR—Form of detector used in f.m. radios.

REACTANCE—Opposition offered by inductors and capacitors to an alternating current. Measured in ohms, it varies with the frequency of the current.

RECTIFICATION—Process of converting an alternating current into a direct current.

RECTIFIER—Device such as a diode which offers an easy path for currents flowing in one direction but a high-resistance path for currents flowing in the opposite direction. It is therefore useful in converting alternating current to one-way current.

RESONANCE—Condition which occurs in a circuit containing inductance and capacitance when, at a particular frequency called the RESONANT FREQUENCY, the current is maximum and the *impedance* is minimum (for a series circuit) or the impedance is maximum and the current is minimum (for a parallel circuit).

S

SCREEN GRID—Grid in a valve immediately outside the *control grid.* Its purpose is to avoid the effective capacitance that normally exists between the anode and the control grid.

SECONDARY ELECTRON—Electron emitted from an atom (or molecule) owing to another electron hitting the atom and being captured by it.

SECONDARY EMISSION—Emission of *secondary electrons.*

SEMICONDUCTOR—Substance such as silicon or germanium which conducts electricity, but not as well as metals. When very pure, these substances are very bad conductors: small quantities of suitable impurities are needed to give the semiconducting properties employed in transistors and rectifiers. Depending on what impurity has been used to 'dope' it, a semiconductor is either *p-type* or *n-type.*

SERIES CONNEXION—Components connected end to end to make a continuous path are said to be connected in series. Components connected across one another so that current is divided between them are said to be connected in parallel or in shunt.

SHUNT CONNEXION—Parallel connexion. The word shunt is frequently applied to the small resistor connected across the terminals of a sensitive meter so that most of the current being measured is 'shunted' through the resistor while only a tiny fraction actually passes through the meter.

SIDEBANDS—Range of frequencies, on either side of the carrier frequency of a radio wave, which contains the frequencies corresponding to the speech or music being broadcast. E.g. if sounds ranging in frequency up to 2,000 c/s (2 kc/s) are to be broadcast on a 1,000 kc/s carrier, the sidebands must extend 2 kc/s on either side of the carrier frequency, i.e. from 998 kc/s to 1,002 kc/s.

SKY WAVE—Radio wave which travels from the Earth's surface out into space, where it is lost unless returned to the ground by charged reflecting layers in the upper atmosphere.

SPACE CHARGE—Cloud of electrons that remains near the cathode of a valve and, since like charges repel, tends to prevent further emission of electrons from the cathode.

SQUARE WAVE—Voltage or current which rises instantly to a fixed level when it remains for a definite time before falling instantly to its initial valve. After a definite time, the pattern is repeated.

SUPERHETERODYNE—Radio receiver employing the *heterodyne* principle. A local oscillator generates a wave slightly different in frequency from the radio wave being received: the two waves are combined and a new wave equal in frequency to the difference between the two is extracted from the combination.

SUPPRESSOR GRID—Electrode situated between the *screen grid* and the *anode* of a valve: it is used to suppress the flow of *secondary electrons.*

T

TANK CIRCUIT—Capacitor connected in parallel with an inductor: at one particular frequency, *resonance* occurs and the tank circuit is said to be tuned to this frequency. Tank circuits form an essential part of valve oscillators.

TETRODE—Valve comprising four electrodes: cathode, anode, control grid and screen grid.

THERMIONIC EMISSION—Emission of electrons from a heated surface, especially the cathode of a valve.

THYRATRON—Gas-filled valve.

THYRISTOR—Transistor equivalent of the gas-filled valve; this device normally presents a very high resistance, but allows current to flow when sufficient voltage is applied to the gate terminal. Formerly called a silicon controlled rectifier.

TICKLER OSCILLATOR—Simple oscillator consisting of a triode with a tank circuit attached to the grid and a 'tickler' coil in the anode circuit. The tank-circuit coil is coupled to the 'tickler' coil, and energy is thus fed back from anode circuit to grid circuit to maintain the oscillating current in the tank circuit.

TRANSCONDUCTANCE—Important *valve constant* defined as the ratio of the small change in anode current to the small change in grid current causing it, when the voltage of the anode is kept constant. Symbol: g_m.

TRANSFORMER—Two inductors (coils) on the same core. An alternating current in one coil induces an alternating voltage in the other. Used for coupling one circuit to another.

TRANSISTOR—Semiconductor equivalent of the valve. For example, an N-P-N transistor consists of a thin slice of p-type semiconductor (the base) sandwiched between two pieces of n-type semiconductor (the emitter and collector).

TRAVELLING-WAVE TUBE—Type of valve in which waves of ultra-high frequency (microwaves) are amplified by making them interact with a beam of electrons. Used in radar.

TUBE—American name for all types of thermionic *valve*.

TUNED-ANODE TUNED-GRID OSCILLATOR—Oscillator having a tank circuit in the anode circuit and another in the grid circuit: the two are coupled only by the capacitor effect which exists in the valve between grid and anode.

TUNED CIRCUIT—See TANK CIRCUIT. A circuit in which *resonance* occurs at a particular frequency. It usually contains a variable capacitor so that the resonant frequency can be altered.

U

U.H.F.—Ultra-high frequency. Frequencies between 300 Mc/s and 3,000 Mc/s, used in radar and some television broadcasting.

UNDAMPED OSCILLATION—Oscillation in which the *amplitude* of successive cycles remains constant.

V

VALVE—Important component in electronics, consisting of a heated cathode, an anode, and sometimes one or more grids, usually in a vacuum. Essential features are: (i) electrons flow from the cathode to the anode, never in the opposite direction; (ii) the strength of the electron flow can be controlled by altering the voltage of the grid.

VALVE CONSTANTS—Numbers which characterize the performance of any type of valve. The three most important valve constants are the *amplification factor*, the a.c. *anode resistance* and the *transconductance*.

V.H.F.—Very-high frequency. Frequencies between 30 Mc/s and 300 Mc/s, used in television and f.m. radio broadcasting.

VIDEO AMPLIFIER—Amplifier capable of handling a wide band of frequencies. Employed in radar for amplifying pulses, and in television for dealing with *video signals*.

VIDEO SIGNAL—Signal in television broadcasting that contains the picture information and the synchronization pulses.

VOLTAGE—Difference in electrical *potential* between two points. The voltage between the ends of conductor governs the size of the current flowing through the conductor. Sometimes used to mean the electromotive force (e.m.f.) which drives electricity in a circuit.

W

WORK FUNCTION—Energy needed to eject an electron from a heated surface. At a given temperature, substances having a low work function (e.g. thoriated tungsten) emit electrons more copiously than substances having a high work function (e.g. pure tungsten).

Z

ZENER DIODE—Diode consisting of a P-N junction in silicon, used for stabilizing voltages within very precise limits.

Index